Analog and Digital Circuits for Electronic Control System

Analog and Digital Circuits for Electronic Control System

Dr. D.K. Chauhan

RANDOM

RANDOM PUBLICATIONS

NEW DELHI (INDIA)

Analog and Digital Circuits for Electronic Control System

ISBN 978-93-5111-570-0

Published in 2015 in India by

RANDOM PUBLICATIONS

Reprinted 2022

4376-A/4B, Gali Murari Lal, Ansari Road
New Delhi-110 002
Phone : +9111-43580356, 011-23289044, 011-43142548
e-mail: sales@randompublications.com,
info@randompublications.com, randomexports@gmail.com

Type Setting by : Friends Media, Delhi-110089
Digitally Printed at : Replika Press Pvt. Ltd.

Preface

Digital electronics, or digital (electronic) circuits, represent signals by discrete bands of analog levels, rather than by a continuous range. All levels within a band represent the same signal state. Relatively small changes to the analog signal levels due to manufacturing tolerance, signal attenuation or parasitic noise do not leave the discrete envelope, and as a result are ignored by signal state sensing circuitry.

Digital circuits are made from analog components. The design must assure that the analog nature of the components doesn't dominate the desired digital behavior. Digital systems must manage noise and timing margins, parasitic inductances and capacitances, and filter power connections.

Digital electronics normally based on 'logic circuits'. These circuits depend on pulses of electricity to make the circuit work. For instance, if current is present - this is represented as '1'. If current is not present, this is represented as '0'. Digital electronics is based on a series of 1s and 0s.

Analog circuits can be very elegant designs with many components, or they can be very simple, like two resistors combining to make a voltage divider. In general, though, analog circuits are much more difficult to design than those which accomplish the same task digitally. It takes a special kind of analog circuit wizard to design an analog radio receiver, or an analog battery charger; digital components exist to make those designs much simpler.

This book is designed for advanced undergraduates and graduate students with background knowledge in basic electronics including biasing, modeling, circuit, analysis, and frequency response.

I would like to thank my team for standing beside me throughout my career and writing this book. My special thanks go to "Random Publications" who have published the book.

–Dr. D.K. Chauhan

Contents

1

Digital and Analog Signals

Signals carry information and are defined as any physical quantity that varies with time, space, or any other independent variable. For example, a sine wave whose amplitude varies with respect to time or the motion of a particle with respect to space can be considered as signals. A system can be defined as a physical device that performs an operation on a signal. For example, an amplifier is used to amplify the input signal amplitude. In this case, the amplifier performs some operation(s) on the signal, which has the effect of increasing the amplitude of the desired information-bearing signal.

Signals can be categorised in various ways; for example discrete and continuous time domains. Discrete-time signals are defined only on a discrete set of times. Continuous-time signals are often referred to as continuous signals even when the signal functions are not continuous; an example is a square-wave signal.

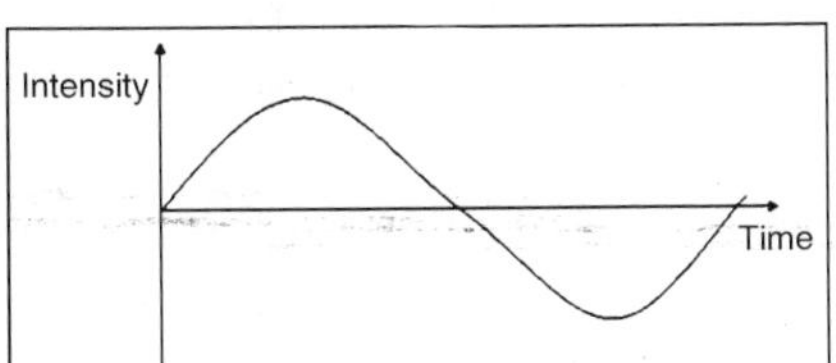

Fig. Analog Signal

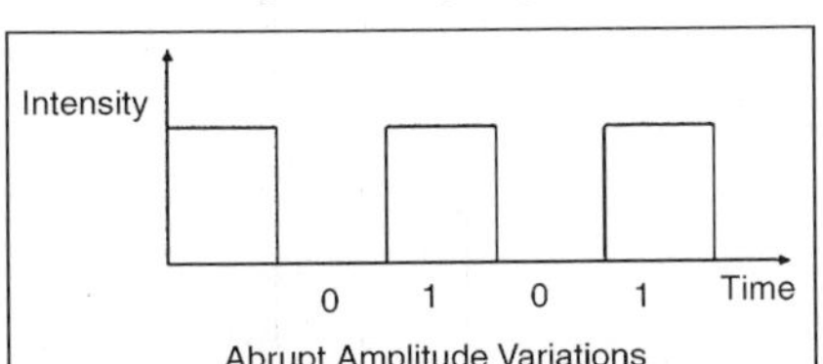

Fig. Digital Signal

Another category of signals is discrete-valued and continuous-valued or otherwise known as digital and analog signals. Digital signals are discrete-valued and analog signals are continuous electrical signals that vary in time as shown in Figures. Analog devices and systems process signals whose voltages or other quantities vary in a continuous manner. They can take on any value across a continuous range of voltage, current, or other metric.

The analog signals can have an infinite number of values. Analog systems can be called wave systems. They have a value that changes steadily over time and can have any one of an infinite set of values in a range. Analog signals represent some physical quantity and they can be a model of the real quantity. Most of the time, the variations corresponds to that of the non-electric (original) signal. For example, the telephone transmitter converts the sounds into an electrical voltage signal. The intensity of the voice causes electric current variations. Therefore, the two are analogous hence the name analog. At the receiving end, the signal is reproduced in the same proportion. Hence the electric current is a model and is an electrical representation of one's voice. Not all analog signals vary as smoothly as the waveform shown in Figure. Digital signals are non-continuous, they change in individual steps. They consist of pulses or digits with discrete levels or values. The value of each pulse is constant, but there is an abrupt change from one digit to the next. Digital signals have two amplitude levels. The value of which are specified as one of two possibilities such as 1 or 0, HIGH or LOW, TRUE or FALSE and so on. In reality, the values are anywhere within specific ranges and we define values within a given range.

A digital system is the one that handles only discrete values or signals. Any set that is restricted to a finite number of elements contains discrete information. The word digital describes any system based on discontinuous data or events. Digital is the method of storing, processing and transmitting information through the use of distinct electronic pulses that represent the binary digits 0 and 1. Examples of discrete sets are the 10 decimal digits, the 26 letters of the alphabet, etc. A digital system would be to flick the light switch on and off. There's no 'in between' values.

ADVANTAGES OF DIGITAL SIGNALS

The usual advantages of digital circuits when compared to analog circuits are:

- *Noise Margin (resistance to noise/robustness):* Digital circuits are less affected by noise. If the noise is below a certain level (the noise margin), a digital circuit behaves as if there was no noise at all. The stream of bits can be reconstructed into a perfect replica of the original source. However, if the noise exceeds this level, the digital circuit cannot give correct results.
- *Error Correction and Detection:* Digital signals can be regenerated to achieve lossless data transmission, within certain limits. Analog signal transmission and processing, by contrast, always introduces noise.

- *Easily Programmable:* Digital systems interface well with computers and are easy to control with software. It is often possible to add new features to a digital system without changing hardware, and to do this remotely, just by uploading new software. Design errors or bugs can be worked-around with a software upgrade, after the product is in customer hands. A digital system is often preferred because of (re-)programmability and ease of upgrading without requiring hardware changes.
- Cheap Electronic Circuits: More digital circuitry can be fabricated per square millimeter of integrated-circuit material. Information storage can be much easier in digital systems than in analog ones. In particular, the great noise-immunity of digital systems makes it possible to store data and retrieve it later without degradation. In an analog system, aging and wear and tear will degrade the information in storage, but in a digital system, as long as the wear and tear is below a certain level, the information can be recovered perfectly. Theoretically, there is no data-loss when copying digital data. This is a great advantage over analog systems, which faithfully reproduce every bit of noise that makes its way into the signal.

DISADVANTAGES OF DIGITAL SIGNALS

The world in which we live is analog, and signals from this world such as light, temperature, sound, electrical conductivity, electric and magnetic fields, and phenomena such as the flow of time, are for most practical purposes continuous and thus analog quantities rather than discrete digital ones. For a digital system to do useful things in the real world, translation from the continuous realm to the discrete digital realm must occur, resulting in quantisation errors. This problem can usually be mitigated by designing the system to store enough digital data to represent the signal to the desired degree of fidelity. The Nyquist-Shannon sampling theorem provides an important guideline as to how much digital data is needed to accurately portray a given analog signal.

Digital systems can be fragile, in that if a single piece of digital data is lost or misinterpreted, the meaning of large blocks of related data can completely change. This problem can be diminished by designing the digital system for robustness. For example, a parity bit or other error-detecting or error-correcting code can be inserted into the signal path so that minor data corruptions can be detected and possibly corrected. Digital circuits use more energy than analog circuits to accomplish the same calculations and signal processing tasks, thus producing more heat as well. In portable or battery-powered systems this can be a major limiting factor.

Digital circuits are made from analog components, and care has to be taken to all noise and timing margins, to parasitic inductances and capacitances, to proper filtering of power and ground connections, to electromagnetic coupling amongst data lines. Inattention to these can cause problems such as "glitches", pulses do not reach valid switching (threshold) voltages, or unexpected ("undecoded") combinations of logic states.

A corollary of the fact that digital circuits are made from analog components is the fact that digital circuits are slower to perform calculations than analog circuits that occupy a similar amount of physical space and consume the same amount of power. However, the digital circuit will perform the calculation with much better repeatability, due to the high noise immunity of digital circuitry.

ANALOG ISSUES IN DIGITAL CIRCUITS

Digital circuits are made from analog components. The design must assure that the analog nature of the components doesn't dominate the desired digital behaviour. Digital systems must manage noise and timing margins, parasitic inductances and capacitances, and filter power connections.

Bad designs have intermittent problems such as "glitches", vanishingly-fast pulses that may trigger some logic but not others, "runt pulses" that do not reach valid "threshold" voltages, or unexpected ("undecoded") combinations of logic states. Since digital circuits are made from analog components, digital circuits calculate more slowly than low-precision analog circuits that use a similar amount of space and power.

However, the digital circuit will calculate more repeatably, because of its high noise immunity. On the other hand, in the high-precision domain (for example, where 14 or more bits of precision are needed), analog circuits require much more power and area than digital equivalents.A digital circuit is often constructed from small electronic circuits called logic gates. Each logic gate represents a function of boolean logic. A logic gate is an arrangement of electrically controlled switches. The output of a logic gate is an electrical flow or voltage, that can, in turn, control more logic gates.

Logic gates often use the fewest number of transistors in order to reduce their size, power consumption and cost, and increase their reliability. Integrated circuits, are the least expensive way to make logic gates in large volumes. Integrated circuits are usually designed by engineers using electronic design automation software.

Another form of digital circuit is constructed from lookup tables, (many sold as "programmable logic devices", though other kinds of PLDs exist). Lookup tables can perform the same functions as machines based on logic gates, but can be easily reprogrammed without changing the wiring. This means that a designer can often repair design errors without changing the arrangement of wires. Therefore, in small volume products, programmable logic devices are often the preferred solution. They are usually designed by engineers using electronic design automation software.

When the volumes are medium to large, and the logic can be slow, or involves complex algorithms or sequences, often a small microcontroller is programmed to make an embedded system. These are usually programmed by software engineers. When only one digital circuit is needed, and its design is totally customized, as for a factory production line controller, the conventional solution is a programmable logic controller, or PLC. These are usually programmed by electricians, using ladder logic.

DIGITAL ELECTRONICS

HALF ADDER CIRCUIT

To understand what is a half adder you need to know what is an adder first. Adder circuit is a combinational digital circuit that is used for adding two numbers. A typical adder circuit produces a sum bit (denoted by *S*) and a carry bit (denoted by *C*) as the output. Typically adders are realised for adding binary numbers but they can be also realised for adding other formats like BCD (binary coded decimal, XS-3, etc. Besides addition, adder circuits can be used for a lot of other applications in digital electronics like address decoding, table index calculation, etc. Adder circuits are of two types: Half adder ad Full adder. Full adder s have been already explained in a previous article and in this topic I am giving stress to half adders.

Half adder is a combinational arithmetic circuit that adds two numbers and produces a sum bit (*S*) and carry bit (*C*) as the output. If *A* and *B* are the input bits, then sum bit (*S*) is the X-OR of *A* and *B* and the carry bit (*C*) will be the AND of A and B. From this it is clear that a half adder circuit can be easily constructed using one X-OR gate and one AND gate. Half adder is the simplest of all adder circuit, but it has a major disadvantage. The half adder can add only two input bits (*A* and *B*) and has nothing to do with the carry if there is any in the input. So if the input to a half adder have a carry, then it will be neglected it and adds only the *A* and B bits. That means the binary addition process is not complete and that's why it is called a half adder. The truth table, schematic representation and XOR//AND realisation of a half adder are shown in the figure below.

Inputs		Outputs	
A	B	s	c
0	0	0	0
1	0	1	0
0	1	1	0
1	1	0	1

Truth Table

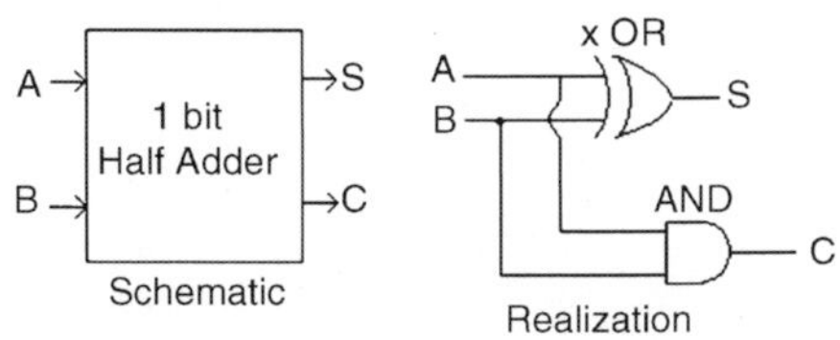

Fig. Truth Table, Schematic and Realisation of Half Adder

NAND gates or NOR gates can be used for realising the half adder in universal logic and the relevant circuit diagrams are shown in the figure below.

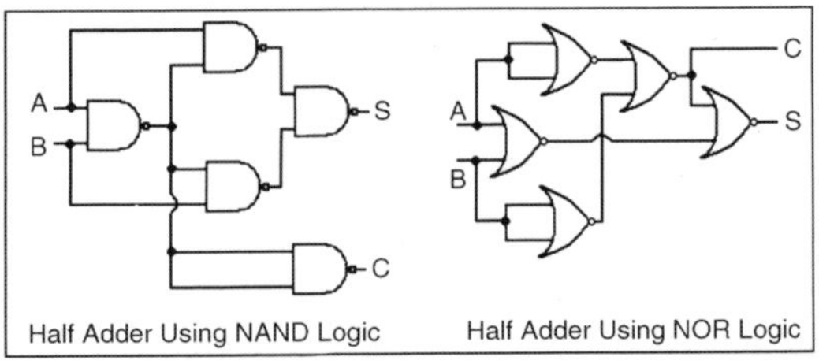

Fig. Half Adder Using NAND and NOR Logic

RIPPLE CARRY ADDER CIRCUIT

Multiple full adder circuits can be cascaded in parallel to add an *N*-bit number. For an *N*- bit parallel adder, there must be *N* number of full adder circuits. *A* ripple carry adder is a logic circuit in which the carry-out of each full adder is the carry in of the succeeding next most significant full adder. It is called a ripple carry adder because each carry bit gets rippled into the next stage. In a ripple carry adder the sum and carry out bits of any half adder stage is not valid until the carry in of that stage occurs.Propagation delays inside the logic circuitry is the reason behind this. Propagation delay is time elapsed between the application of an input and occurance of the corresponding output.

Consider a NOT gate, When the input is "03 the output will be "13 and vice versa. The time taken for the NOT gate's output to become "03 after the application of logic "13 to the NOT gate's input is the propagation delay here. Similarly the carry propagation delay is the time elapsed between the application of the carry in signal and the occurance of the carry out (Cout) signal. Circuit diagram of a 4-bit ripple carry adder is shown below.

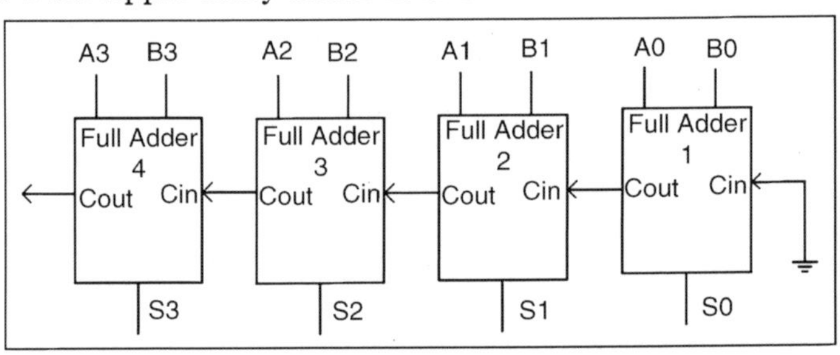

Fig. Ripple Carry Adder

Sum out S0 and carry out Cout of the Full Adder 1 is valid only after the propagation delay of Full Adder 1. In the same way, Sum out S3 of the Full Adder 4 is valid only after the joint propagation delays of Full Adder 1 to Full Adder 4. In simple words, the final result of the ripple carry adder is valid only after the joint propogation delays of all full adder circuits inside it.

Full Adder

To understand the working of a ripple carry adder completely, you need to have a look at the full adder too. Full adder is a logic circuit that adds two input operand bits plus a Carry in bit and outputs a Carry out bit and a sum bit.. The Sum out (Sout) of a full adder is the XOR of input operand bits *A*, *B* and the Carry in (Cin) bit. Truth table and schematic of a 1 bit Full adder is shown below There is a simple trick to find results of a full adder. Consider the second last row of the truth table, here the operands are 1, 1, 0 ie (*A, B, C*in). Add them together ie 1+1+0 = 10.

In binary system, the number order is 0, 1, 10, 11....... and so the result of 1+1+0 is 10 just like we get 1+1+0 =2 in decimal system. 2 in the decimal system corresponds to 10 in the binary system. Swapping the result "103 will give S=0 and Cout = 1 and the second last row is justified. This can be applied to any row in the table.

Inputs			Outputs	
A	B	Cin	Cout	S
0	0	0	0	0
1	0	0	0	1
0	1	0	0	1
1	1	0	1	0
0	0	1	0	1
1	0	1	1	0
0	1	1	1	0
1	1	1	1	1

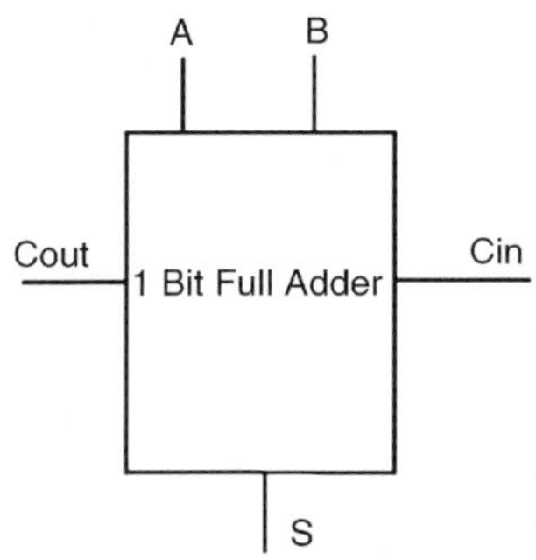

Fig. Bit Full Adder Schematic and Truth Table

A Full adder can be made by combining two half adder circuits together *(a half adder is a circuit that adds two input bits and outputs a sum bit and a carry bit).*

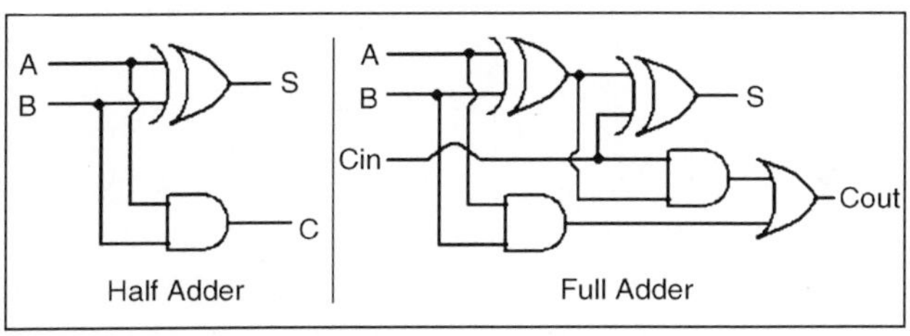

Fig. Full Adder and Half Adder Circuit

Full Adder Using NAND or NOR Logic

Alternatively the full adder can be made using NAND or NOR logic. These schemes are universally accepted and their circuit diagrams are shown below.

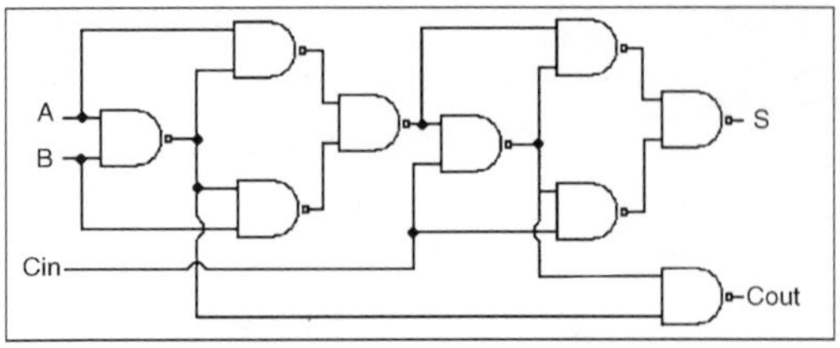

Fig. Full Adder Using NAND Logic

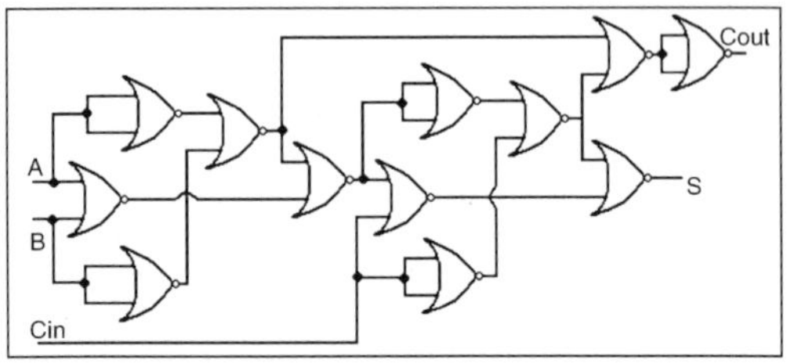

Fig. Full Adder Using NOR Logic

FLIP FLOP – TRIGGERING METHODS

After going through my post on flip flop, you must have understood the importance of triggering a flip flop. The output of a flip flop can be changed by bring a small change in the input signal. This small change can be brought with the help of a clock pulse or commonly known as a trigger pulse.

When such a trigger pulse is applied to the input, the output changes and thus the flip flop is said to be triggered. Flip flops are applicable in designing counters or registers which stores data in the form of multi-bit numbers.But such registers need a group of flip flops connected to each other as sequential circuits. And these sequential circuits require trigger pulses.

The number of trigger pulses that is applied to the input of the circuit determines the number in a counter. A single pulse makes the bit move one position, when it is applied onto a register that stores multi-bit data.

In the case of SR Flip Flops, the change in signal level decides the type of trigger that is to be given to the input. But the original level must be regained before giving a second pulse to the circuit. If a clock pulse is given to the input of the flip flop at the same time when the output of the flip flop is changing, it may cause instability to the circuit. The reason for this instability is the feedback that is given from the output combinational circuit to the memory elements. This problem can be solved to a certain level by making the flip flop more sensitive to the pulse transition rather than the pulse duration.

There are mainly four types of pulse-triggering methods. They differ in the manner in which the electronic circuits respond to the pulse. They are

High Level Triggering

When a flip flop is required to respond at its HIGH state, a HIGH level triggering method is used. It is mainly identified from the straight lead from the clock input. Take a look at the symbolic representation shown below.

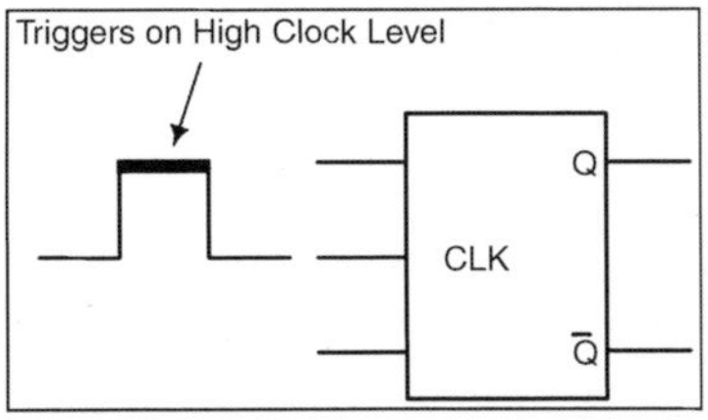

Fig. High Level Triggering

Low Level Triggering

When a flip flop is required to respond at its LOW state, a LOW level triggering method is used.. It is mainly identified from the clock input lead along with a low state indicator bubble. Take a look at the symbolic representation shown below.

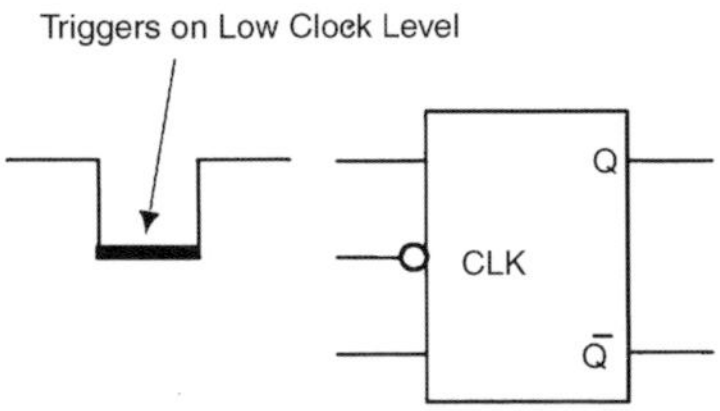

Fig. Low Level Triggering

Positive Edge Triggering

When a flip flop is required to respond at a LOW to HIGH transition state, POSITIVE edge triggering method is used. It is mainly identified from the clock input lead along with a triangle. Take a look at the symbolic representation shown below.

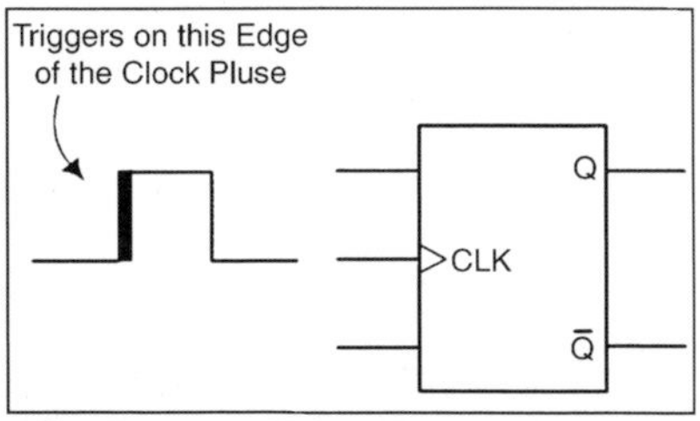

Fig. Positive Edge Triggering

Negative Edge Triggering

When a flip flop is required to respond during the HIGH to LOW transition state, a NEGATIVE edge triggering method is used.. It is mainly identified from the clock input lead along with a low-state indicator and a triangle. Take a look at the symbolic representation shown below.

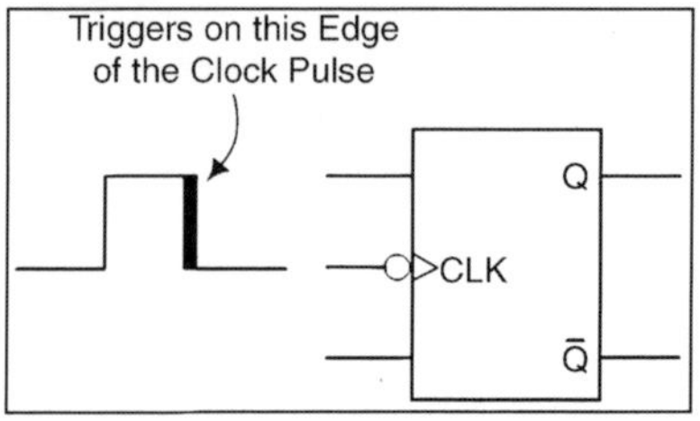

Fig. Negative Edge Triggering

CLOCK PULSE TRANSITION

The movement of a trigger pulse is always from a 0 to 1 and then 1 to 0 of a signal. Thus it takes two transitions in a single signal. When it moves from 0 to 1 it is called a positive transition and when it moves from 1 to 0 it is called a negative transition. To understand more take a look at the images below.

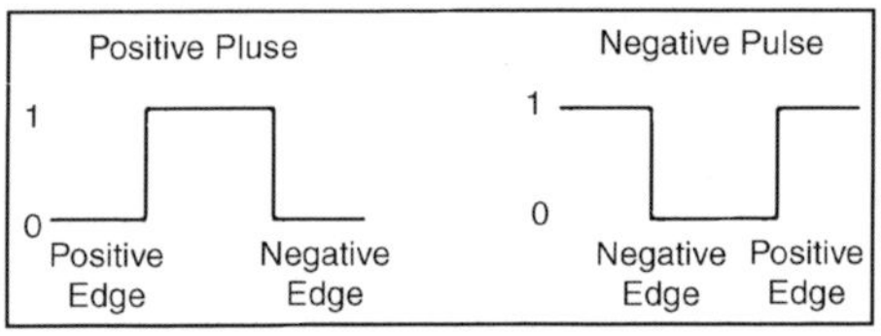

Fig. Clock Pulse Transition

The clocked flip-flops already introduced are triggered during the 0 to 1 transition of the pulse, and the state transition starts as soon as the pulse reaches the HIGH level. If the other inputs change while the clock is still 1, a new output state may occur. If the flip-flop is made to then the multiple-transition problem can be eliminated.

The multi-transition problem can be stopped is the flip flop is made to respond to the positive or negative edge transition only, other than responding to the entire pulse duration.

ANALOG AND DIGITAL CIRCUITS

The quantities that are to be measured, monitored, recorded, processed and controlled are analog and digital, depending on the type of system used. It is important when dealing with various quantities that we be able to represent their values efficiently and accurately. There are basically two ways of representing the numerical value of quantities, analog and digital.

ANALOG REPRESENTATION

Systems which are capable of processing a continuous range of values varying with respect to time are called analog systems. In analog representation a quantity is represented by a voltage, current, or meter movement that is proportional to the value of that quantity. Analog quantities such as those cited above have an important characteristic: They can vary over a continuous range of values.

Analog Voltage vs Time

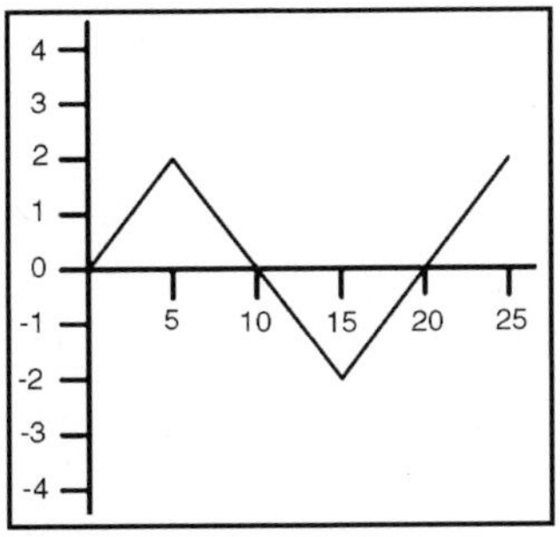

DIGITAL REPRESENTATION

Systems which process discrete values are called digital systems. In digital representation the quantities are represented not by proportional quantities but by symbols called digits.

As an example, consider the digital watch, which provides the time of the day in the form of decimal digits representing hours and minutes (and sometimes seconds). As we know, time of day changes continuously, but the digital watch reading does not change continuously; rather, it changes in steps of one per minute (or per second).

In other words, time of day digital representation changes in discrete steps, as compared to the representation of time provided by an analog watch, where the dial reading changes continuously. *Below is a diagram of digital voltage vs time*: Here input voltage changes from +4 Volts to -4 Volts; it can be converted to digital form by Analog to Digital converters (ADC). An ADC converts continuous signals into samples per second. Well, this is an entirely different theory.

Digital Voltage vs Time

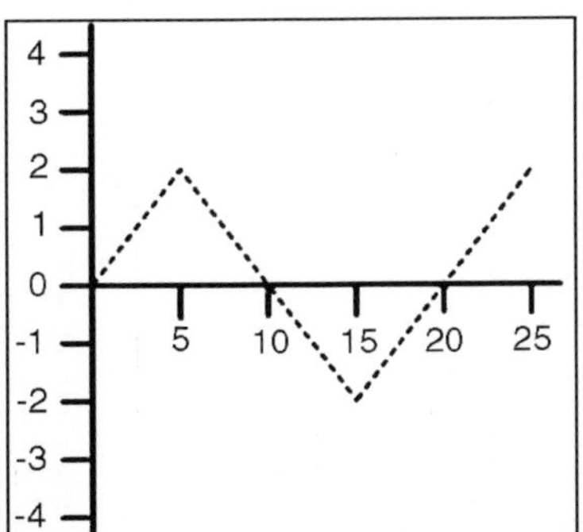

The major difference between analog and digital quantities, then, can be stated

simply as follows:

- Analog = Continuous
- Digital = Discrete (step by step)

ADVANTAGES OF DIGITAL TECHNIQUES

- Easier to design. Exact values of voltage or current are not important, only the range (HIGH or LOW) in which they fall.
- Information storage is easy.
- Accuracy and precision are greater.
- Operations can be programmed. Analog systems can also be programmed, but the available operations variety and complexity is severely limited.
- Digital circuits are less affected by noise, as long as the noise is not large enough to prevent us from distinguishing HIGH from LOW (we discuss this in detail in an advanced digital tutorial section).
- More digital circuitry can be fabricated on IC chips.

LIMITATIONS OF DIGITAL TECHNIQUES

- Most physical quantities in real world are analog in nature, and these quantities are often the inputs and outputs that are being monitored, operated on, and controlled by a system. Thus conversion to digital format and re-conversion to analog format is needed.

2

Digital Electronics System

DIGITAL SYSTEM

While the binary numeration system is an interesting mathematical abstraction, we haven't yet seen its practical application to electronics. What makes binary numeration so important to the application of digital electronics is the ease in which bits may be represented in physical terms. Because a binary bit can only have one of two different values, either 0 or 1, any physical medium capable of switching between two saturated states may be used to represent a bit. Consequently, any physical system capable of representing binary bits is able to represent numerical quantities, and potentially has the ability to manipulate those numbers. Electronic circuits are physical systems that lend themselves well to the representation of binary numbers. Transistors, when operated at their bias limits, may be in one of two different states: either cutoff.

If a transistor circuit is designed to maximize the probability of falling into either one of these states (and not operating in the linear, or active, mode), it can serve as a physical representation of a binary bit. A voltage signal measured at the output of such a circuit may also serve as a representation of a single bit, a low voltage representing a binary "0" and a (relatively) high voltage representing a binary "1." Consider the following transistor circuit:

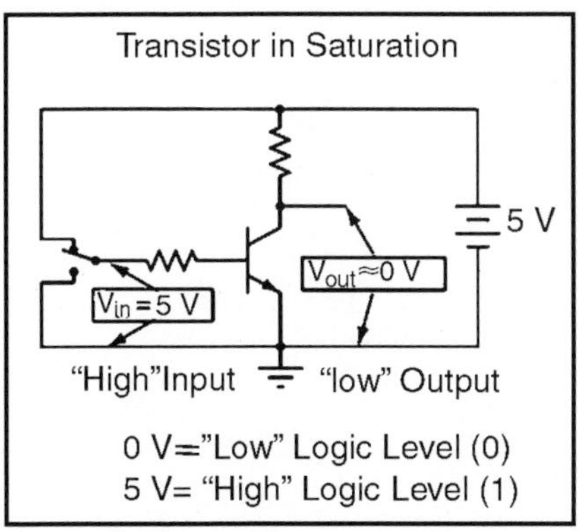

In this circuit, the transistor is in a state of saturation by virtue of the applied input voltage (5 volts) through the two-position switch. Because it's saturated, the transistor drops very little voltage between collector and emitter, resulting in an output voltage of (practically) 0 volts.

If we were using this circuit to represent binary bits, we would say that the input signal is a binary "1" and that the output signal is a binary "0." Any voltage close to full supply voltage (measured in reference to ground, of course) is considered a "1" and a lack of voltage is considered a "0." Alternative terms for these voltage levels are high (same as a binary "1") and low (same as a binary "0").

A general term for the representation of a binary bit by a circuit voltage is logic level. Moving the switch to the other position, we apply a binary "0" to the input and receive a binary "1" at the output:

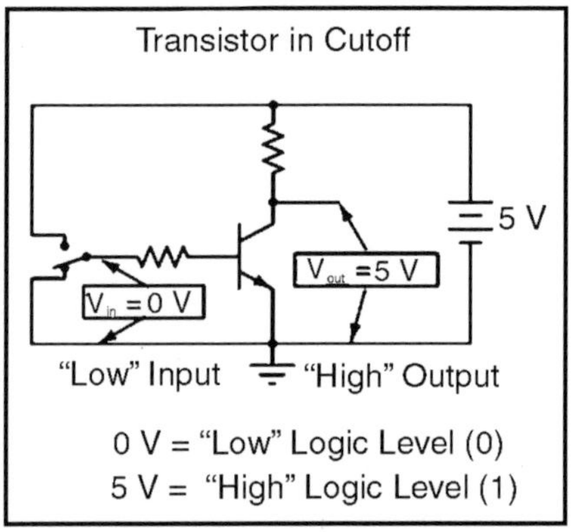

What we've created here with a single transistor is a circuit generally known as a logic gate, or simply gate. A gate is a special type of amplifier circuit designed to accept and generate voltage signals corresponding to binary 1's and 0's. As such, gates are not intended to be used for amplifying analog signals (voltage signals between 0 and full voltage). Used together, multiple gates may be applied to the task of binary number storage (memory circuits) or manipulation (computing circuits), each gate's output representing one bit of a multi-bit binary number.

The gate shown here with the single transistor is known as an inverter, or NOT gate, because it outputs the exact opposite digital signal as what is input. For convenience, gate circuits are generally represented by their own symbols rather than by their constituent transistors and resistors.

The following is the symbol for an inverter:

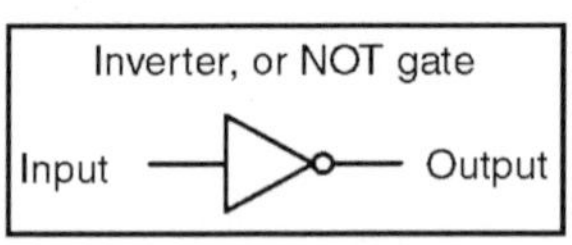

An alternative symbol for an inverter is shown here:

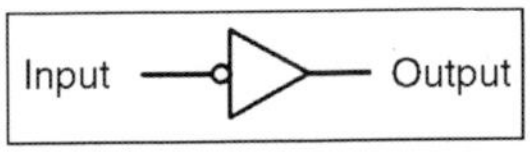

Notice the triangular shape of the gate symbol, much like that of an operational amplifier. As was stated before, gate circuits actually are amplifiers. The small circle, or "bubble" shown on either the input or output terminal is standard for representing the inversion function. As you might suspect, if we were to remove the bubble from the gate symbol, leaving only a triangle, the resulting symbol would no longer indicate inversion, but merely direct amplification. Such a symbol and such a gate actually do exist, and it is called a buffer, the subject of the next section. Like an operational amplifier symbol, input and output connections are shown as single wires, the implied reference point for each voltage signal being "ground." In digital gate circuits, ground is almost always the negative connection of a single voltage source (power supply).

Dual, or "split," power supplies are seldom used in gate circuitry. Because gate circuits are amplifiers, they require a source of power to operate. Like operational amplifiers, the power supply connections for digital gates are often omitted from the symbol for simplicity's sake. If we were to show all the necessary connections needed for operating this gate, the schematic would look something like this:

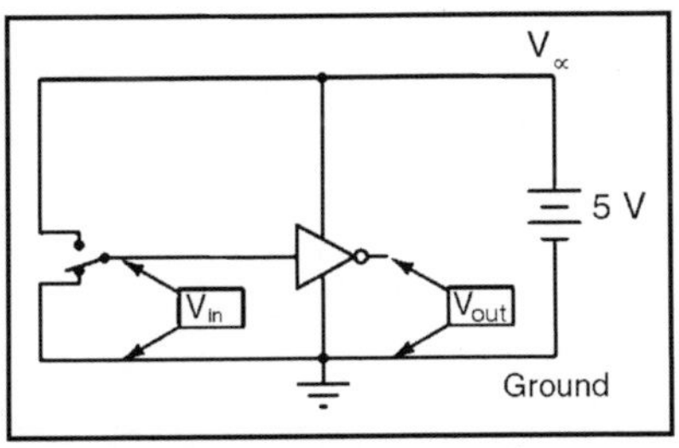

Power supply conductors are rarely shown in gate circuit schematics, even if the power supply connections at each gate are. Minimizing lines in our schematic, we get this:

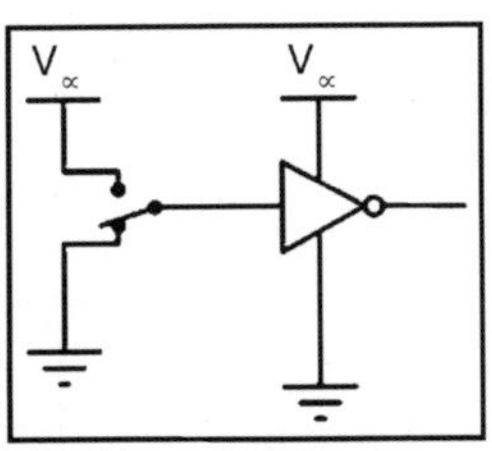

"V_{cc}" stands for the constant voltage supplied to the collector of a bipolar junction transistor circuit, in reference to ground. Those points in a gate circuit marked by the label "V_{cc}" are all connected to the same point, and that point is the positive terminal of a DC voltage source, usually 5 volts. As we will see in other sections of this chapter, there are quite a few different types of logic gates, most of which have multiple input terminals for accepting more than one signal.

The output of any gate is dependent on the state of its input(s) and its logical function.

One common way to express the particular function of a gate circuit is called a truth table. Truth tables show all combinations of input conditions in terms of logic level states (either "high" or "low," "1" or "0," for each input terminal of the gate), along with the corresponding output logic level, either "high" or "low." For the inverter, or NOT, circuit just illustrated, the truth table is very simple indeed:

NOT Gate Truth Table

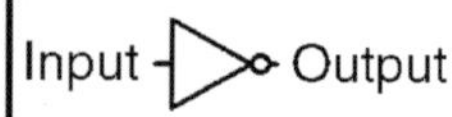

Input	Output
0	1
1	0

Truth tables for more complex gates are, of course, larger than the one shown for the NOT gate. A gate's truth table must have as many rows as there are possibilities for unique input combinations. For a single-input gate like the NOT gate, there are only two possibilities, 0 and 1. For a two input gate, there are four possibilities (00, 01, 10, and 11), and thus four rows to the corresponding truth table. For a three-input gate, there are eight possibilities (000, 001, 010, 011, 100, 101, 110, and 111), and thus a truth table with eight rows are needed. The mathematically inclined will realise that the number of truth table rows needed for a gate is equal to 2 raised to the power of the number of input terminals.

STRUCTURE OF DIGITAL SYSTEMS

Engineers use many methods to minimize logic functions, in order to reduce the circuit's complexity. When the complexity is less, the circuit also has fewer errors and less electronics, and is therefore less expensive. The most widely used simplification is a minimization algorithm like the Espresso heuristic logic minimizer within a CAD system, although historically, binary decision diagrams, an automatedQuine–McCluskey algorithm, truth tables, Karnaugh maps, and Boolean algebra have been used.

Representations are crucial to an engineer's design of digital circuits. Some analysis methods only work with particular representations. The classical way to represent a digital circuit is with an equivalent set of logic gates. Another way, often with the least electronics, is to construct an equivalent system of electronic switches (usually transistors). One of the easiest ways is to simply have a memory containing a truth table. The inputs are fed into the address of the memory, and the data outputs of the memory become the outputs.

For automated analysis, these representations have digital file formats that can be processed by computer programmes. Most digital engineers are very careful to select computer programmes ("tools") with compatible file formats. To choose representations, engineers consider types of digital systems. Most digital systems divide into "combinational systems" and "sequential systems." A combinational system always presents the same output when given the same inputs. It is basically a representation of a set of logic functions. A sequential system is a combinational system with some of the outputs fed back as inputs. This makes the digital machine perform a "sequence" of operations. The simplest sequential system is probably a flip flop, a mechanism that represents a binary digit or "bit". Sequential systems are often designed as state machines. In this way, engineers can design a system's gross behaviour, and even test it in a simulation, without considering all the details of the logic functions. Sequential systems divide into two further subcategories. "Synchronous" sequential systems change state all at once, when a "clock" signal changes state. "Asynchronous" sequential systemspropagate changes whenever inputs change. Synchronous sequential systems are made of well-characterised asynchronous circuits such as flip-flops, that change only when the clock changes, and which have carefully designed timing margins.

The usual way to implement a synchronous sequential state machine is to divide it into a piece of combinational logic and a set of flip flops called a "state register." Each time a clock signal ticks, the state register captures the feedback generated from the previous state of the combinational logic, and feeds it back as an unchanging input to the combinational part of the state machine. The fastest rate of the clock is set by the most time-consuming logic calculation in the combinational logic. The state register is just a representation of a binary number. If the states in the state machine are numbered (easy to arrange), the logic function is some combinational logic that produces the number of the next state.

In comparison, asynchronous systems are very hard to design because all possible states, in all possible timings must be considered. The usual method is to construct a table of the minimum and maximum time that each such state can exist, and then adjust the circuit to minimize the number of such states, and force the circuit to periodically wait for all of its parts to enter a compatible state (this is called "self-resynchronisation"). Without such careful design, it is easy to accidentally produce asynchronous logic that is "unstable", that is, real electronics will have unpredictable results because of the cumulative delays caused by small variations in the values of the electronic components. Certain circuits (such as the synchroniser flip-flops, switchdebouncers, arbiters, and the like which allow external unsynchronised signals to enter synchronous logic circuits) are inherently asynchronous in their design and must be analysed as such.

As of 2005, almost all digital machines are synchronous designs because it is much easier to create and verify a synchronous design—the software currently used to simulate digital machines does not yet handle asynchronous designs. However, asynchronous logic is thought to be superior, if it can be made to work, because its speed is not constrained by an arbitrary clock; instead, it runs at the maximum speed of its logic gates. Building an asynchronous circuit using faster parts makes the circuit faster.

Many digital systems are data flow machines. These are usually designed using synchronous register transfer logic, using hardware description languages such as VHDL or Verilog. In register transfer logic, binary numbers are stored in groups of flip flops called registers.

The outputs of each register are a bundle of wires called a "bus" that carries that number to other calculations. A calculation is simply a piece of combinational logic. Each calculation also has an output bus, and these may be connected to the inputs of several registers. Sometimes a register will have a multiplexer on its input, so that it can store a number from any one of several buses. Alternatively, the outputs of several items may be connected to a bus through buffers that can turn off the output of all of the devices except one. A sequential state machine controls when each register accepts new data from its input. In the 1980s, some researchers discovered that almost all synchronous register-transfer machines could be converted to asynchronous designs by using first-in-first-out synchronisation logic. In this scheme, the digital machine is characterised as a set of data flows. In each step of the flow, an asynchronous "synchronisation circuit" determines when the outputs of that step are valid, and presents a signal that says, "grab the data" to the stages that use that stage's inputs. It turns out that just a few relatively simple synchronisation circuits are needed.

The most general-purpose register-transfer logic machine is a computer. This is basically an automatic binary abacus. The control unit of a computer is usually designed as a microprogram run by a microsequencer. A microprogram is much like a player-piano roll. Each table entry or "word" of the microprogram commands the state of every bit that controls the computer. The sequencer then counts, and the count addresses the memory or combinational logic machine that contains the microprogram. The bits from the microprogram control the arithmetic logic unit, memory and other parts of the computer, including the microsequencer itself. In this way, the complex task of designing the controls of a computer is reduced to a simpler task of programming a collection of much simpler logic machines.

Computer architecture is a specialised engineering activity that tries to arrange the registers, calculation logic, buses and other parts of the computer in the best way for some purpose. Computer architects have applied large amounts of ingenuity to computer design to reduce the cost and increase the speed and immunity to programming errors of computers. An increasingly common goal is to reduce the power used in a battery-powered computer system, such as a cell-phone. Many computer architects serve an extended apprenticeship as microprogrammers. "Specialised computers" are usually a conventional computer with a special-purpose microprogram.

GATES, DECODERS, MULTIPLEXERS

GATES

Logic gates (or simply gates) are the fundamental building blocks of digital circuitry. As their name implies, they function by "opening" or "closing" to admit or reject the flow of digital information. Gates implement electronically simple logical opera-tions on boolean (Bool's algebra) variables, *i.e.*, variables that can have only one of two states (0/1, low/high, false/true). From an electrical point of view and for the TTL (transistor-transistor-logic) family of digital electronics, any voltage in the range 0-0,7 V and in the range 2,5-5 V, represent logic states 0 and 1, respectively. In the following figure the accepted electronic symbols for different gates are shown, along with their corresponding "truth tables" and their symbolic logical expressions. All variables (X, A, B, ...) are booleans.

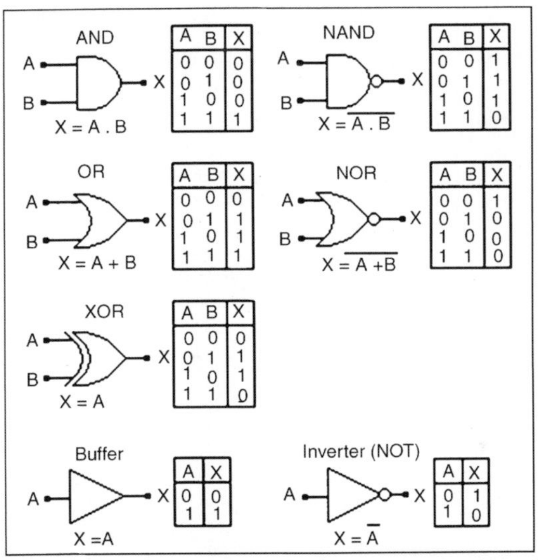

A	B	X (AND)
0	0	0
0	1	0
1	0	0
1	1	1

A	B	X (NAND)
0	0	1
0	1	1
1	0	1
1	1	0

A	B	X (OR)
0	0	0
0	1	1
1	0	1
1	1	1

A	B	X (NOR)
0	0	1
0	1	0
1	0	0
1	1	0

A	B	X (XOR)
0	0	0
0	1	1
1	0	1
1	1	0

A	X (Buffer)
0	0
1	1

A	X (Inverter)
0	1
1	0

The most typical logical operations are implemented by AND and OR gates. The logical ex-pres-sion for the AND operation is "if *A* is true AND *B* is true then *X* is true", and for the OR operation is "if A is true OR B is true then *X* is true". The inverted logic AND and OR gates are com-mon-ly known as NAND (Not AND) and NOR (Not OR) gates. *A* XOR (Exclusive-OR) gate im-plements the logical expression "if *A* is different than B then *X* is true", hence sometimes this ga-te is called "inequality comparator".

The buffer and the inverter are not gates but their use is closely associated with them. A buffer doesn't change the logic state of its input. It is only occasionally used for increasing the fan-out, *i.e.*, the capability of the output of one gate to drive a number of other gates. The inverter is much more important and it is used for inverting a logic state, *i.e.*, for performing the logical operation of negation (NOT). The logical expressions for a buffer and an inverter are "*X* is *A*" and "*X* is NOT *A*", respectively. AND, OR, NAND and NOR gates can have more than 2 inputs. In this case their truth tables are extended to all inputs combinations and their corresponding expressions as well. For example, the logical expression for a 4-input AND is "if *A* is true AND B is true AND *C* is true AND *D* is true then *X* is true". The corresponding expression for a 3-input NOR gate is "if *A* is true OR *B* is true OR *C* is true then *X* is false"

DECODERS

Decoders are circuits with two or more inputs and one or more outputs, resulting by combining various types of gates. Their basic function is to accept a binary word (code) as an input and crea-te a different binary word as an output. A typical decoder is the so-called full adder (3 inputs-2 outputs) implementing the addition of two one-digit numbers (A_i, B_i) taking into consideration the status of any previous carry (C_{i-1}), resulting into the sum (S_i), and generating a new carry (C_i). The addition of two 1-digits numbers and the correspon-ding truth table of full adder are shown below:

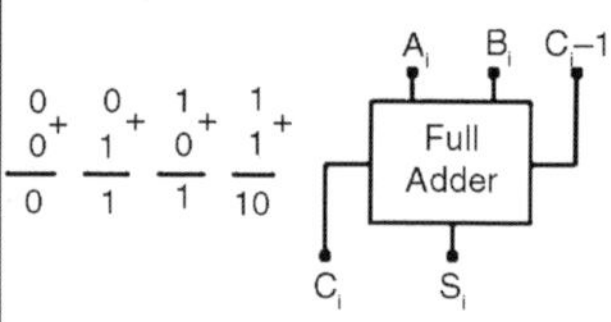

A_i	B_i	C_i-1	S_i	C_i
0	0	0	0	0
0		0	1	0
1	0	0	1	0
1	1	0	0	1
0	0	1	1	0
0		1	0	1
1	0	1	0	1
1	1	1	1	1

N full adders can be cascaded to form a unit for the addition of two *N*-digits binary numbers. Decoders with any type of truth table can be constructed by using simple or complicated combi-nations of gates. Implementation of Bool's algebra rules generally simplifies the overall design. Simple and useful decoders are the so-called "2-to-4" and "3-to-8" decoders.

MULTIPLEXERS

Generally, multiplexers are circuits behaving like a controlled rotary switch, *i.e.*, any one of a number of inputs may be selected as output. In digital electronics, a multiplexer is a combination of logic gates resulting into circuits with two or more inputs (data inputs) and one output. The selection of the channel to be read into the output is controlled by supplying a specific digital word to a different set of inputs (select inputs). A typical 4 input channels (D3-D0) digital multi-plexer, and its corresponding truth table is shown below:

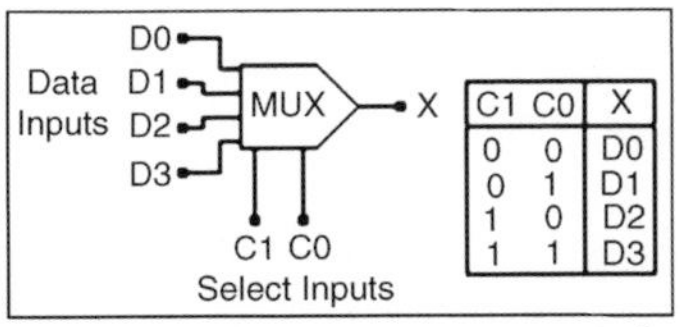

The active input channel is selected by supplying the appropriate code to select inputs (*C1*, *C0*).

APPLET

With this easy to use applet you can train yourself with some simple simulated circuits of digital gates.

You can select one out of five (sets of) circuits by clicking on the corresponding radiobutton:

- Circuit "Gates 1" contains all 2-input gates including a buffer and an inverter.
- Circuit "Gates 2" contains some typical examples of gates with more than 2 inputs.
- Circuit "Full adder" contains a combination of gates implementing the function of a full adder. In the same circuit a 4-bit adder is implemented by cascading 4 full adder circuits.
- Circuit "Decoders" contains a 2-to-4 and a 3-to-8 decoder.
- Circuit "Multiplexer" contains a 4-input line multiplexer.

With all circuits you can change the logic state of any input by clicking on the corresponding small buttons, of which, each one is acting as logic state generator. Observe how the logic state of the out-put(s) and of some "test points" of their internal circuitry is affected by these changes and verify the validity of the truth tables of individual gates. It is of interest to note how a 2-input AND, OR, NAND or NOR gate can control the "flow" of digital data supplied to one of their inputs, by applying different logic states to the other input (control input). Observe also how a XOR gate can act as a buffer or an inverter by applying 0 or 1 to one of its inputs.

DIGITAL INTEGRATED CIRCUITS

Digital integrated circuits, sometimes called an IC, chip or microchip, are the building blocks for nearly every type of modern electronics and electrical devices, whether for home or office, work or pleasure.

Origin

Digital integrated circuit development was necessary to resolve the fundamental issues of vacuum tubes, the leading technology of the first half of the 20th century. Vacuum tubes are bulky, fragile, power hungry, hot and failed frequently.

History

On Sept. 12, 1958, Jack Kilby successfully demonstrated the first IC with a single transistor and all its components on a single piece of germanium semiconductor material. In the first 50 years, ICs have advanced from single transistors to complex microprocessors with multiple CPU cores. On Dec. 3, 2009, the Helsinki University of Technology announced that a team of researchers had succeeded in building the first single atom transistor, the theoretical minimum component size. Technological alternatives such as quantum computing evolved as semiconductor technology approached theoretical limits.

Features

Digital integrated circuits operate based on binary mathematics using electrical signals to represent "ones" and "zeros." Digital integrated circuits can perform any number of operations, from a simple switch to a complex microprocessor.

HISTORY OF INTEGRATED CIRCUITS

The experiments performed on semi conductors revealed that they can perform the variety of vacuum tube functions. The incorporation of bulk of puny transistors on a small chip was a major leap in manually assembling electronic components on a really small circuit. Mass production and efficiency of Integrated Circuits greatly replaced the discrete transistors from many devices.

The two main features of integrated circuits like low cost and high performance gave it edge over the discrete transistors. The low cost is associated to the bulk production and printing using photolithography. The material used in constructing an IC is less than that used in discrete circuits. The high performance can be contributed to the quick switching and low power consumption.

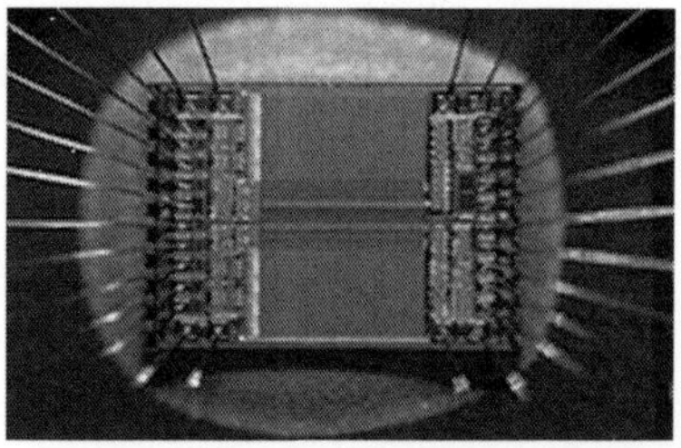

The idea of IC was developed by the scientists working for Royal Radar. The idea propagated further when Jack Kilby and Robert Noyce working separately brought a remarkable invention. The chips made by both of them were meant to perform same functions whereas the material used was not the same. Noyce made a silicone chip and the Kilby manufactured a germanium chip. The development in this field was not new it was in 1949 when the first idea related to integrated chip came to view. From then onwards the development and growth is continued.

Applications of Integrated Circuts

Integrated circuits today have become ubiquitous. It seems that it was this invention which has led to the growth in technology. This remarkable invention has helped companies and manufacturers of electronic devices to design and introduce new and advance products. It can be said that it was due to the micro chips that we can use portable digital devices today.

The integrated chip is an important component of laptops, palm tops, MP3s, playstaions, mobile phones, net books and an unlimited array of electronic devices. The existence of all modern communication and electronic devices is based today on the use of integrated circuits. This chip is small but really efficient and swift when it comes to the performance. It is due to this desirable feature that it is part of almost all cell phones and we carry them anywhere in the world without being worrying about performance decline.

Taxonomy of Integrated Circuits

The integrated circuits can be divided into three broad categories like digital ICs, analog ICs and mixed signal ICs. Digital integrated circuits contain many flip flops, multiplexers and logic gates in a single chip comprising only few millimeters. This small size has provided its edge over all other technologies. Therefore, the manufactures of digital devices now prefer it over discrete transistors. These small chips consume less power and they are efficient. The manufacturing of these circuits help in reducing manufacturing cost due to the wide scale integration.

The working of the digital microchips is guided by the binary mathematical process in the form of 1 and 0 signals. On the contrary the analog micro circuits work by continuous flow of signals. The various examples of analog circuits include power management circuits, sensors and operational amplifiers.

These integrated circuits are used for performing functions like active filtering, demodulation, amplification and mixing. The expertly designed analog circuits remove the need for manufacturing an entirely new circuit from abrasion. This provides the facility to the mobile phone manufactures to buy and install the IC instead of preparing one. The digital and analog circuits can be combined in on one chip to enhance the performance.

3

Digital Signals and Systems

DIGITAL SIGNALS

INTRODUCTION

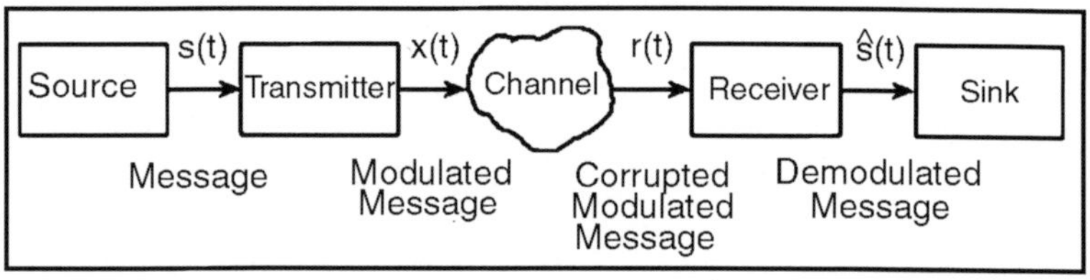

Definition of a system

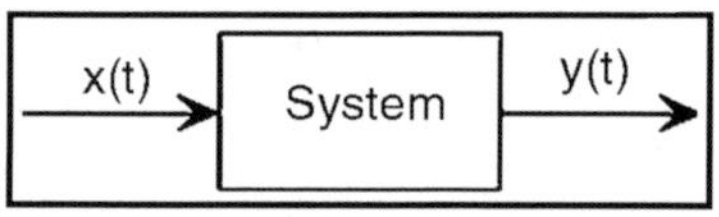

The fundamental model of communications is portrayed in Figure 1. In this fundamental model, each message-bearing signal, exemplified by s(t), is analog and is a function of time. A system operates on zero, one, or several signals to produce more signals or to simply absorb them (Figure above). In electrical engineering, we represent a system as a box, receiving input signals (usually coming from the left) and producing from them new output signals.

This graphical representation is known as a block diagram. We denote input signals by lines having arrows pointing into the box, output signals by arrows pointing away. As typified by the communications model, how information flows, how it is corrupted and manipulated, and how it is ultimately received is summarized by interconnecting block diagrams: The outputs of one or more systems serve as the inputs to others.

In the communications model, the source produces a signal that will be absorbed by the sink. Examples of time-domain signals produced by a source are music, speech, and characters typed on a keyboard. Signals can also be functions of two variables—an image is a signal that depends on two spatial variables—or more—television pictures (video signals) are functions of two spatial variables and time.

Thus, information sources produce signals. In physical systems, each signal corresponds to an electrical voltage or current. To be able to design systems, we must understand electrical science and technology. However, we first need to understand the big picture to appreciate the context in which the electrical engineer works.

In communication systems, messages—signals produced by sources—must be recast for transmission. The block diagram has the message s(t) passing through a block labeled transmitter that produces the signal x(t). In the case of a radio transmitter, it accepts an input audio signal and produces a signal that physically is an electromagnetic wave radiated by an antenna and propagating as Maxwell's equations predict.

In the case of a computer network, typed characters are encapsulated in packets, attached with a destination address, and launched into the Internet. From the communication systems "big picture" perspective, the same block diagram applies although the systems can be very different.

In any case, the transmitter should not operate in such a way that the message s(t) cannot be recovered from x(t). In the mathematical sense, the inverse system must exist, else the communication system cannot be considered reliable. (It is ridiculous to transmit a signal in such a way that no one can recover the original. However, clever systems exist that transmit signals so that only the "in crowd" can recover them. Such crytographic systems underlie secret communications.)

Transmitted signals next pass through the next stage, the evil channel. Nothing good happens to a signal in a channel: It can become corrupted by noise, distorted, and attenuated among many possibilities. The channel cannot be escaped (the real world is cruel), and transmitter design and receiver design focus on how best to jointly fend off the channel's effects on signals. The channel is another system in our block diagram, and produces r(t), the signal received by the receiver. If the channel were benign (good luck finding such a channel in the real world), the receiver would serve as the inverse system to the transmitter, and yield the message with no distortion.

However, because of the channel, the receiver must do its best to produce a received message s(t) that resembles s(t) as much as possible. Shannon showed in his 1948 paper that reliable—for the moment, take this word to mean error-free—digital communication was possible over arbitrarily noisy channels. It is this result that modern communications systems exploit, and why many communications systems are going"digital. Finally, the received message is passed to the information sink that somehow makes use of the message. In the communications model, the source is a system having no input but producing an output; a sink has an input and no output.

Understanding signal generation and how systems work amounts to understanding signals, the nature of the information they represent, how information is transformed between analog and digital forms, and how information can be processed by systems operating on information-bearing signals.

This understanding demands two different fields of knowledge. One is electrical science: How are signals represented and manipulated electrically? The second is signal science: What is the structure of signals, no matter what their source, what is their information content, and what capabilities does this structure force upon communication systems?

LOGIC SIGNAL VOLTAGE LEVELS

Logic gate circuits are designed to input and output only two types of signals: "high" (1) and "low" (0), as represented by a variable voltage: full power supply voltage for a "high" state and zero voltage for a "low" state. In a perfect world, all logic circuit signals would exist at these extreme voltage limits, and never deviate from them (*i.e.*, less than full voltage for a "high," or more than zero voltage for a "low"). However, in reality, logic signal voltage levels rarely attain these perfect limits due to stray voltage drops in the transistor circuitry, and so we must understand the signal level limitations of gate circuits as they try to interpret signal voltages lying somewhere between full supply voltage and zero.

TTL gates operate on a nominal power supply voltage of 5 volts, +/- 0.25 volts. Ideally, a TTL "high" signal would be 5.00 volts exactly, and a TTL "low" signal 0.00 volts exactly. However, real TTL gate circuits cannot output such perfect voltage levels, and are designed to accept "high" and "low" signals deviating substantially from these ideal values. "Acceptable" input signal voltages range from 0 volts to 0.8 volts for a "low" logic state, and 2 volts to 5 volts for a "high" logic state. "Acceptable" output signal voltages (voltage levels guaranteed by the gate manufacturer over a specified range of load conditions) range from 0 volts to 0.5 volts for a "low" logic state, and 2.7 volts to 5 volts for a "high" logic state:

If a voltage signal ranging between 0.8 volts and 2 volts were to be sent into the input of a TTL gate, there would be no certain response from the gate. Such a signal would be considered uncertain, and no logic gate manufacturer would guarantee how their gate circuit would interpret such a signal.

As you can see, the tolerable ranges for output signal levels are narrower than for input signal levels, to ensure that any TTL gate outputting a digital signal into the input of another TTL gate will transmit voltages acceptable to the receiving gate. The difference between the tolerable output and input ranges is called the noise margin of the gate. For TTL gates, the low-level noise margin is the difference between 0.8 volts and 0.5 volts (0.3 volts), while the high-level noise margin is the difference between 2.7 volts and 2 volts (0.7 volts).

Simply put, the noise margin is the peak amount of spurious or "noise" voltage that may be superimposed on a weak gate output voltage signal before the receiving gate might interpret it wrongly:

CMOS gate circuits have input and output signal specifications that are quite different from TTL. For a CMOS gate operating at a power supply voltage of 5 volts, the acceptable input signal voltages range from 0 volts to 1.5 volts for a "low" logic state, and 3.5 volts to 5 volts for a "high" logic state. "Acceptable" output signal voltages (voltage levels guaranteed by the gate manufacturer over a specified range of load conditions) range from 0 volts to 0.05 volts for a "low" logic state, and 4.95 volts to 5 volts for a "high" logic state:

It should be obvious from these figures that CMOS gate circuits have far greater noise margins than TTL: 1.45 volts for CMOS low-level and high-level margins, versus a maximum of 0.7 volts for TTL. In other words, CMOS circuits can tolerate over twice the amount of superimposed "noise" voltage on their input lines before signal interpretation errors will result.

CMOS noise margins widen even further with higher operating voltages. Unlike TTL, which is restricted to a power supply voltage of 5 volts, CMOS may be powered by voltages as high as 15 volts (some CMOS circuits as high as 18 volts). Shown here are the acceptable "high" and "low" states, for both input and output, of CMOS integrated circuits operating at 10 volts and 15 volts, respectively:

The margins for acceptable "high" and "low" signals may be greater than what is shown in the previous illustrations. What is shown represents "worst-case" input signal performance, based on manufacturer's specifications. In practice, it may be found that a gate circuit will tolerate "high" signals of considerably less voltage and "low" signals of considerably greater voltage than those specified here.

Conversely, the extremely small output margins shown — guaranteeing output states for "high" and "low" signals to within 0.05 volts of the power supply "rails" — are optimistic. Such "solid" output voltage levels will be true only for conditions of minimum loading. If the gate is sourcing or sinking substantial current to a load, the output voltage will not be able to maintain these optimum levels, due to internal channel resistance of the gate's final output MOSFETs.

Within the "uncertain" range for any gate input, there will be some point of demarcation dividing the gate's actual "low" input signal range from its actual "high" input signal range. That is, somewhere between the lowest "high" signal voltage level and the highest "low" signal voltage level guaranteed by the gate manufacturer, there is a threshold voltage at which the gate will actually switch its interpretation of a signal from "low" or "high" or visa-versa. For most gate circuits, this unspecified voltage is a single point:

In the presence of AC "noise" voltage superimposed on the DC input signal, a single threshold point at which the gate alters its interpretation of logic level will result in an erratic output:

If this scenario looks familiar to you, it's because you remember a similar problem with (analog) voltage comparator op-amp circuits. With a single threshold point at which an input causes the output to switch between "high" and "low" states, the presence of significant noise will cause erratic changes in the output:

The solution to this problem is a bit of positive feedback introduced into the amplifier circuit. With an op-amp, this is done by connecting the output back around to the non-inverting (+) input through a resistor.

In a gate circuit, this entails redesigning the internal gate circuitry, establishing the feedback inside the gate package rather than through external connections. A gate so designed is called a Schmitt trigger. Schmitt triggers interpret varying input voltages according to two threshold voltages: a positive-going threshold (V_{T+}), and a negative-going threshold (V_{T-}):

Schmitt trigger gates are distinguished in schematic diagrams by the small "hysteresis" symbol drawn within them, reminiscent of the B-H curve for a ferromagnetic material. Hysteresis engendered by positive feedback within the gate circuitry adds an additional level of noise immunity to the gate's performance.

Schmitt trigger gates are frequently used in applications where noise is expected on the input signal line(s), and/or where an erratic output would be very detrimental to system performance. The differing voltage level requirements of TTL and CMOS technology present problems when the two types of gates are used in the same system.

Although operating CMOS gates on the same 5.00 volt power supply voltage required by the TTL gates is no problem, TTL output voltage levels will not be compatible with CMOS input voltage requirements. Take for instance a TTL NAND gate outputting a signal into the input of a CMOS inverter gate. Both gates are powered by the same 5.00 volt supply (V_{cc}). If the TTL gate outputs a "low" signal (guaranteed to be between 0 volts and 0.5 volts), it will be properly interpreted by the CMOS gate's input as a "low" (expecting a voltage between 0 volts and 1.5 volts):

However, if the TTL gate outputs a "high" signal (guaranteed to be between 5 volts and 2.7 volts), it might not be properly interpreted by the CMOS gate's input as a "high" (expecting a voltage between 5 volts and 3.5 volts):

Given this mismatch, it is entirely possible for the TTL gate to output a valid "high" signal (valid, that is, according to the standards for TTL) that lies within the "uncertain" range for the CMOS input, and may be (falsely) interpreted as a "low" by the receiving gate. An easy "fix" for this problem is to augment the TTL gate's "high" signal voltage level by means of a pullup resistor:

Something more than this, though, is required to interface a TTL output with a CMOS input, if the receiving CMOS gate is powered by a greater power supply voltage:

There will be no problem with the CMOS gate interpreting the TTL gate's "low" output, of course, but a "high" signal from the TTL gate is another matter entirely. The guaranteed output voltage range of 2.7 volts to 5 volts from the TTL gate output is nowhere near the CMOS gate's acceptable range of 7 volts to 10 volts for a "high" signal.

If we use an open-collector TTL gate instead of a totem-pole output gate, though, a pullup resistor to the 10 volt V_{dd} supply rail will raise the TTL gate's "high" output voltage to the full power supply voltage supplying the CMOS gate. Since an open-collector gate can only sink current, not source current, the "high" state voltage level is entirely determined by the power supply to which the pullup resistor is attached, thus neatly solving the mismatch problem:

Due to the excellent output voltage characteristics of CMOS gates, there is typically no problem connecting a CMOS output to a TTL input. The only significant issue is the current loading presented by the TTL inputs, since the CMOS output must sink current for each of the TTL inputs while in the "low" state.

When the CMOS gate in question is powered by a voltage source in excess of 5 volts (V_{cc}), though, a problem will result. The "high" output state of the CMOS gate, being greater than 5 volts, will exceed the TTL gate's acceptable input limits for a "high" signal. A solution to this problem is to create an "open-collector" inverter circuit using a discrete NPN transistor, and use it to interface the two gates together:

The "R_{pullup}" resistor is optional, since TTL inputs automatically assume a "high" state when left floating, which is what will happen when the CMOS gate output is "low" and the transistor cuts off. Of course, one very important consequence of implementing this solution is the logical inversion created by the transistor: when the CMOS gate outputs a "low" signal, the TTL gate sees a "high" input; and when the CMOS gate outputs a "high" signal, the transistor saturates and the TTL gate sees a "low" input. So long as this inversion is accounted for in the logical scheme of the system, all will be well.

CONCEPT OF INFORMATION

The concept of information is central to communication. There is no precise definition of the word "information". So, instead of information, we deal with "message". Message is defined as the physical manifestation of information as produced by the source

ANALOG

An analog message is a physical quantity that varies with time usually in a smooth and continuous fashion.

Examples of analog messages are:

- The acoustic pressure produced when you speak
- The angular position of an aircraft gyro
- The light intensity at some point in a television image.

DIGITAL

A digital message is an ordered sequence of symbols selected from a finite set of discrete elements.

Examples of digital messages are:

- Letters printed on this page
- Listing of hourly temperature readings
- The keys you press at a computer terminal.

Whether the message is analog or digital, it needs to be converted into an electrical signal. (There are only few messages, which are inherently electrical.)

INPUT AND OUTPUT

The input transducer converts the message to an electrical signal say a voltage or current. Another transducer at the destination converts the output signal to the desired message form.

Illustration

The transducers in a voice communication system could be microphone at the input and a loudspeaker at the output.

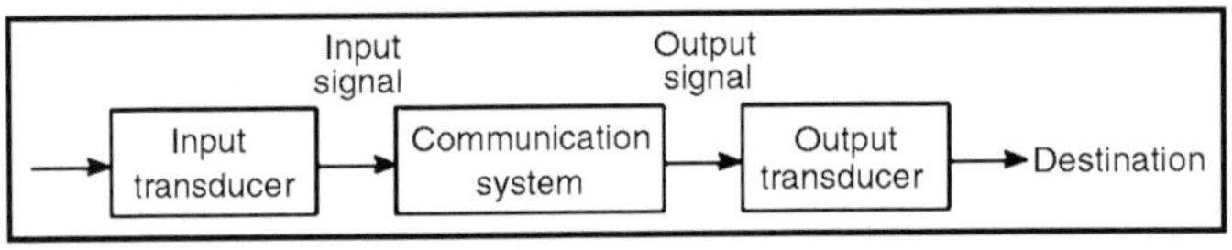

Communication system with input and output transducers

STEP,RAMP, AND IMPULSE FUNCTIONS

Because of the very high switching rate and relatively low signal strength found on data, address, and other buses within a computer, direct extension of the buses beyond the confines of the main circuit board or plug-in boards would pose serious problems. First, long runs of electrical conductors, either on printed circuit boards or through cables, act like receiving antennas for electrical noise radiated by motors, switches, and electronic circuits:

Such noise becomes progressively worse as the length increases, and may eventually impose an unacceptable error rate on the bus signals. Just a single bit error in transferring an instruction code from memory to a microprocessor chip may cause an invalid instruction to be introduced into the instruction stream, in turn causing the computer to totally cease operation.

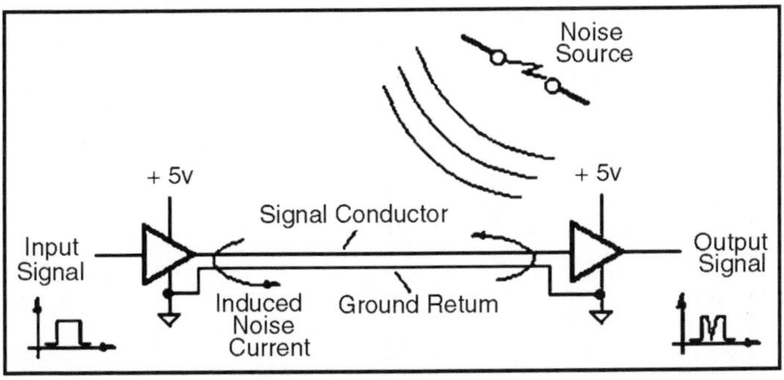

A second problem involves the distortion of electrical signals as they pass through metallic conductors. Signals that start at the source as clean, rectangular pulses may be received as rounded pulses with ringing at the rising and falling edges:

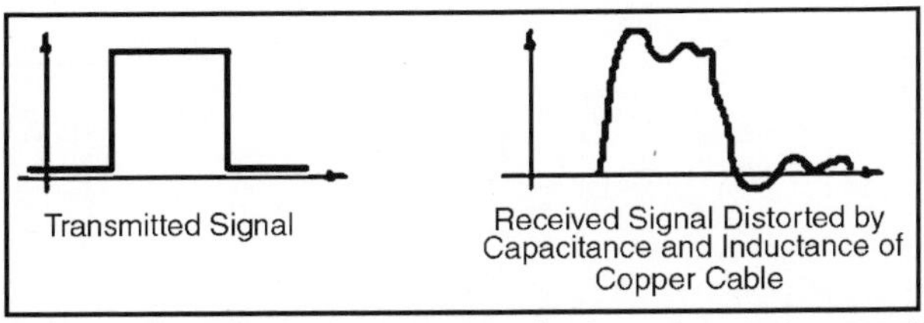

These effects are properties of transmission through metallic conductors, and become more pronounced as the conductor length increases. To compensate for distortion, signal power must be increased or the transmission rate decreased.

Special amplifier circuits are designed for transmitting direct (unmodulated) digital signals through cables. For the relatively short distances between components on a printed circuit board or along a computer backplane, the amplifiers are in simple IC chips that operate from standard +5v power. The normal output voltage from the amplifier for logic '1' is slightly higher than the minimum needed to pass the logic '1' threshold. Correspondingly for logic '0', it is slightly lower. The difference between the actual output voltage and the threshold value is referred to as the noise margin, and represents the amount of noise voltage that can be added to the signal without creating an error:

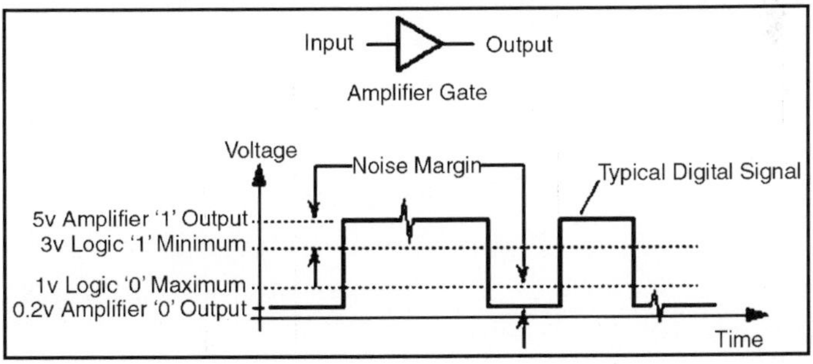

TRANSMISSION OVER MEDIUM DISTANCES

Computer peripherals such as a printer or scanner generally include mechanisms that cannot be situated within the computer itself. Our first thought might be just to extend the computer's internal buses with a cable of sufficient length to reach the peripheral. Doing so, however, would expose all bus transactions to external noise and distortion even though only a very small percentage of these transactions concern the distant peripheral to which the bus is connected.

If a peripheral can be located within 20 feet of the computer, however, relatively simple electronics may be added to make data transfer through a cable efficient and reliable. To accomplish this, a bus interface circuit is installed in the computer:

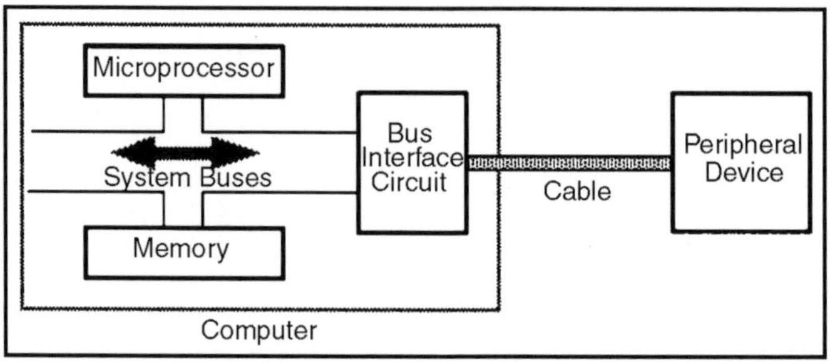

It consists of a holding register for peripheral data, timing and formatting circuitry for external data transmission, and signal amplifiers to boost the signal sufficiently for transmission through a cable. When communication with the peripheral is necessary, data is first deposited in the holding register by the microprocessor. This data will then be reformatted, sent with error-detecting codes, and transmitted at a relatively slow rate by digital hardware in the bus interface circuit.

In addition, the signal power is greatly boosted before transmission through the cable. These steps ensure that the data will not be corrupted by noise or distortion during its passage through the cable. In addition, because only data destined for the peripheral is sent, the party-line transactions taking place on the computer's buses are not unnecessarily exposed to noise.

Data sent in this manner may be transmitted in byte-serial format if the cable has eight parallel channels (at least 10 conductors for half-duplex operation), or in bit-serial format if only a single channel is available.

TRANSMISSION OVER LONG DISTANCES

When relatively long distances are involved in reaching a peripheral device, driver circuits must be inserted after the bus interface unit to compensate for the electrical effects of long cables:

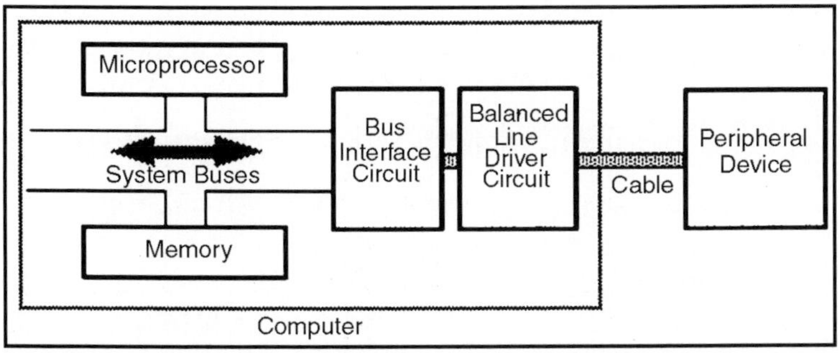

This is the only change needed if a single peripheral is used. However, if many peripherals are connected, or if other computer stations are to be linked, a local area network (LAN) is required, and it becomes necessary to drastically change both the electrical drivers and the protocol to send messages through the cable. Because multiconductor cable is expensive, bit-serial transmission is almost always used when the distance exceeds 20 feet.

In either a simple extension cable or a LAN, a balanced electrical system is used for transmitting digital data through the channel. This type of system involves at least two wires per channel, neither of which is a ground. Note that a common ground return cannot be shared by multiple channels in the same cable as would be possible in an unbalanced system.

The basic idea behind a balanced circuit is that a digital signal is sent on two wires simultaneously, one wire expressing a positive voltage image of the signal and the other a negative voltage image. When both wires reach the destination, the signals are subtracted by a summing amplifier, producing a signal swing of twice the value found on either incoming line. If the cable is exposed to radiated electrical noise, a small voltage of the same polarity is added to both wires in the cable. When the signals are subtracted by the summing amplifier, the noise cancels and the signal emerges from the cable without noise:

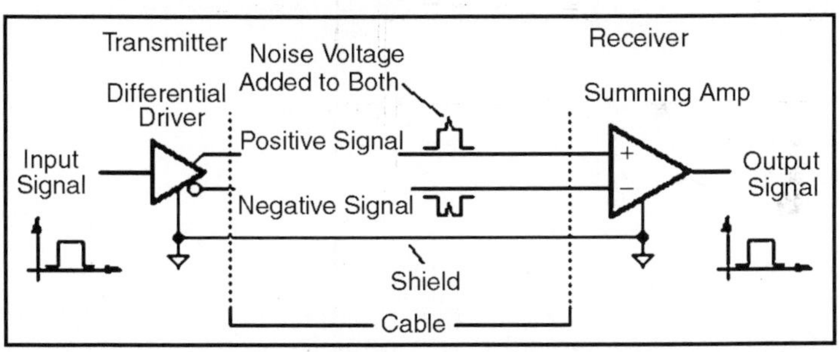

A great deal of technology has been developed for LAN systems to minimize the amount of cable required and maximize the throughput. The costs of a LAN have been concentrated in the electrical interface card that would be installed in PCs or peripherals to drive the cable, and in the communications software, not in the cable itself (whose cost has been minimized). Thus, the cost and complexity of a LAN are not particularly affected by the distance between stations.

TRANSMISSION OVER DISTANCES (GREATER THAN 4000 FEET)

Data communications through the telephone network can reach any point in the world. The volume of overseas fax transmissions is increasing constantly, and computer networks that link thousands of businesses, governments, and universities are pervasive. Transmissions over such distances are not generally accomplished with a direct-wire digital link, but rather with digitally-modulated analog carrier signals. This technique makes it possible to use existing analog telephone voice channels for digital data, although at considerably reduced data rates compared to a direct digital link.

Transmission of data from your personal computer to a timesharing service over phone lines requires that data signals be converted to audible tones by a modem. An audio sine wave carrier is used, and, depending on the baud rate and protocol, will encode data by varying the frequency, phase, or amplitude of the carrier. The receiver's modem accepts the modulated sine wave and extracts the digital data from it. Several modulation techniques typically used in encoding digital data for analog transmission are shown below:

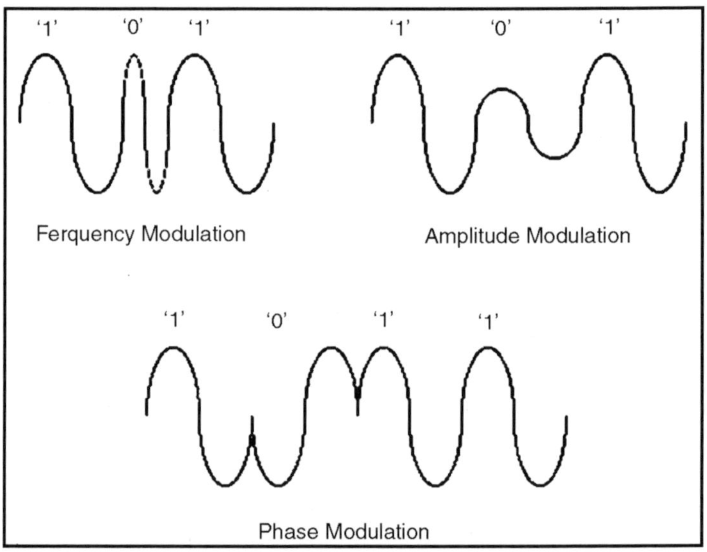

Similar techniques may be used in digital storage devices such as hard disk drives to encode data for storage using an analog medium.

PASSIVE AND ACTIVE NETWORKS

"Passive DRM protects its content from onlookers who do not have a DRM-enabled client. Encryption is generally used for Passive DRM, so that the content is meaningless garbage unless you have the right bits in your client. I consider this 'passive' protection, because the data is inaccessible by default and only becomes accessible if you have the right kind of client, with the right key.

Active DRM, then, would be a scheme where protection is only provided if the client in use is one that is correctly coded to block access where it has not been specifically granted. This is a scheme in which the data is readily accessible to most normal viewers/ players, but has a special code that tells a DRM-enabled viewer/player to hide the content from people who haven't been approved."

The whole problem is his two categories are a false distinction. You can't arbitrarily draw a line through a system and say "this is passive, this is active." For your CSS example, if you consider a given player's decryption code along with an arbitrary encrypted DVD, you have a system with both active and passive elements. If you leave out either of those elements, you have a disc that won't play or a player with no disc, the only perfectly secure system (assuming your cryptography is good.) When judging the efficiency of new compression schemes, the size of the decoder is added to the size of the compressed data to get a fair assessment of its efficiency. Otherwise you could win contests with a one-byte file and a 10 GB decoder programme that simply contains all the actual data.

Whichever way you design a system, complexity is being pushed from one party to another but never eliminated. For DVD, where most of the complexity is in the player, there is a huge variety of player implementations that each have their own bugs. The author of every disc needs to test against many combinations of players because of that problem.

Likewise, if you push the complexity onto the disc by including executable code there, the player gets simpler but the disc could be buggy. However, in that case, the content author will get a bad reputation for the buggy disc (see the Sony rootkit fiasco he mentions).

This doesn't just apply to DRM. While he might consider a MPEG4-AVC video file as "passive" in his terminology, it is really a complex series of instructions to the decoder. Look at the number of different but valid ways to encode video and you'll see it's closer to a programme than to "passive" data.

Now in his definition for "Active DRM", he is not actually describing the general class of software protection techniques. He is describing a system that is poorly-designed, often due to an attempt to retrofit DRM onto an existing system without it. Ofcourse it makes sense that if you have two ways to access the content, one with DRM and the other without, the additional complexity makes no sense to the end-user or mass copiers. It may make economic sense to the content author, but they have to weigh the potential risks to their business also (annoying users vs. stopping some casual copying.)

Even assuming his terminology makes sense, the Windows Media Center system he references is actually a combination of "active" and "passive". The cable video stream is encrypted ("passive"), and the Windows DRM component is "active". In particular, it has a "black box" DLL that checks the host environment and hashes various items to derive a key, hence the problem.

All I can distill from what Alun says is "an unprotected system is made more complex by adding DRM." I agree, but this doesn't say anything larger about "active" versus "passive" DRM.

Full disclosure: I was previously one of the designers of the Blu-ray protection layer (BD+), a unique approach to disc protection that involves both cryptography and software protection. You can consider me biased, but my analysis should be able to stand on its own.

So, if you encrypt a file, and hand me the encrypted bits on a disk, that's what I would call "passive DRM", because its protection does not require me to be running the DRM client code. Only its unlocking requires me to run the DRM client code (or a reasonable facsimile thereof) and get a key.

OSI MODEL CONCEPTS

The standard model for networking protocols and distributed applications is the International Standard Organization's Open System Interconnect (ISO/OSI) model. It defines seven network layers. Short for Open System Interconnection, an ISO standard for worldwide communications that defines a networking framework for implementing protocols in seven layers. Control is passed from one layer to the next, starting at the application layer in one station, proceeding to the bottom layer, over the channel to the next station and back up the hierarchy. At one time, most vendors agreed to support OSI in one form or another, but OSI was too loosely defined and proprietary standards were too entrenched. Except for the OSI-compliant X.400 and X.500 e-mail and directory standards, which are widely used, what was once thought to become the universal communications standard now serves as the teaching model for all other protocols. Understanding how the OSI Model works is not only useful for taking certification exams, but also for real life scenarios. Read How to use the OSI Model to Troubleshoot Networks for more info.

LAYER 1 - PHYSICAL

Physical layer defines the cable or physical medium itself, *e.g.*, thinnet, thicknet, unshielded twisted pairs (UTP). All media are functionally equivalent. The main difference is in convenience and cost of installation and maintenance. Converters from one media to another operate at this level.

LAYER 2 - DATA LINK

Data Link layer defines the format of data on the network. A network data frame, aka packet, includes checksum, source and destination address, and data.

The largest packet that can be sent through a data link layer defines the Maximum Transmission Unit (MTU). The data link layer handles the physical and logical connections to the packet's destination, using a network interface. A host connected to an Ethernet would have an Ethernet interface to handle connections to the outside world, and a loopback interface to send packets to itself. Ethernet addresses a host using a unique, 48-bit address called its Ethernet address or Media Access Control (MAC) address. MAC addresses are usually represented as six colon-separated pairs of hex digits, *e.g.*, 8:0:20:11:ac:85. This number is unique and is associated with a particular Ethernet device. Hosts with multiple network interfaces should use the same MAC address on each. The data link layer's protocol-specific header specifies the MAC address of the packet's source and destination.

LAYER 3 - NETWORK

NFS uses Internetwork Protocol (IP) as its network layer interface. IP is responsible for routing, directing datagrams from one network to another. The network layer may have to break large datagrams, larger than MTU, into smaller packets and host receiving the packet will have to reassemble the fragmented datagram. The Internetwork Protocol identifies each host with a 32-bit IP address. IP addresses are written as four dot-separated decimal numbers between 0 and 255, *e.g.*, 129.79.16.40. The leading 1-3 bytes of the IP identify the network and the remaining bytes identifies the host on that network. The network portion of the IP is assigned by InterNIC Registration Services, under the contract to the National Science Foundation, and the host portion of the IP is assigned by the local network administrators. For large sites, the first two bytes represents the network portion of the IP, and the third and fourth bytes identify the subnet and host respectively.

LAYER 4 - TRANSPORT

Transport layer subdivides user-buffer into network-buffer sized datagrams and enforces desired transmission control. Two transport protocols, Transmission Control Protocol (TCP) and User Datagram Protocol (UDP), sits at the transport layer. Reliability and speed are the primary difference between these two protocols. TCP establishes connections between two hosts on the network through 'sockets' which are determined by the IP address and port number. TCP keeps track of the packet delivery order and the packets that must be resent. Maintaining this information for each connection makes TCP a stateful protocol. UDP on the other hand provides a low overhead transmission service, but with less error checking. NFS is built on top of UDP because of its speed and statelessness. Statelessness simplifies the crash recovery.

LAYER 5 - SESSION

The session protocol defines the format of the data sent over the connections. The NFS uses the Remote Procedure Call (RPC) for its session protocol. RPC may be built on either TCP or UDP. Login sessions uses TCP whereas NFS and broadcast use UDP.

LAYER 6 - PRESENTATION

External Data Representation (XDR) sits at the presentation level. It converts local representation of data to its canonical form and vice versa. The canonical uses a standard byte ordering and structure packing convention, independent of the host.

LAYER 7 - APPLICATION

The OSI Reference Model is founded on a suggestion developed by the International Organization for Standardization (ISO). The model is known as ISO OSI (Open Systems Interconnection) Reference Model because it relates with connecting open systems - that is, systems that are open for communication with other systems.

OSI Model is a set of protocols that try to identify and homogenize the data communication practices. The OSI Model has the support of most computer and network vendors, many big customers, and most governments, including the United States.

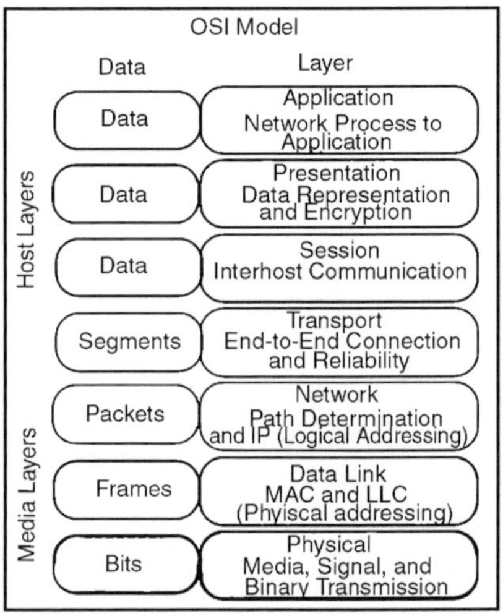

The OSI Model is a model that illustrates how data communications should take place. It segregates the process into seven groups, called layers. Into these layers are integrated the protocol standards developed by the ISO and other standards organization, including the Institute of Electrical and Electronic Engineers (IEEE), American National Standards Institute (ANSI), and the International Telecommunications Union (ITU), formerly known as the CCITT (Comite Consultatif Internationale de Telegraphique et Telephone). The OSI Model affirms what protocols and standards should be used at each layer. It is modular, each layer of the OSI Model functions with the one above and below it.

The short form used to memorize the layer names of the OSI Model is "All People SeemTo Need Data Processing". The lower two layers are normally put into practice with hardware and software. The remaining five layers are only implemented with software.

The layered approach to network communications gives the subsequent advantages: Reduced intricacy, enhanced teaching/learning, modular engineering, accelerated advancement, interoperable technology, and standard interfaces.

TCP / IP MODEL

TCP/IP Model
Application Layer
Transport Layer
Internet Layer
Network Interface Layer

The most significant downside with the TCP/IP model is that if you reference it, fewer people will know what you are talking about!

For a better description of why the OSI model should go the way of the dodo, disco, and DivX, read Kill the Beast: Why the Seven-Layer Model Must Die.

OSI 7 LAYERS REFERENCE MODEL

Open Systems Interconnection (OSI) model is a reference model developed by ISO (International Organization for Standardization) in 1984, as a conceptual framework of standards for communication in the network across different equipment and applications by different vendors. It is now considered the primary architectural model for inter-computing and internetworking communications. Most of the network communication protocols used today have a structure based on the OSI model.

The OSI model defines the communications process into 7 layers, which divides the tasks involved with moving information between networked computers into seven smaller, more manageable task groups. A task or group of tasks is then assigned to each of the seven OSI layers. Each layer is reasonably self-contained so that the tasks assigned to each layer can be implemented independently. This enables the solutions offered by one layer to be updated without adversely affecting the other layers.

The OSI 7 layers model has clear characteristics. Layers 7 through 4 deal with end to end communications between data source and destinations. Layers 3 to 1 deal with communications between network devices.

On the other hand, the seven layers of the OSI model can be divided into two groups: upper layers (layers 7, 6 & 5) and lower layers (layers 4, 3, 2, 1).

The upper layers of the OSI model deal with application issues and generally are implemented only in software. The highest layer, the application layer, is closest to the end user. The lower layers of the OSI model handle data transport issues. The physical layer and the data link layer are implemented in hardware and software. The lowest layer, the physical layer, is closest to the physical network medium (the wires, for example) and is responsible for placing data on the medium.

LINEAR AND NON- LINEAR NETWORK

LINEARITY

The growth in areas of application of electric circuits has led to an evolution from simple to complex circuits. To handle the complexity, engineers over the years have developed some theorems to simplify circuit analysis. Such theorems include Thevenin's and Norton's theorems. Since these theorems are applicable to linear circuits, we first discuss the concept of circuit linearity. In addition to circuit theorems, we discuss the concepts of superposition, source transformation, and maximum power transfer in this chapter. The concepts we develop are applied in the last section to source modeling and resistance measurement.

Linearity is the property of an element describing a linear relationship between cause and effect. Although the property applies to many circuit elements, we shall limit its applicability to resistors in this chapter. The property is a combination of both the homogeneity (scaling) property and the additive property.

The homogeneity property requires that if the input (also called the excitation) is multiplied by a constant, then the output (also called response) is multiplied by the same constant. For a resistor, for example, Ohm's law relates the input I to the output v,

$$v=iR$$

If the current is increased by constant k, then the voltage increases correspondingly by k, that is,

$$kiR=kv$$

The additive property requires that the response to a sum of inputs is the sum of the responses to each input applied separately. Using the voltage current relationship of a resistor, if,

$$v_1=i_1R$$

and

$$v_2=i_2R$$

then, applying (i_1+i_2) gives,

$$v=(i_1+i_2)R=i_1R+i_2R=v_1+v_2$$

We say that a resistor is a linear element because the voltage-current relationship satisfies both the homogeneity and the additive properties. In general, a circuit is linear if it is both additive and homogeneous. A linear circuit consists of only linear elements, linear dependent sources, and independent sources.

A linear circuit is one whose output is linearly related (or directly proportional) to its input.note that since $p=i^2R=v^2/R$ (making it a quadratic function rather than a linear one), the relationship between power and voltage (or current) is non-linear. Therefore, the theorems covered in this chapter are not applicable to power.

To illustrate the linearity principle, consider the linear circuit shown in Figure. The linear circuit has no independent sources inside it. It is excited by a voltage source vs, which serves as the input. The circuit is terminated by a load R. We may take the current i through R as the output. Suppose $v_S=10V$ gives i = 2 A. According to the linearity principle, $v_S=1$V will give i = 0.2 A. By the same token, i = 1 mA must be due to $v_S=5$mV.

SUPERPOSITION

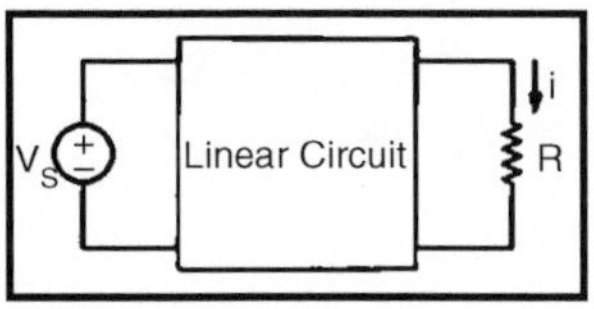

Fig. A linear circuit with input Vs and output

If a circuit has two or more independent sources, one way to determine the value of a specific variable (voltage or current) is to use nodal or mesh analysis as in Chapter. Another way is to determine the contribution of each independent source to the variable and then add them up. The latter approach is known as the superposition.

The idea of superposition rests on the linearity property. The superposition principle states that the voltage across (or current through) an element in a linear circuit is the algebraic sum of the voltages across (or currents through) that elements due to each independent source acting alone. The principle of superposition helps us to analyze a linear circuit with more than one independent source by calculating the contribution of each independent source separately.

However, to apply the superposition principle, we must keep two things in mind:

- We consider one independent source at a time while all other independent sources are turned off. This implies that we replace every voltage source by 0 V (or a short circuit), and every current source by 0 A (or an open circuit). This way we obtain a simpler and more manageable circuit.
- Dependent sources are left intact because they are controlled by circuit variables.

With this in mind, we apply the superposition principle in three steps:

Steps to Apply Superposition Principle

- Turn off all independent sources except one source. Find the output (voltage or current) due to that active source using the techniques covered in Chapters.
- Repeat step 1 for each of the other independent sources.

Find the total contribution by adding algebraically all the contributions due to the independent sources.

Analyzing a circuit using superposition has one major disadvantage: it may very likely involve more work. If the circuit has three independent sources, we may have to analyze three simpler circuits each providing the contribution due to the respectively individual source. However, superposition does help reduce a complex circuit to simpler circuits through replacement of voltage sources by short circuits and of current sources by open circuits.

Keep in mind that superposition is based on linearity. For this reason, it not applicable to the effect on power due to each source, because the power absorbed by a resistor depends on the square of the voltage or current. If the power value is needed, the current through (or voltage across) the element must be calculated first using superposition.

SOURCE TRANSFORMATION

We have noticed that series-parallel combination and wye-delta transformation help simplify circuits. Source transformation is another tool for simplifying circuits. Basic to these tools is concept of equivalence. We recall that an equivalent circuit is one whose v-i characteristics are identical with the original circuit.

In Section, we saw that node-voltage (or mesh current) equations can be obtained by mere inspection of a circuit when the sources are all independent current (or all independent voltage) sources. It is therefore expedient in circuit analysis to be able to substitute a voltage source in series with a resistor for a current source in parallel with a resistor, or vice versa, as shown in Figure Either substitution is known as source transformation.

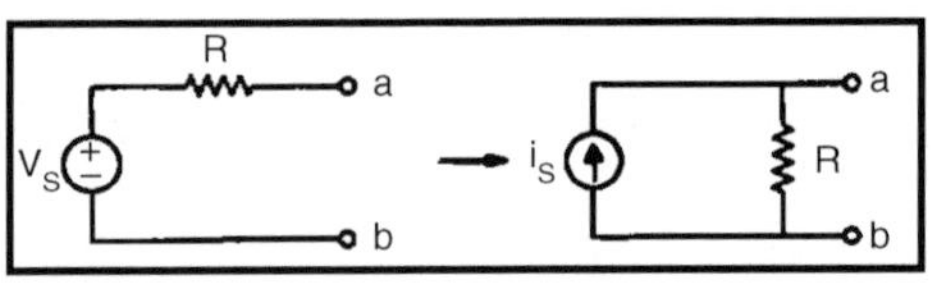

Fig. Transformation of independent sources.

A source transformation is the process of replacing a voltage source vs in series with a resistor R by a current source is in parallel with a resistor R, or vice versa. The two circuits in Figure are equivalent – provided they have the same voltage–current relation at terminals a-b. It is easy to show that they are indeed equivalent. If the sources are turned off, the equivalent resistance at terminals a-b in both circuits is R. Also, when terminals a-b are short circuited, the short circuit current flowing from a to b is $i_{SG}=v_S/R$ in the circuit in the left hand side and $i_{SG}=i_S$ for the circuit on right hand side. Thus, $v_S/R=i_S$ in order for the two circuits to be equivalent. Hence, source transformation requires that

$$v_S = i_S R (6)$$

or

$$i_S = v_S R$$

Source transformation also applies to dependent sources, provided we carefully handle the dependent variable. As shown in Figure, a dependent voltage source in series with a resistor can be transformed to a dependent current source in parallel with the resistor or vice versa where we make sure that Equation is satisfied.

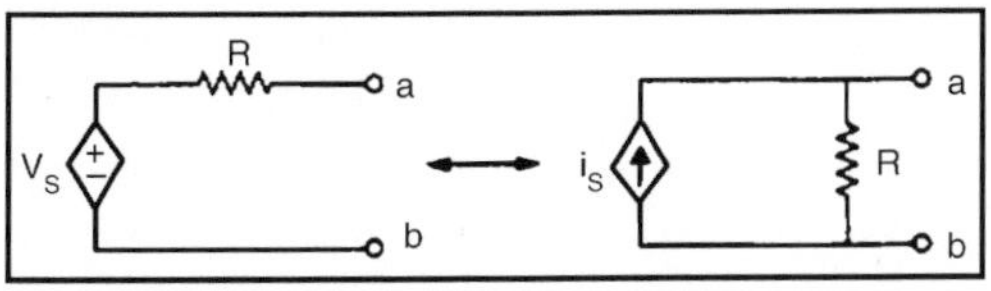

Fig. Transformation of dependent sources.

Like wye-delta transformation, a source transformation does not affect the remaining part of the circuit. When applicable, source transformation is a powerful tool that allows circuit manipulations to ease circuit analysis.

However, we should keep the following points in mind when dealing with source transformation:

- Note from Figure that the arrow of the current source is directed towards the positive terminal of the voltage source.
- Note from Equation that source transformation is not possible when R = 0, which is the case with an ideal voltage source. However, for a practical, non-ideal voltage source, $R \neq 0$. Similarly, an ideal current source with $R = \infty$ cannot be replaced by a finite voltage source.

REQUIREMENTS FOR LINEARITY

A system is called linear, if it has two mathematical properties: homogeneity (hôma-gen-â-ity) and additivity. If you can show that a system has both properties, then you have proven that the system is linear. Likewise, if you can show that a system doesn't have one or both properties, you have proven that it isn't linear. A third property, shift invariance, is not a strict requirement for linearity, but it is a mandatory property for most DSP techniques.

When you see the term linear system used in DSP, you should assume it includes shift invariance unless you have reason to believe otherwise. These three properties form the mathematics of how linear system theory is defined and used. Later in this chapter we will look at more intuitive ways of understanding linearity. For now, let's go through these formal mathematical properties.

As illustrated in Figure; homogeneity means that a change in the input signal's amplitude results in a corresponding change in the output signal's amplitude. In mathematical terms, if an input signal of *x[n]* results in an output signal of *y[n]*, an input of *kx[n]* results in an output of *ky[n]*, for any input signal and constant, *k*.

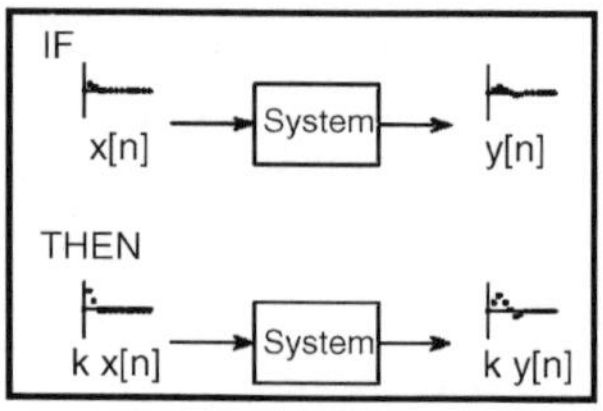

A simple resistor provides a good example of both homogenous and non-homogeneous systems. If, the input to the system is the voltage across the resistor, *v(t)*, and the output from the system is the current through the resistor, *i(t)*, the system is homogeneous. Ohm's law guarantees this; if the voltage is increased or decreased, there will be a corresponding increase or decrease in the current.

Now, consider another system where the input signal is the voltage across the resistor, *v(t)* but the output signal is the power being dissipated in the resistor, *p(t)*. Since power is proportional to the square of the voltage, if, the input signal is increased by a factor of two, the output signal is increase by a factor of four. This system is not homogeneous and therefore cannot be linear. The property of additivity is illustrated in Figure; Consider a system where an input of $x_1[n]$ produces an output of $y_1[n]$. Further suppose that a different input, $x_2[n]$, produces another output, $y_2[n]$. The system is said to be additive, if an input of $x_1[n] + x_2[n]$ results in an output of $y_1[n] + y_2[n]$, for all possible input signals. In words, signals added at the input produce signals that are added at the output.

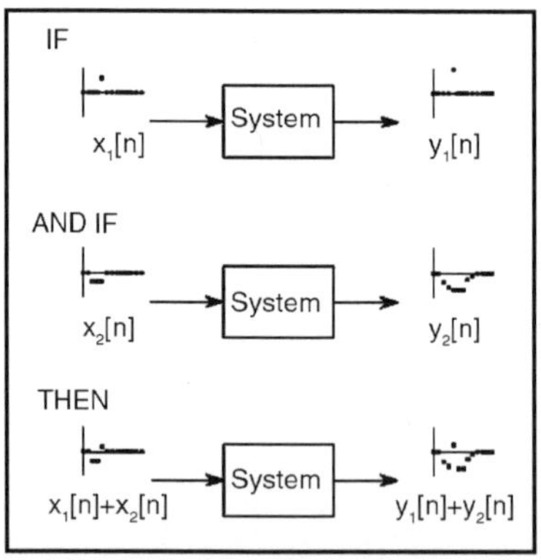

The important point is that added signals pass through the system without interacting. As an example, think about a telephone conversation with your Aunt Edna and Uncle Bernie. Aunt Edna begins a rather lengthy story about how well her radishes are doing this year. In the background, Uncle Bernie is yelling at the dog for having an accident in his favourite chair. The two voice signals are added and electronically transmitted through the telephone network. Since this system is additive, the sound you hear is the sum of the two voices as they would sound if transmitted individually. You hear Edna and Bernie, not the creature, Ednabernie.

A good example of a non-additive circuit is the mixer stage in a radio transmitter. Two signals are present: an audio signal that contains the voice or music, and a carrier wave that can propagate through space when applied to an antenna. The two signals are added and applied to a non-linearity, such as a pn junction diode. This results in the signals merging to form a third signal, a modulated radio wave capable of carrying the information over great distances.

As shown in Figure; shift invariance means that a shift in the input signal will result in nothing more than an identical shift in the output signal. In more formal terms, if an input signal of *x[n]* results in an output of *y[n]*, an input signal of *x[n + s]* results in an output of *y[n + s]*, for any input signal and any constant's.

Pay particular notice to how the mathematics of this shift is written, it will be used in upcoming chapters. By adding a constant's to the independent variable, *n*, the waveform can be advanced or retarded in the horizontal direction. For example, when $s = 2$, the signal is shifted left by two samples; when $s = -2$, the signal is shifted right by two samples.

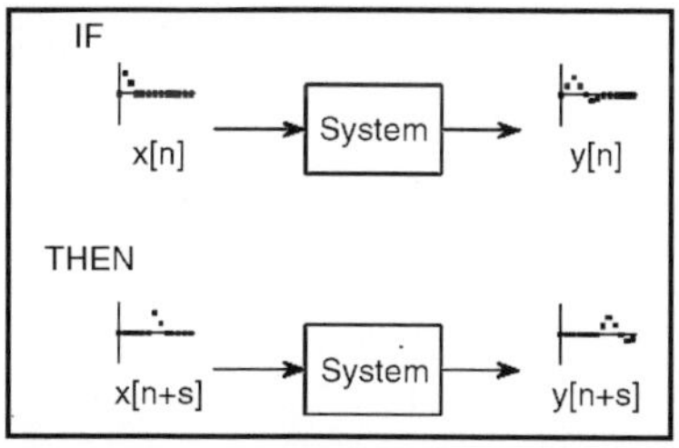

Shift invariance is important because it means the characteristics of the system do not change with time (or whatever the independent variable happens to be). If a blip in the input causes a blop in the output, you can be assured that another blipwill cause an identical blop. Most of the systems you encounter will be shift invariant. This is fortunate, because it is difficult to deal with systems that change their characteristics while in operation.

For example, imagine that you have designed a digital filter to compensate for the degrading effects of a telephone transmission line. Your filter makes the voices sound more natural and easier to understand. Much to your surprise, along comes winter and you find the characteristics of the telephone line have changed with temperature. Your compensation filter is now mismatched and doesn't work especially well. This situation may require1a more sophisticated algorithm that can adapt to changing conditions.

Why do homogeneity and additivity play a critical role in linearity, while shift invariance is something on the side? This is because linearity is a very broad concept, encompassing much more than just signals and systems. For example, consider a farm er selling oranges for $2 per crate and apples for $5 per crate. If the farmer sells only oranges, he will receive $20 for 10 crates, and $40 for 20 crates, making the exchange homogenous. If he sells 20 crates of oranges and 10 crates of apples, the farmer will receive: This is the same amount as if the two had been sold individually, making the transaction additive.

Being both homogenous and additive, this sale of goods is a linear process. However, since there are no signals involved, this is not a system and shift invariance has no meaning. Shift invariance can be thought of as an additional aspect of linearity needed when signals and systems are involved.

NON- LINEARITY

Homogeneity, additivity, and shift invariance are important because they provide the mathematical basis for defining linear systems. Unfortunately, these properties alone don't provide most scientists and engineers with an intuitive feeling of what linear systems are about. The properties of static linearity and sinusoidal fidelityare often of help here. These are not especially important from a mathematical standpoint, but relate to how humans think about and understand linear systems.

You should pay special attention to this section. Static linearity defines how a linear system reacts when the signals aren't changing, *i.e.*, when they are *DC* or static.

The output is the input multiplied by a constant. That is a graph of the possible input values plotted against the corresponding output values is a straight line that passes through the origin. This is shown in Figure for two common linear systems: Ohm's law for resistors and Hooke's law for springs. For comparison, Figure shows the static relationship for two non-linear systems: a pn junction diode and the magnetic properties of iron.

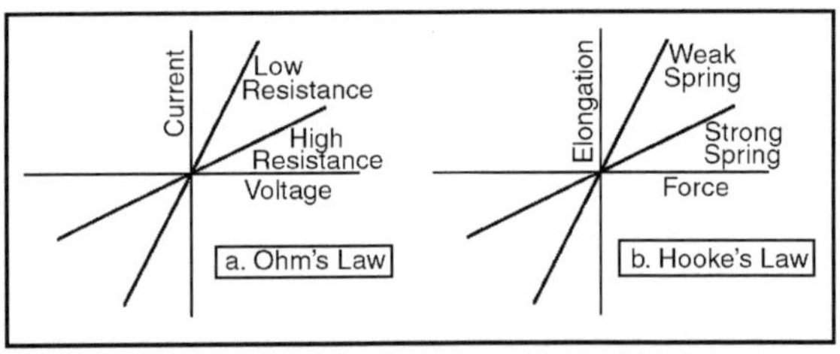

All linear systems have the property of static linearity. The opposite is usually true, but not always. There are systems that show static linearity, but are not linear with respect to changing signals. However, a very common class of systems can be completely understood with static linearity alone. In these systems it doesn't matter if the input signal is static or changing. These are called memoryless systems because the output depends only on the present state of the input, and not on its history.

For example, the instantaneous current in a resistor depends only on the instantaneous voltage across it, and not on how the signals came to be the value they are. If a system has static linearity and is memoryless then the system must be linear. This provides an important way to understand (and prove) the linearity of these simple systems.

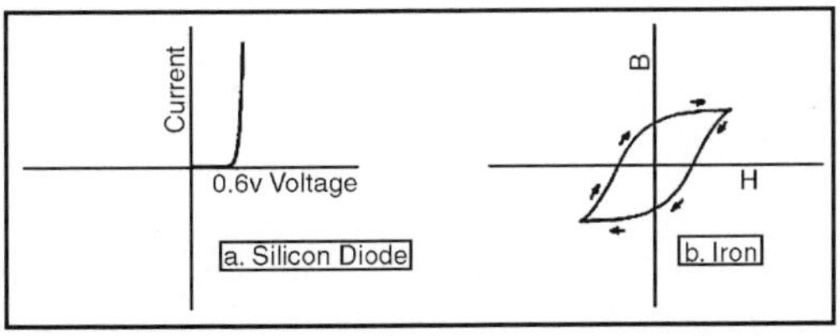

An important characteristic of linear systems is how they behave with sinusoids a property we will call sinusoidal fidelity: If the input to a linear system is a sinusoidal wave, the output will also be a sinusoidal wave, and at exactly the same frequency as the input. Sinusoids are the only waveform that have this property.

For instance, there is no reason to expect that a square wave entering a linear system will produce a square wave on the output. Although a sinusoid on the input guarantees a sinusoid on the output, the two may be different in amplitude and phase. This should be familiar from your knowledge of electronics: a circuit can be described by its frequency response, graphs of how the circuit's gain and phase vary with frequency.

Now for the reverse question: If a system always produces a sinusoidal output in response to a sinusoidal input, is the system guaranteed to be linear? The answer is no, but the exceptions are rare and usually obvious. For example, imagine an evil demon hiding inside a system, with the goal of trying to mislead you. The demon has an oscilloscope to observe the input signal and a sine wave generator to produce an output signal. When you feed a sine wave into the input, the demon quickly measures the frequency and adjusts his signal generator to produce a corresponding output.

Of course, this system is not linear because it is not additive. To show this, place the sum of two sine waves into the system. The demon can only respond with a single sine wave for the output. This example is not as contrived as you might think phase lock loops operate in much this way.

To get a better feeling for linearity, think about a technician trying to determine if an electronic device is linear. The technician would attach a sine wave generator to the input of the device and an oscilloscope to the output. With a sine wave input, the technician would look to see if the output is also a sine wave.

For example, the output cannot be clipped on the top or bottom, the top half cannot look different from the bottom half, there must be no distortion where the signal crosses zero, etc. Next, the technician would vary the amplitude of the input and observe the effect on the output signal. If the system is linear, the amplitude of the output must track the amplitude of the input. Lastly, the technician would vary the input signal's frequency, and verify that the output signal's frequency changes accordingly.

As the frequency is changed, there will likely be amplitude and phase changes seen in the output, but these are perfectly permissible in a linear system. At some frequencies, the output may even be zero, that is, a sinusoid with zero amplitude. If the technician sees all these things, he will conclude that the system is linear. While this conclusion is not a rigourous mathematical proof, the level of confidence is justifiably high.

4

Boolean Functions and Binary Numbers System

BOOLEAN FUNCTIONS

The most obvious way to simplify Boolean expressions is to manipulate them in the same way as normal algebraic expressions are manipulated. With regards to logic relations in digital forms, a set of rules for symbolic manipulation is needed in order to solve for the unknowns.

A set of rules formulated by the English mathematician George Boole describe certain propositions whose outcome would be either true or false. With regard to digital logic, these rules are used to describe circuits whose state can be either, 1 (true) or 0 (false). In order to fully understand this, the relation between the AND gate, OR gate and NOT gate operations should be appreciated. A number of rules can be derived from these relations as demonstrates.

Boolean Postulates:

- *P*1: X = 0 or X = 1
- *P*2: 0. 0 = 0
- *P*3: 1 + 1 = 1
- *P*4: 0 + 0 = 0
- *P*5: 1. 1 = 1
- *P*6: 1. 0 = 0. 1 = 0
- *P*7: 1 + 0 = 0 + 1 = 1

LAWS OF BOOLEAN ALGEBRA

Table shows the basic Boolean laws. Note that every law has two expressions, (a) and (b). This is known as *duality*. These are obtained by changing every AND(.) to OR(+), every OR(+) to AND(.) and all 1's to 0's and vice-versa. It has become conventional to drop the . (AND symbol) *i.e.* A.B is written as AB.

BOOLEAN LAWS

*T*1: Commutative Law
(a) $A + B = B + A$
(b) $A\,B = B\,A$

*T*2: Associate Law
(a) $(A + B) + C = A + (B + C)$
(b) $(A\,B)\,C = A\,(B\,C)$

*T*3: Distributive Law
(a) $A\,(B + C) = A\,B + A\,C$
(b) $A + (B\,C) = (A + B)\,(A + C)$

*T*4: Identity Law
(a) $A + A = A$
(b) $A\,A = A$

*T*5: (a) $AB + A\overline{B} = A$
(b) $(A + B)(A + \overline{B}) = A$

*T*6: Redundance Law
(a) $A + A\,B = A$
(b) $A\,(A + B) = A$

*T*7: (a) $0 + A = A$
(b) $0\,A = 0$

*T*8: (a) $1 + A = 1$
(b) $1\,A = A$

*T*9: (a) $\overline{A} + A = 1$
(b) $\overline{A}\,A = 0$

*T*10: (a) $A + \overline{A}B = A + B$
(b) $A(\overline{A} + B) = AB$

*T*11: De Morgan's Theorem
(a) $(\overline{A + B}) = \overline{A}\,\overline{B}$
(b) $(\overline{AB}) = \overline{A} + \overline{B}$

Examples

*Prove T*10: (a) $A + \overline{A}B = A + B$

- *Algebraically:*

$$
\begin{aligned}
A + \overline{A}B &= A1 + \overline{A}B && T7(a)\\
&= A(1 + B) + \overline{A}B && T7(c)\\
&= A + AB + \overline{A}B && T3(a)\\
&= A + B\left(A + \overline{A}\right) && T3(a)\\
&= A + B && T(8)
\end{aligned}
$$

- *Using the truth table*:

A	B	A+B	$\bar{A}B$	$A+\bar{A}B$
0	0	0	0	0
0	1	1	1	1
1	0	1	0	1
1	1	1	0	1

Using the laws given above, complicated expressions can be simplified.

$$Z=\left(A+\bar{B}+\bar{C}\right)\left(A+\bar{B}C\right)$$

$$Z=AA+A\bar{B}C+A\bar{B}+\bar{B}\bar{B}C+A\bar{C}+\bar{B}C\bar{C}$$

$$Z=A\left(1+\bar{B}C+\bar{B}+\bar{C}\right)+\bar{B}C+\bar{B}C\bar{C}\text{ From laws TBb and Tb}$$

$$Z=A+\bar{B}C\text{ From laws T8b and T9b}$$

KARNAUGH MAPS

So far we can see that applying Boolean algebra can be awkward in order to simplify expressions. Apart from being laborious (and requiring the remembering all the laws) the method can lead to solutions which, though they appear minimal, are not.

The Karnaugh map provides a simple and straight-forward method of minimising boolean expressions. With the Karnaugh map Boolean expressions having up to four and even six variables can be simplified. .

The Karnaugh map (K map) provides a systematic method for simplifying a Boolean expression or a truth table function. When used properly, the K map will produce the simplest SOP or POS expression possible. Familiarity with the law and rules of Boolean algebra is not required. Instead, simplification is done graphically using the K mapping technique.

The K map is a table consisting of $N = 2^n$ cells, where n is the number of input variables. For a SOP expression each cell represents one particular combination of the variables in product form. The table format is such that there is a single variable change between any adjacent cells.

This is the characteristic the will determine adjacency. To illustrate the above points let us consider an example where n = 2 and N = 4. Assuming the input variable are A and B then the K map illustrating the four (4) possible variable combinations $\bar{A}\bar{B}, \bar{A}B, A\bar{B}$ and AB is show below in Table.

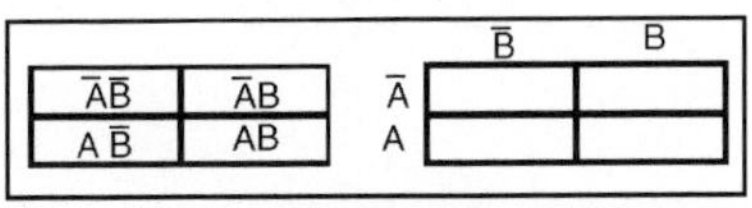

Fig. Two Variable K Map Format.

If we extend our example to consider the case where n = 3 and N = 8 then assuming that our input variables are A, B and C the associated K Map is shown in Table.

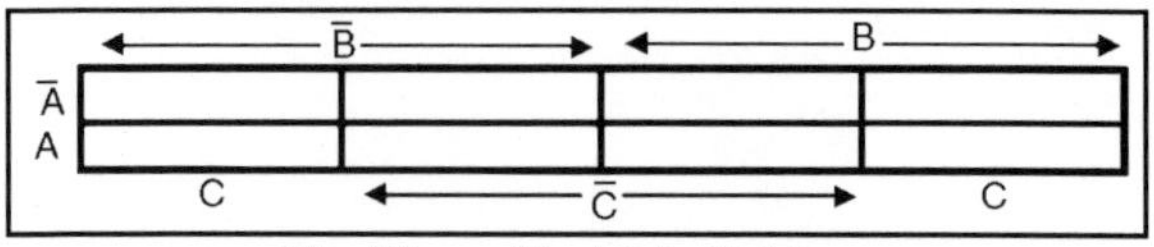

Fig. Three Variable K-Map.

For n>5 the K map technique becomes impractical unless implemented on computer.

Simplifying Using the K-Map

To simplify a SOP for of a Boolean expression using a K map, first identify all the input combinations that produce an output of logic level 1 and place them in their appropriate K map cell. Consequently, all other cells must contain zero (0). Second, group the adjacent cells that contain 1 in a manner that maximises the size of the groups but also minimises the total number of groups. All 1's in the output must be included in a group even if the group is only one cell. Third, as each SOP term represents an AND expression, each (AND) grouping is written with only the input variables that are common to the group. Finally, the simplified expression is formed by ORing each of the (AND) groups. To illustrate lets consider the function, $X = \bar{A}BC + AB\bar{C} + ABC$ whose truth table and K map are illustrated in Tables.

Truth Table of $X = \bar{A}BC + AB\bar{C} + ABC$

Input			Output
A	**B**	**C**	**X**
0	0	0	0
0	0	1	0
0	1	0	0
0	1	1	1
1	0	0	0
1	0	1	0
1	1	0	1
1	1	1	1

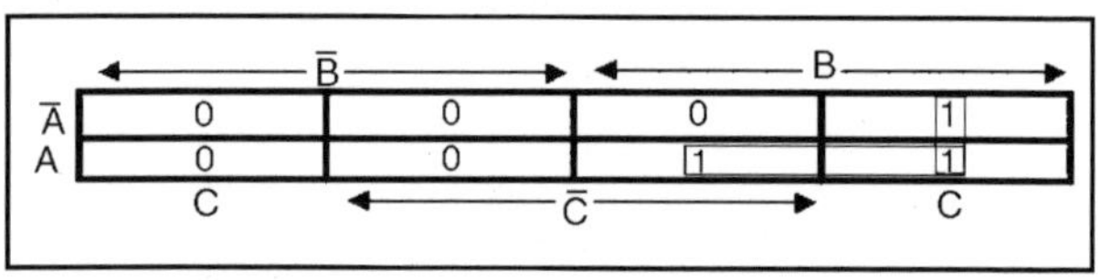

K-Map for $X = \bar{A}BC + AB\bar{C} + ABC$

By examination of the K map, the simplified expression for $X = \bar{A}BC + AB\bar{C} + ABC$ is $X = BC + AB$

Let us now use the K map technique to simplify the SOP Boolean expression $X = \bar{A}\bar{B}\bar{C} + A\bar{B}\bar{C} + \bar{A}\bar{B}C + A\bar{B}C + ABC$. The associated truth table and K map are presented in Table 1-19 and Table, respectively.

Table. Truth $X=\bar{A}\bar{B}\bar{C}+A\bar{B}\bar{C}+\bar{A}\bar{B}C+A\bar{B}C+ABC$

Input			Output
A	**B**	**C**	**X**
0	0	0	1
0	0	1	1
0	1	0	0
0	1	1	0
1	0	0	1
1	0	1	1
1	1	0	0
1	1	1	1

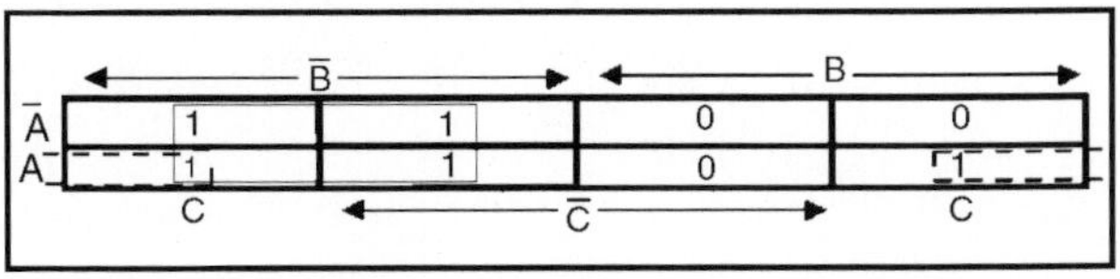

K-Map for $X=\bar{A}\bar{B}\bar{C}+A\bar{B}\bar{C}+\bar{A}\bar{B}C+A\bar{B}C+ABC$

By examination of the K map, the simplified expression for $X=\bar{A}\bar{B}\bar{C}+A\bar{B}\bar{C}+\bar{A}\bar{B}C+A\bar{B}C+ABC$ is $X=\bar{B}+AC$.

Suppressed Variables

Sometimes a SOP expression does not have a complete set of variables in each of its terms. For example, the expression $X=A\bar{B}\bar{C}+B\bar{C}+ABC$ contains a term that does not include the variable A or its complement. A missing variable in a SOP expression is called a *suppressed* variable.

Before plotting an expression with suppressed variables on a K-MAP we must first expand out all of the shortened terms to include the missing variables and its complement.

To illustrate let us consider use our example above:

$$B\bar{C}=B\bar{C}(1)$$

$$B\bar{C}=B\bar{C}\left(A+\bar{A}\right)$$

$$B\bar{C}=AB\bar{C}+\bar{A}B\bar{C}$$

Then our expanded expression becomes,

$$X=A\bar{B}\bar{C}+AB\bar{C}+\bar{A}B\bar{C}+ABC$$

and we may now use the truth table and the associated K map, as illu,strated in Tables to simplify.

Truth Table of $X=A\bar{B}\bar{C}+AB\bar{C}+\bar{A}B\bar{C}+ABC$

Input			Output
A	**B**	**C**	**X**
0	0	0	0
0	0	1	0

0	1	0	1
0	1	1	0
1	0	0	1
1	0	1	0
1	1	0	1
1	1	1	1

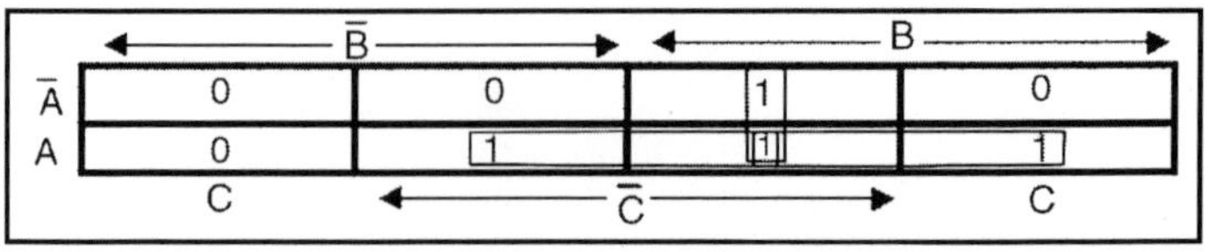

K-Map for $X = A\bar{B}\bar{C} + AB\bar{C} + \bar{A}B\bar{C} + ABC$

From the three groupings shown in the K map we form the expression $X = B\bar{C} + A\bar{C} + AB$ ·

BOOLEAN ALGEBRA VS ORDINARY ALGEBRA

Mathematical rules are based on the defining limits we place on the particular numerical quantities dealt with. When we say that, 1 + 1 = 2 or 3 + 4 = 7, we are implying the use of integer quantities: the same types of numbers we all learned to count in elementary education. What most people assume to be self-evident rules of arithmetic valid at all times and for all purposes — actually depend on what we define a number to be. For instance, when calculating quantities in AC circuits, we find that the "real" number quantities which served us so well in DC circuit analysis are inadequate for the task of representing AC quantities. We know that voltages add when connected in series, but we also know that it is possible to connect a 3-volt AC source in series with a 4-volt AC source and end up with 5 volts total voltage (3 + 4 = 5)! Does this mean the inviolable and self-evident rules of arithmetic have been violated?

No, it just means that the rules of "real" numbers do not apply to the kinds of quantities encountered in AC circuits, where every variable has both a magnitude and a phase. Consequently, we must use a different kind of numerical quantity, or object, for AC circuits (*complex* numbers, rather than *real* numbers), and along with this different system of numbers comes a different set of rules telling us how they relate to one another.

An expression such as (3 + 4 = 5) is nonsense within the scope and definition of real numbers, but it fits nicely within the scope and definition of complex numbers (think of a right triangle with opposite and adjacent sides of 3 and 4, with a hypotenuse of 5). Because complex numbers are two-dimensional, they are able to "add" with one another trigonometrically as single-dimension "real" numbers cannot. Logic is much like mathematics in this respect: the so-called "Laws" of logic depend on how we define what a proposition is. The Greek philosopher Aristotle founded a system of logic based on only two types of propositions: true and false. His bivalent (two-mode) definition of truth led to the four foundational laws of logic: the Law of Identity (A is A); the Law of Non-contradiction (A is not non-A); the Law of the Excluded Middle (either A or non-A); and the Law of Rational Inference.

These so-called Laws function within the scope of logic where a proposition is limited to one of two possible values, but may not apply in cases where propositions can hold values other than "true" or "false." In fact, much work has been done and continues to be done on "multivalued," or *fuzzy* logic, where propositions may be true or false *to a limited degree*.

In such a system of logic, "Laws" such as the Law of the Excluded Middle simply do not apply, because they are founded on the assumption of bivalence. Likewise, many premises which would violate the Law of Non-contradiction in Aristotelian logic have validity in "fuzzy" logic. Again, the defining limits of propositional values determine the Laws describing their functions and relations. Boole wrote a treatise on the subject in 1854, titled *An Investigation of the Laws of Thought, on Which Are Founded the Mathematical Theories of Logic and Probabilities*, which codified several rules of relationship between mathematical quantities limited to one of two possible values: true or false, 1 or 0. His mathematical system became known as Boolean algebra.

All arithmetic operations performed with Boolean quantities have but one of two possible outcomes: either 1 or 0. There is no such thing as "2" or "–1" or "1/2" in the Boolean world. It is a world in which all other possibilities are invalid by fiat. As one might guess, this is not the kind of math you want to use when balancing a checkbook or calculating current through a resistor. However, Claude Shannon of MIT fame recognized how Boolean algebra could be applied to on-and-off circuits, where all signals are characterized as either "high" (1) or "low" (0). The thesis, titled *A Symbolic Analysis of Relay and Switching Circuits*, put Boole's theoretical work to use in a way Boole never could have imagined, giving us a powerful mathematical tool for designing and analysing digital circuits.

In this chapter, you will find a lot of similarities between Boolean algebra and "normal" algebra, the kind of algebra involving so-called real numbers. Just bear in mind that the system of numbers defining Boolean algebra is severely limited in terms of scope, and that there can only be one of two possible values for any Boolean variable: 1 or 0.

Consequently, the "Laws" of Boolean algebra often differ from the "Laws" of real-number algebra, making possible such statements as $1 + 1 = 1$, which would normally be considered absurd. Once you comprehend the premise of all quantities in Boolean algebra being limited to the two possibilities of 1 and 0, and the general philosophical principle of Laws depending on quantitative definitions, the "non-sense" of Boolean algebra disappears.It should be clearly understood that Boolean numbers are not the same as *binary* numbers. Whereas Boolean numbers represent an entirely different system of mathematics from real numbers, binary is nothing more than an alternative *notation* for real numbers. The two are often confused because both Boolean math and binary notation use the same two ciphers: 1 and 0.The difference is that Boolean quantities are restricted to a single bit (either 1 or 0), whereas binary numbers may be composed of many bits adding up in place-weighted form to a value of any finite size. The binary number 10011_2("nineteen") has no more place in the Boolean world than the decimal number 2_{10} ("two") or the octal number 32_8 ("twenty-six").

BOOLEAN EXPRESSIONS

In mathematics, an *identity* is a statement true for all possible values of its variable or variables. The algebraic identity of $x + 0 = x$ tells us that anything (x) added to zero equals the original "anything," no matter what value that "anything" (x) may be. Like ordinary algebra, Boolean algebra has its own unique identities based on the bivalent states of Boolean variables. The first Boolean identity is that the sum of anything and zero is the same as the original "anything." This identity is no different from its real-number algebraic equivalent:

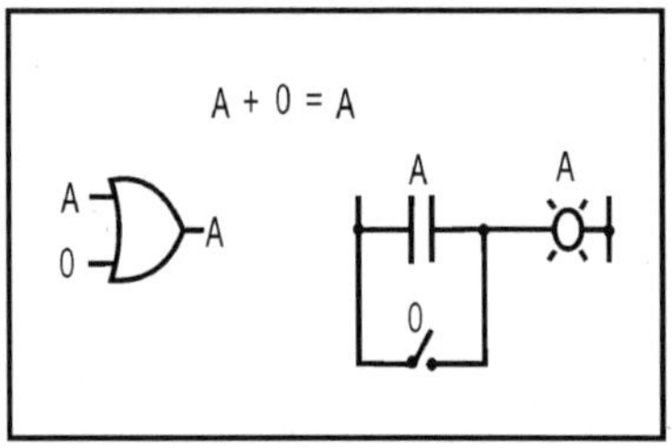

No matter what the value of A, the output will always be the same: when A=1, the output will also be 1; when A=0, the output will also be 0.

The next identity is most definitely *different* from any seen in normal algebra. Here we discover that the sum of anything and one is one:

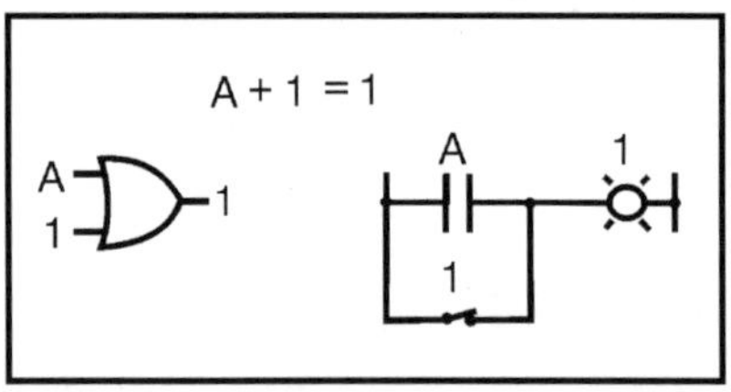

No matter what the value of A, the sum of A and 1 will always be 1. In a sense, the "1" signal *overrides*the effect of A on the logic circuit, leaving the output fixed at a logic level of 1.Next, we examine the effect of adding A and A together, which is the same as connecting both inputs of an OR gate to each other and activating them with the same signal:

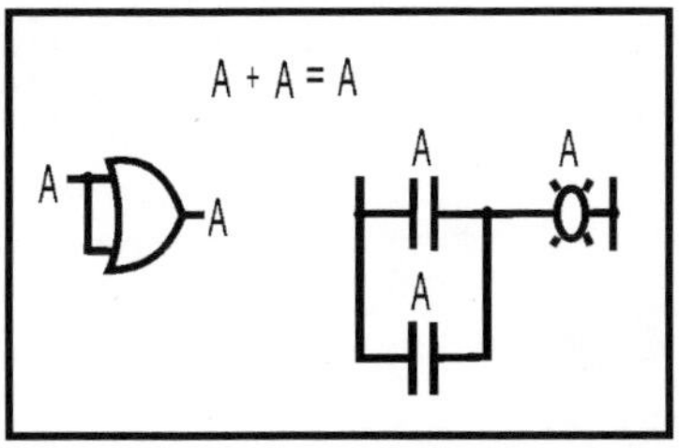

In real-number algebra, the sum of two identical variables is twice the original variable's value (x + x = 2x), but remember that there is no concept of "2" in the world of Boolean math, only 1 and 0, so we cannot say that A + A = 2A. Thus, when we add a Boolean quantity to itself, the sum is equal to the original quantity: 0 + 0 = 0, and 1 + 1 = 1.

Introducing the uniquely Boolean concept of complementation into an additive identity, we find an interesting effect. Since there must be one "1" value between any variable and its complement, and since the sum of any Boolean quantity and 1 is 1, the sum of a variable and its complement must be 1:

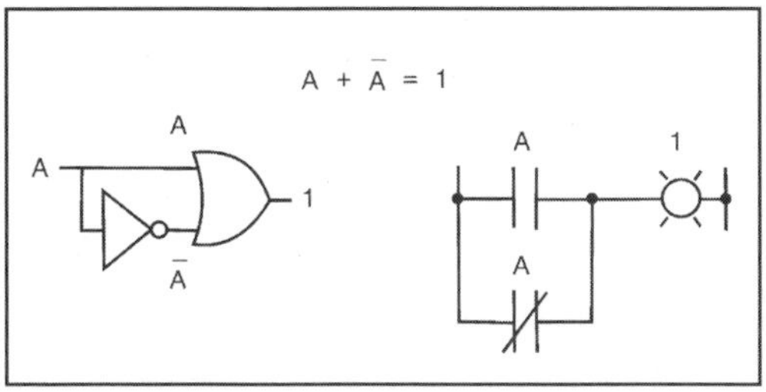

Just as there are four Boolean additive identities (A+0, A+1, A+A, and A+A'), so there are also four multiplicative identities: A × 0, A × 1, A × A, and A × A'. Of these, the first two are no different from their equivalent expressions in regular algebra:

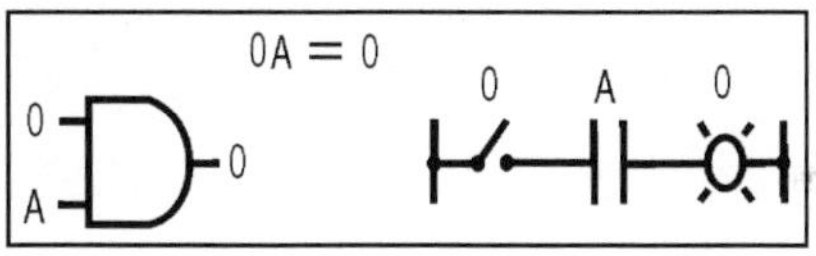

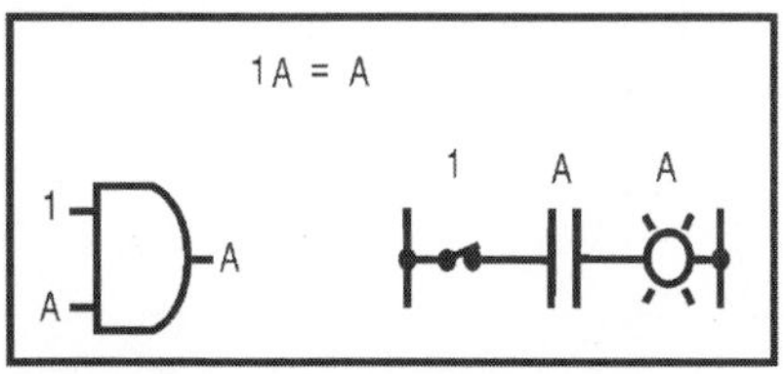

The third multiplicative identity expresses the result of a Boolean quantity multiplied by itself. In normal algebra, the product of a variable and itself is the *square* of that variable ($3 \times 3 = 3^2 = 9$). However, the concept of "square" implies a quantity of 2, which has no meaning in Boolean algebra, so we cannot say that A × A = A^2. Instead, we find that the product of a Boolean quantity and itself is the original quantity, since 0 × 0 = 0 and 1 × 1 = 1:

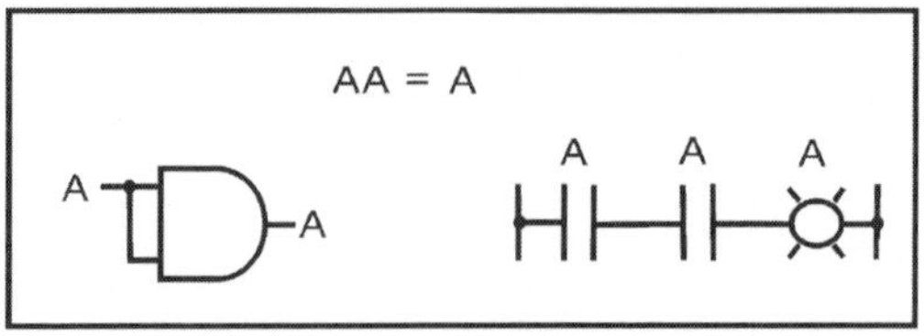

The fourth multiplicative identity has no equivalent in regular algebra because it uses the complement of a variable, a concept unique to Boolean mathematics. Since there must be one "0" value between any variable and its complement, and since the product of any Boolean quantity and 0 is 0, the product of a variable and its complement must be 0:

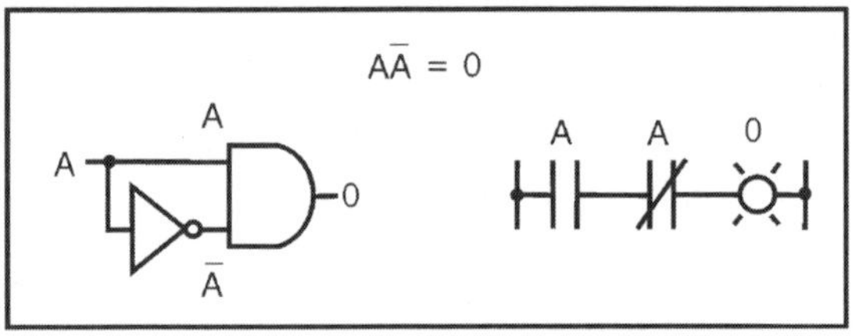

To summarize, then, we have four basic Boolean identities for addition and four for multiplication:

Table. Basic Boolean Algebraic Identities

Additive	Multiplicative
$A + 0 = A$	$0A = 0$
$A + 1 = 1$	$1A = A$
$A + A = A$	$AA = A$
$A + \overline{A} = 1$	$A\overline{A} = 0$

Another identity having to do with complementation is that of the *double complement*: a variable inverted twice.

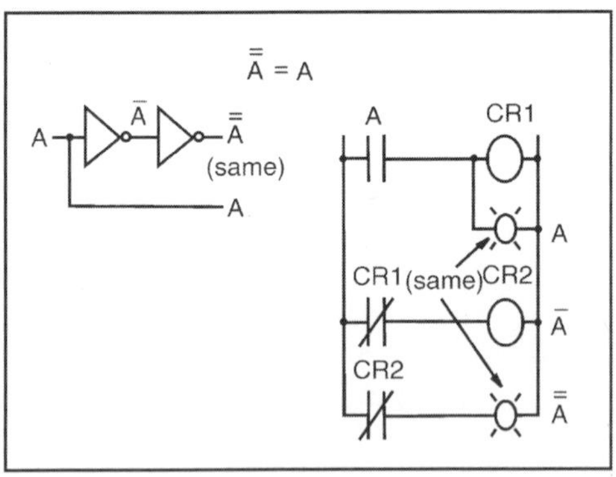

Complementing a variable twice (or any even number of times) results in the original Boolean value. This is analogous to negating (multiplying by -1) in real-number algebra: an even number of negations cancel to leave the original value:

BOOLEAN ALGEBRA PROPERTIES

Another type of mathematical identity, called a "property" or a "law," describes how differing variables relate to each other in a system of numbers. One of these properties is known as the *commutative property*, and it applies equally to addition and multiplication. In essence, the commutative property tells us we can reverse the order of variables that are either added together or multiplied together without changing the truth of the expression:

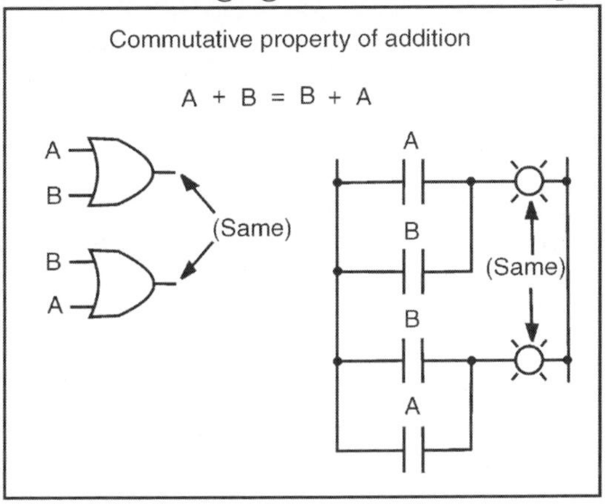

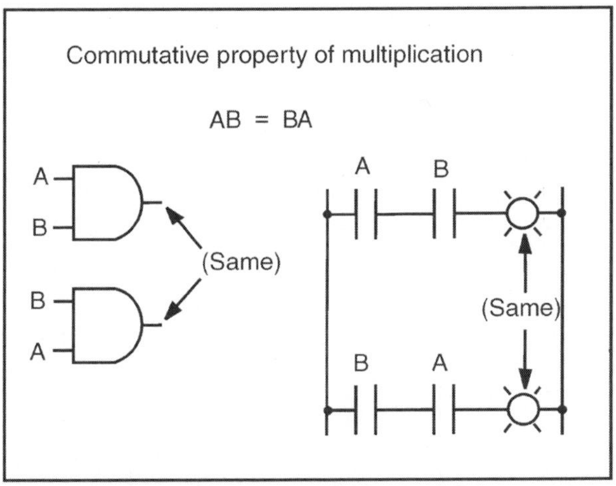

Along with the commutative properties of addition and multiplication, we have the *associative property*, again applying equally well to addition and multiplication. This property tells us we can associate groups of added or multiplied variables together with parentheses without altering the truth of the equations.

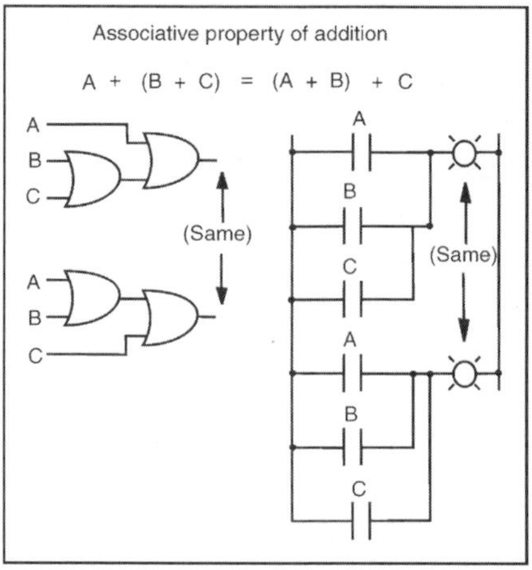

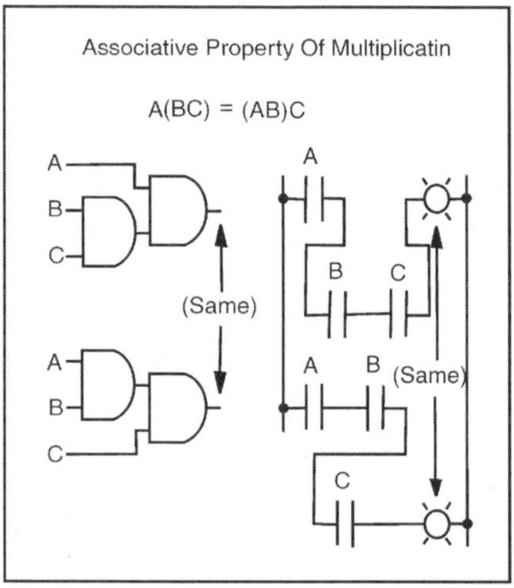

Lastly, we have the *distributive property*, illustrating how to expand a Boolean expression formed by the product of a sum, and in reverse shows us how terms may be factored out of Boolean sums-of-products:

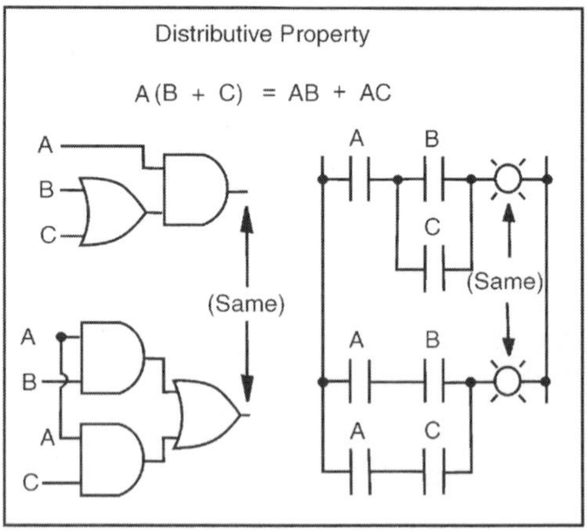

To summarize, here are the three basic properties: commutative, associative, and distributive.

Table. Basic Bookean Algebraic Properties

Additive	Multiplicative
A + B = B + A	AB = BA
A + (B + C) = (A + B) + C	A(BC) = (AB) C
A (B + C) = AB + AC	

SIMPLIFICATION RULES

Boolean algebra finds its most practical use in the simplification of logic circuits. If we translate a logic circuit's function into symbolic (Boolean) form, and apply certain algebraic rules to the resulting equation to reduce the number of terms and/or arithmetic operations, the simplified equation may be translated back into circuit form for a logic circuit performing the same function with fewer components.

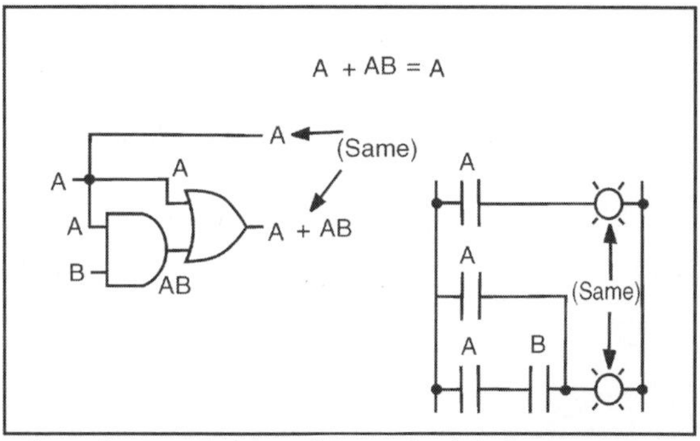

If equivalent function may be achieved with fewer components, the result will be increased reliability and decreased cost of manufacture.To this end, there are several rules of Boolean algebra presented in this section for use in reducing expressions to their simplest forms.

The identities and properties already reviewed in this chapter are very useful in Boolean simplification, and for the most part bear similarity to many identities and properties of "normal" algebra.

A + AB

↓ Factoring A Out of Both Terms

A (1 + B)

↓ Applying identity A + 1 = 1

A (1)

↓ Applying identity 1A = A

A

However, the rules shown in this section are all unique to Boolean mathematics. This rule may be proven symbolically by factoring an "A" out of the two terms, then applying the rules of A + 1 = 1 and 1A = A to achieve the final result:

Please note how the rule A + 1 = 1 was used to reduce the (B + 1) term to 1. When a rule like "A + 1 = 1" is expressed using the letter "A", it doesn't mean it only applies to expressions containing "A". What the "A" stands for in a rule like A + 1 = 1 is *any* Boolean variable or collection of variables. This is perhaps the most difficult concept for new students to master in Boolean simplification: applying standardized identities, properties, and rules to expressions not in standard form.

For instance, the Boolean expression ABC + 1 also reduces to 1 by means of the "A + 1 = 1" identity. In this case, we recognize that the "A" term in the identity's standard form can represent the entire "ABC" term in the original expression. The next rule looks similar to the first one shown in this section, but is actually quite different and requires a more clever proof:

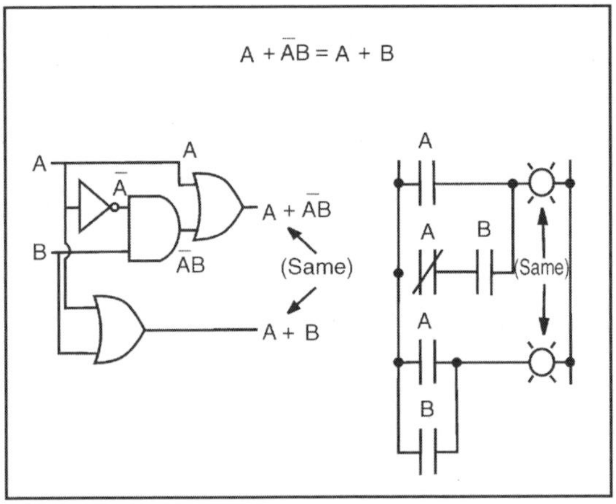

Note how the last rule (A + AB = A) is used to "un-simplify" the first "A" term in the expression, changing the "A" into an "A + AB".

$A + \bar{A}B$

↓ Applying the Previous Rule to Expand A Term $A + AB = A$

$A + AB + \bar{A}B$

↓ Factoring B Out of 2nd and 3rd Terms

$A + B(A + \bar{A})$

↓ Applying Identity $A + \bar{A} = 1$

$A + B(1)$

↓ Applying Identity $1A = A$

$A + B$

While this may seem like a backward step, it certainly helped to reduce the expression to something simpler. Sometimes in mathematics we must take "backward" steps to achieve the most elegant solution. Knowing when to take such a step and when not to is part of the art-form of algebra, just as a victory in a game of chess almost always requires calculated sacrifices.

Another rule involves the simplification of a product-of-sums expression:

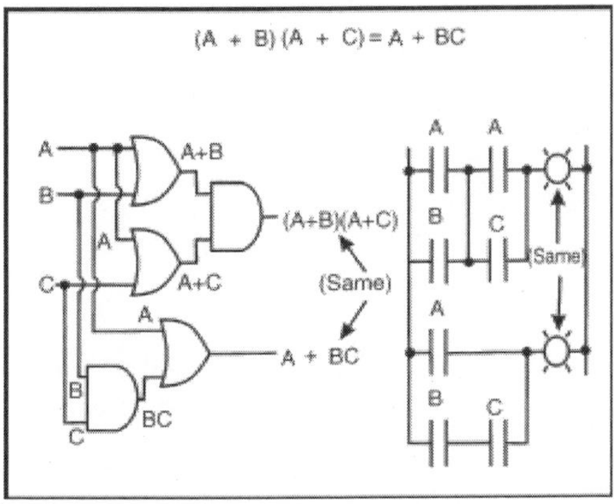

And, the corresponding proof:

(B + C) (A + C)

↓ Distributing Terms

AA + AC + AB + BC

↓ Applying Identity AA = A

A + AC + AB + BC

↓ Applying Rule A + AB = A to the A + AC term

A + AB + BC

↓ Applying Rule A + AB = A to the A + AB term

A + BC

Useful Boolean Rules:

$$A + AB = A$$

$$A + \overline{A}B = A + B$$

$$(A + B)(A + C) = A + BC$$

CIRCUIT SIMPLIFICATION

Let's begin with a semiconductor gate circuit in need of simplification. The "A," "B," and "C" input signals are assumed to be provided from switches, sensors, or perhaps other gate circuits. Where these signals originate is of no concern in the task of gate reduction.

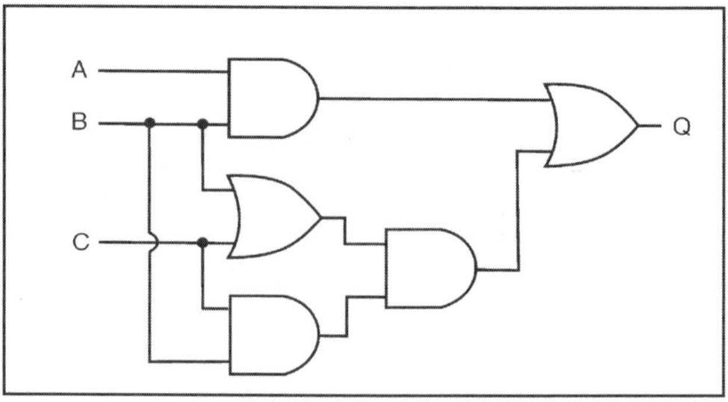

Our first step in simplification must be to write a Boolean expression for this circuit.

This task is easily performed step by step if we start by writing sub-expressions at the output of each gate, corresponding to the respective input signals for each gate. Remember that OR gates are equivalent to Boolean addition, while AND gates are equivalent to Boolean multiplication.

For example, I'll write sub-expressions at the outputs of the first three gates:

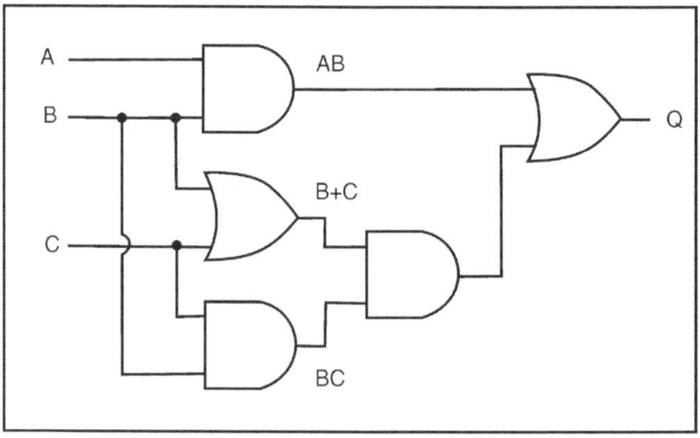

... then another sub-expression for the next gate:

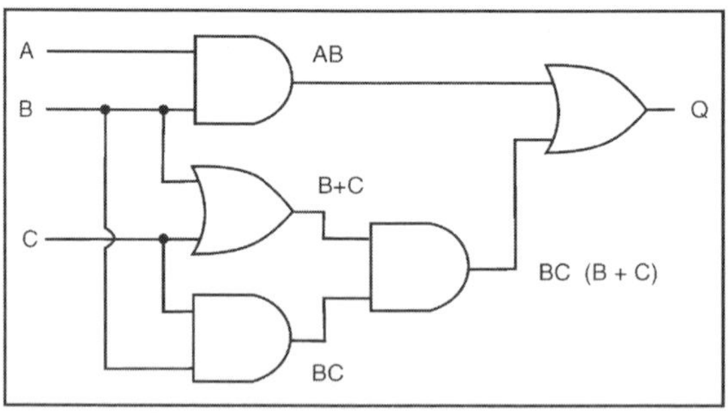

Finally, the output ("Q") is seen to be equal to the expression AB + BC(B + C):

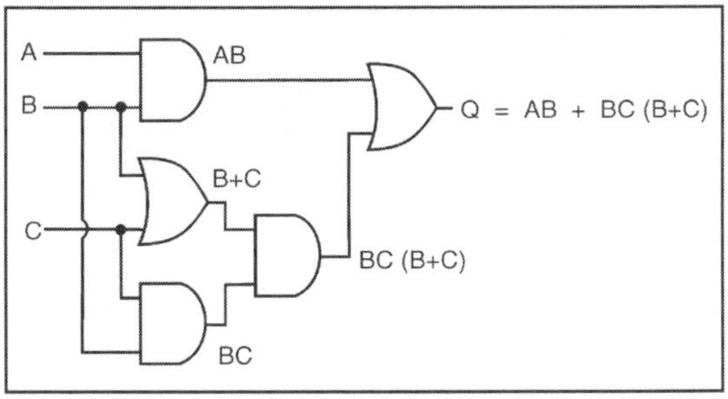

Now that we have a Boolean expression to work with, we need to apply the rules of Boolean algebra to reduce the expression to its simplest form (simplest defined as requiring the fewest gates to implement):

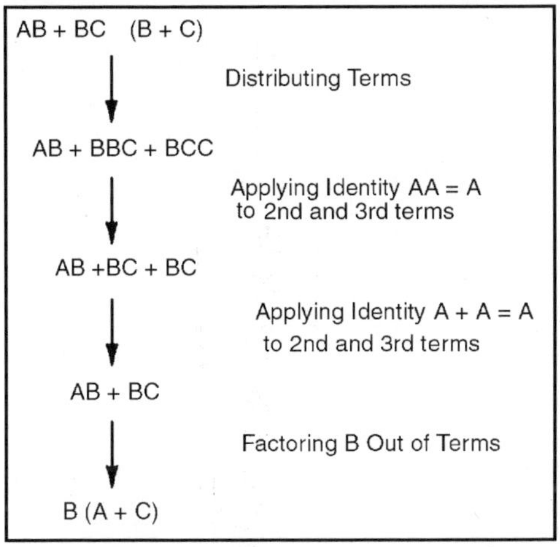

The final expression, B(A + C), is much simpler than the original, yet performs the same function. If you would like to verify this, you may generate a truth table for both expressions and determine Q's status (the circuits' output) for all eight logic-state combinations of A, B, and C, for both circuits. The two truth tables should be identical.

Now, we must generate a schematic diagram from this Boolean expression. To do this, evaluate the expression, following proper mathematical order of operations (multiplication before addition, operations inside parentheses before anything else), and draw gates for each step.

Remember again that OR gates are equivalent to Boolean addition, while AND gates are equivalent to Boolean multiplication. In this case, we would begin with the sub-expression "A + C", which is an OR gate:

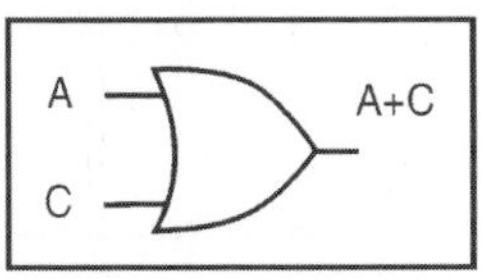

The next step in evaluating the expression "B(A + C)" is to multiply (AND gate) the signal B by the output of the previous gate (A + C):

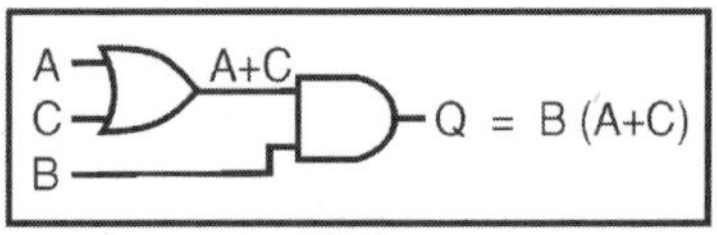

Obviously, this circuit is much simpler than the original, having only two logic gates instead of five. Such component reduction results in higher operating speed (less delay time from input signal transition to output signal transition), less power consumption, less cost, and greater reliability. Electromechanical relay circuits, typically being slower, consuming more electrical power to operate, costing more, and having a shorter average life than their semiconductor counterparts, benefit dramatically from Boolean simplification. Let's consider an example circuit:

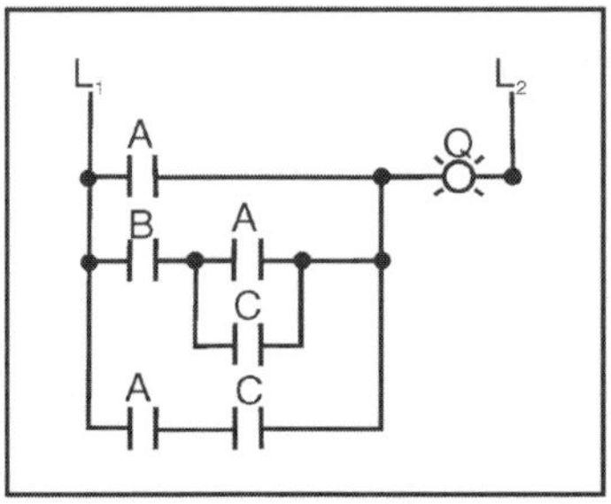

As before, our first step in reducing this circuit to its simplest form must be to develop a Boolean expression from the schematic. For example, examine the following resistor network with its resistors arranged in the same connection pattern as the relay contacts in the former circuit, and corresponding total resistance formula:

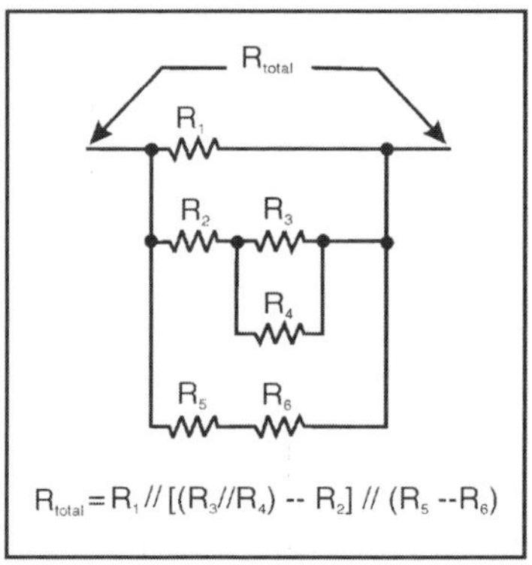

Remember that parallel contacts are equivalent to Boolean addition, while series contacts are equivalent to Boolean multiplication. Write a Boolean expression for this relay contact circuit, following the same order of precedence that you would follow in reducing a series-parallel resistor network to a total resistance. It may be helpful to write a Boolean sub-expression to the left of each ladder "rung," to help organize your expression-writing:

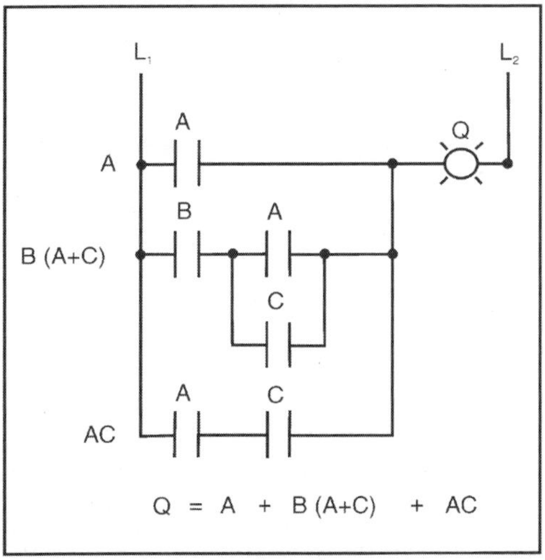

Now that we have a Boolean expression to work with, we need to apply the rules of Boolean algebra to reduce the expression to its simplest form (simplest defined as requiring the fewest relay contacts to implement):

A + B (A + C) + AC

↓ Distributing Terms

A + AB + BC + AC

↓ Applying Rule A + AB = A to 1st and 2nd terms

A + BC + AC

↓ Applying Rule A + AB = A to 1st and 3rd terms

A + BC

The more mathematically inclined should be able to see that the two steps employing the rule "A + AB = A" may be combined into a single step, the rule being expandable to:

"A + AB + AC + AD +... = A"

A + B (A + C) + AC

↓ Distributing Terms

A + AB + BC + AC

↓ Applying (Expanded) Rule A + AB = A to 1st, 2nd, and 4th terms

A + BC

As you can see, the reduced circuit is much simpler than the original, yet performs the same logical function:

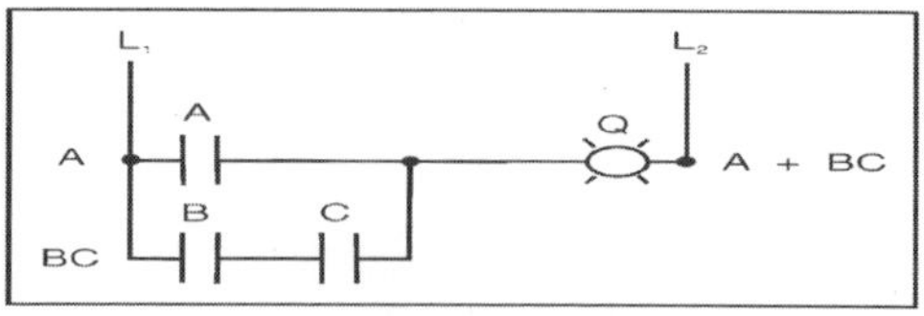

Review: To convert a gate circuit to a Boolean expression, label each gate output with a Boolean sub-expression corresponding to the gates' input signals, until a final expression is reached at the last gate.

To convert a Boolean expression to a gate circuit, evaluate the expression using standard order of operations: multiplication before addition, and operations within parentheses before anything else. To convert a ladder logic circuit to a Boolean expression, label each rung with a Boolean sub-expression corresponding to the contacts' input signals, until a final expression is reached at the last coil or light.

To determine proper order of evaluation, treat the contacts as though they were resistors, and as if you were determining total resistance of the series-parallel network formed by them.

In other words, look for contacts that are either *directly* in series or *directly* in parallel with each other first, then "collapse" them into equivalent Boolean sub-expressions before proceeding to other contacts. To convert a Boolean expression to a ladder logic circuit, evaluate the expression using standard order of operations: multiplication before addition, and operations within parentheses before anything else.

LOGIC CIRCUITS

Any logic circuit, no matter how complex, may be completely described using the Boolean operations, because the OR gate, AND gate, and NOT circuit are the basic building blocks of digital systems.

This is an example of the circuit using Boolean expression:

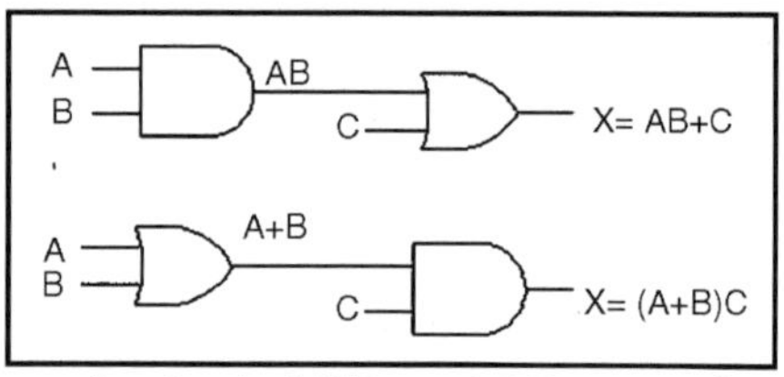

If an expression contains both AND and OR operations, the AND operations are performed first (X=AB+C: AB is performed first), unless there are parentheses in the expression, in which case the operation inside the parentheses is to be performed first (X=(A+B)+C: A+B is performed first).

CIRCUITS CONTAINING INVERTERS

Whenever an INVERTER is present in a logic-circuit diagram, its output expression is simply equal to the input expression with a prime (') over it.

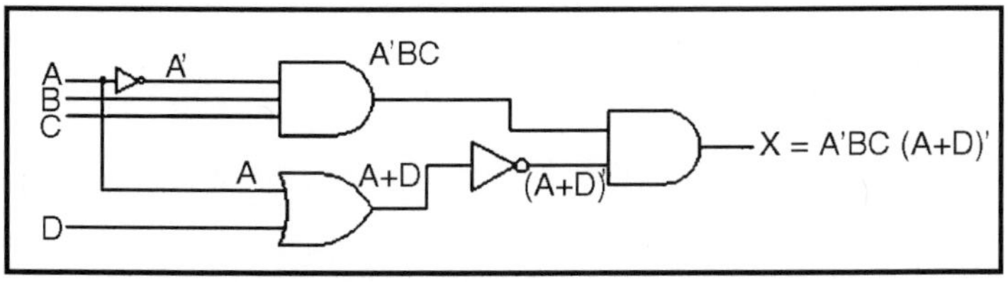

Evaluating Logic Circuit Outputs

Once the Boolean expression for a circuit output has been obtained, the output logic level can be determined for any set of input levels.

This are two examples of the evaluating logic circuit output:

Let A=0, B=1, C=1, D=1

X = A'BC (A+D)'
 = 0'*1*1* (0+1)'
 = 1 *1*1* (1)'
 = 1 *1*1* 0
 = 0

Let A=0, B=0, C=1, D=1, E=1

X = [D+ ((A+B)C)'] * E
 = [1 + ((0+0)1)'] * 1
 = [1 + (0*1)'] * 1
 = [1+ 0'] *1
 = [1+ 1] * 1
 = 1

In general, the following rules must always be followed when evaluating a Boolean expression:

- First, perform all inversions of single terms; that is, 0 = 1 or 1 = 0.
- Then perform all operations within parentheses.
- Perform an AND operation before an OR operation unless parentheses indicate otherwise.
- If an expression has a bar over it, perform the operations of the expression first and then invert the result.

Determining Output Level

The output logic level for given input levels can also be determined directly from the circuit diagram without using the Boolean expression.

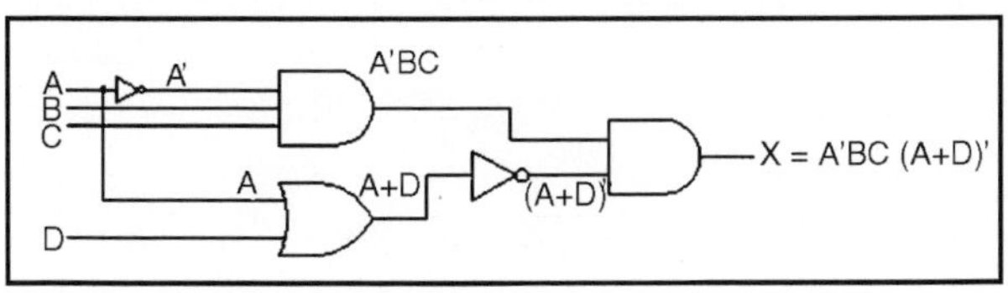

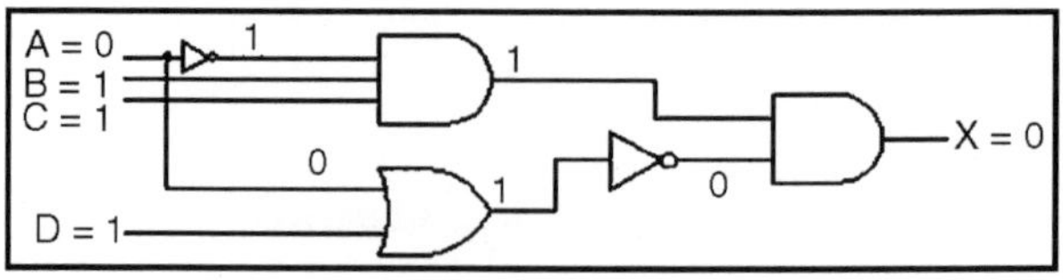

Implementing Circuits From Boolean Expression

If the operation of a circuit is defined by a Boolean expression, a logic-circuit diagram can he implemented directly from that expression. Suppose that we wanted to construct a circuit whose output is y = AC+BC' + A'BC. This Boolean expression contains three terms (AC, BC', A'BC), which are ORed together. This tells us that a three-input OR gate is required with inputs that are equal to AC, BC', and A'BC, respectively.

Each OR-gate input is an AND product term, which means that an AND gate with appropriate inputs can be used to generate each of these terms. Note the use of INVERTERs to produce the A' and C' terms required in the expression.

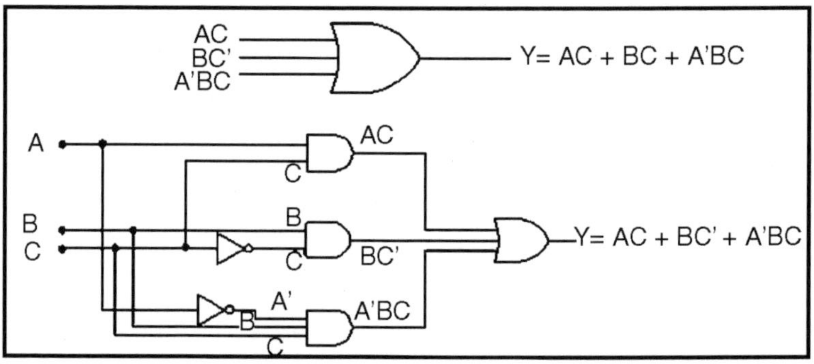

BOOLEAN ARITHMETIC

In computer science, the Boolean or logical data type is a primitive data type having one of two values: true or false, intended to represent the truth values of logic and Boolean algebra. In programming languages that have a built-in Boolean data type, such as Pascal and Java, the comparison operators such as '>' and '?' are usually defined to return a Boolean value.

Also, conditional and iterative commands may be defined to test Boolean-valued expressions. Languages without an explicit Boolean data type, like C and Lisp, may still represent truth values it by some other data type. Lisp uses an empty list for false, and any other value for true. C uses an integer type, with false represented as the zero value, and true as any non-zero value (such as 1 or –1). Indeed, a Boolean variable may be regarded (and be implemented) as a numerical variable with a single binary digit (bit), which can store only two values.

Let us begin our exploration of Boolean algebra by adding numbers together:

$0 + 0 = 0$

$0 + 1 = 1$

$1 + 0 = 1$

$1 + 1 = 1$

The first three sums make perfect sense to anyone familiar with elementary addition.

The last sum, though, is quite possibly responsible for more confusion than any other single statement in digital electronics, because it seems to run contrary to the basic principles of mathematics. Well, it *does* contradict principles of addition for real numbers, but not of elimination. It does not matter how many or few terms we add together, either.

Consider the following sums:

$0 + 1 + 1 = 1$

$1 + 1 + 1 = 1$

$0 + 1 + 1 + 1 = 1$

$1 + 0 + 1 + 1 + 1 = 1$

Take a close look at the two-term sums in the first set of equations. Does that pattern look familiar to you? It should! It is the same pattern of 1's and 0's as seen in the truth table for an OR gate. In other words, Boolean addition corresponds to the logical function of an "OR" gate, as well as to parallel switch contacts:

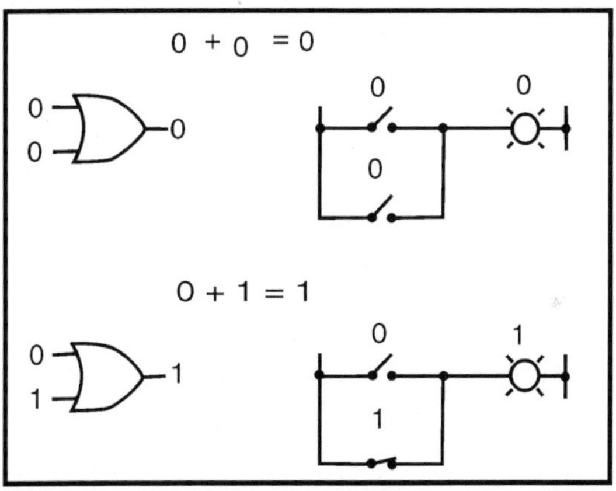

There is no such thing as subtraction in the realm of Boolean mathematics. Subtraction implies the existence of negative numbers: 5–3 is the same thing as 5 + (–3), and in Boolean algebra negative quantities are forbidden. There is no such thing as division in Boolean mathematics, either, since division is really nothing more than compounded subtraction, in the same way that multiplication is compounded addition.

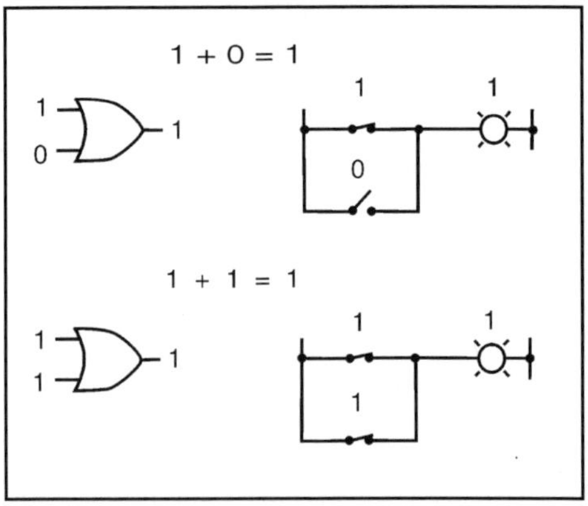

Multiplication is valid in Boolean algebra, and thankfully it is the same as in real-number algebra: anything multiplied by 0 is 0, and anything multiplied by 1 remains unchanged:

$0 \times 0 = 0$

$0 \times 1 = 0$

$1 \times 0 = 0$

$1 \times 1 = 1$

This set of equations should also look familiar to you: it is the same pattern found in the truth table for an AND gate. In other words, Boolean multiplication corresponds to the logical function of an "AND" gate, as well as to series switch contacts:

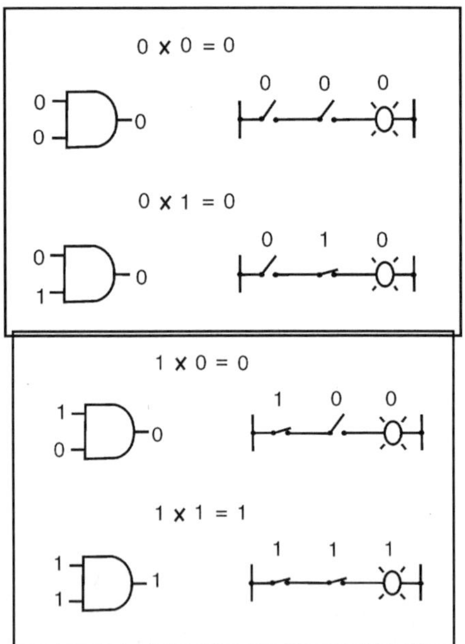

Like "normal" algebra, Boolean algebra uses alphabetical letters to denote variables. Unlike "normal" algebra, though, Boolean variables are always

CAPITAL letters, never lower-case. Because they are allowed to possess only one of two possible values, either 1 or 0, each and every variable has a*complement*: the opposite of its value.

For example, if variable "A" has a value of 0, then the complement of A has a value of 1. Boolean notation uses a bar above the variable character to denote complementation, like this:

If: $A = 0$

Then: $\overline{A} = 1$

If: $A = 1$

Then: $\overline{A} = 1$

In written form, the complement of "A" denoted as "A-not" or "A-bar". Sometimes a "prime" symbol is used to represent complementation. For example, A' would be the complement of A, much the same as using a prime symbol to denote differentiation in calculus rather than the fractional notation d/dt.

Usually, though, the "bar" symbol finds more widespread use than the "prime" symbol, for reasons that will become more apparent later in this chapter. Boolean complementation finds equivalency in the form of the NOT gate, or a normally-closed switch or relay contact:

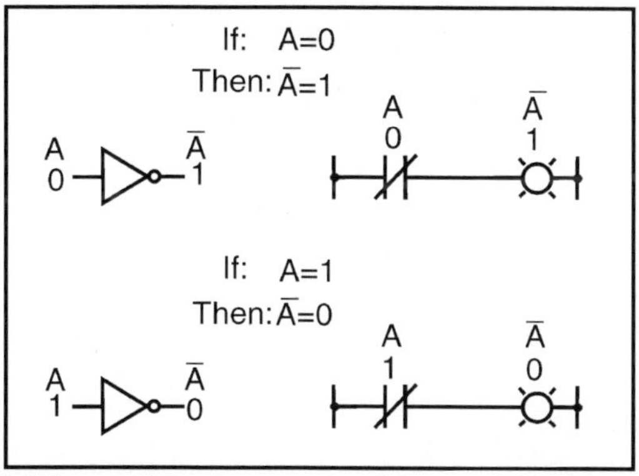

The basic definition of Boolean quantities has led to the simple rules of addition and multiplication, and has excluded both subtraction and division as valid arithmetic operations. We have a symbology for denoting Boolean variables, and their complements. In the next section we will proceed to develop Boolean identities.

Review:

- Boolean addition is equivalent to the *OR* logic function, as well as *parallel* switch contacts.
- Boolean multiplication is equivalent to the *AND* logic function, as well as *series* switch contacts.
- Boolean complementation is equivalent to the *NOT* logic function, as well as *normally-closed*relay contacts.

BINARY NUMBERS SYSTEM

Binary representations of positive can be understood in the same way as their decimal counterparts. For example,

$$86_{10} = 1*64 + 0*32 + 1*16 + 0*8 + 1*4 + 1*2 + 0*1$$

or

$$86_{10} = 1*2^6 + 0*2^5 + 1*2^4 + 0*2^3 + 1*2^2 + 1*2^1 + 0*2^0$$

or

$$86_{10} = 1010110_2$$

The subscript 2 denotes a binary number. Each digit in a binary number is called a bit. The number 1010110 is represented by 7 bits. Any number can be broken down this way, by finding all of the powers of 2 that add up to the number in question (in this case 2^6, 2^4, 2^2 and 2^1). You can see this is exactly analagous to the decimal deconstruction of the number 125 that was done earlier.

Likewise we can make a similar set of observations:

- To multiply a number by 2 you can simply shift it to the left by one digit, and fill in the rightmost digit with a 0. To divide a number by 2, simply shift the number to the right by one digit.
- To see how many digits a number needs, you can simply take the logarithm (base 2) of the number, and add 1 to it. The integer part of the result is the number of digits. For instance,
  $\log_2(86)+1=7.426$.
  The integer part of that is 7, so 7 digits are needed.
- With n digits, 2^n unique numbers (from 0 to 2^n-1) can be represen0ted. If n=8, 256 (=2^8) numbers can be represented 0-255.

It is often convenient to handle groups of bits, rather than individually. The most common grouping is 8 bits, which forms a byte. A single byte can represent 256 (2^8) numbers. Memory capacity is usually referred to in bytes. Two bytes is usually called a word, or short word (though word-length depends on the application).

A two-byte word is also the size that is usually used to represent integers in programming languages. A long word is usually twice as long as a word. A less common unit is the nibble which is 4 bits, or half of a byte. It is cumbersome for humans to deal with writing, reading and remembering individual bits, because it takes many of them to represent even fairly small numbers.

A number of different ways have been developed to make the handling of binary data easier for us. The most common is hexadecimal. In hexadecimal notation, 4 bits (a nibble) are represented by a single digit. There is obviously a problem with this since 4 bits gives 16 possible combinations, and there are only 10 unique decimal digits, 0 to 9. This is solved by using the first 6 letters (A–F) of the alphabet as numbers. The table shows the relationship between decimal, hexadecimal and binary.

Decimal	Hexadecimal	Binary
0	0	0000
1	1	0001
2	2	0010
3	3	0011
4	4	0100
5	5	0101
6	6	0110
7	7	0111
8	8	1000
9	9	1001
10	A	1010
11	B	1011
12	C	1100
13	D	1101
14	E	1110
15	F	1111

There are some significant advantages to using hexadecimal when dealing with electronic representations of numbers (if people had 16 fingers, we wouldn't be saddled with the awkward decimal system). Using hexadecimal makes it very easy to convert back and forth from binary because each hexadecimal digit corresponds to exactly 4 bits ($\log_2(16) = 4$) and each byte is two hexadecimal digit. In contrast, a decimal digit corresponds to $\log_2(10) = 3.322$ bits and a byte is 2.408 decimal digits. Clearly hexadecimal is better suited to the task of representing binary numbers than is decimal.

As an example, the number $CA3_{16} = 1100\ 1010\ 0011_2$ ($1100_2 = C_{16}$, $1010_2 = A_{16}$, $0011_2 = 3_{16}$). It is convenient to write the binary number with spaces after every fourth bit to make it easier to read. Converting back and forth to decimal is more difficult, but can be done in the same way as before.

$$3235_{10} = C_{16}*256 + A_{16}*16 + 3_{16}*1 = C_{16}*16^2 + A_{16}*16^1 + 3_{16}*16^0$$

or

$$3235_{10} = 12*256 + 10*16 + 3*1 = 12*16^2 + 10*16^1 + 3*16^0$$

Octal notation is yet another compact method for writing binary numbers. There are 8 octal characters, 0–7. Obviously this can be represented by exactly 3 bits. Two octal digits can represent numbers up to 64, and three octal digits up to 512. A byte requires 2.667 octal digits. Octal used to be quiete common, it was the primary way of doing low level I/O on some old DEC computers. It is much less common today but is still used occasionally (*e.g.,* to set read, write and execute permissions on Unix systems)

In summary:

- *Bit*: A single binary digit, either zero or one.
- *Byte*: 8 bits, can represent positive numbers from 0 to 255.
- *Hexadecimal*: A representation of 4 bits by a single digit 0..9,A..F. In this way a byte can be represented by two hexadecimal digits

- *Long word*: A long word is usually twice as long as a word.
- *Nibble*: 4 bits, half of a byte.
- *Octal*: A representation of 3 bits by a single digit 0..7. This is used much less commonly than it once was (early DEC computers used octal for much of their I/O)
- *Word*: Usually 16 bits, or two bytes. But a word can be almost any size, depending on the application being considered—32 and 64 bits are common sizes.

BASIC RULES OF BINARY SYSTEM

To understand binary numbers, begin by recalling elementary school math. When we first learned about numbers, we were taught that, in the decimal system, things are organized into columns:

$$\text{H}|\text{T}|\text{O}$$
$$1|9|3$$

Such that "H" is the hundreds column, "T" is the tens column, and "O" is the ones column. So the number "193" is 1-hundreds plus 9-tens plus 3-ones. later, we learned that the ones column meant 10^0, the tens column meant 10^1, the hundreds column 10^2 and so on, such that

$$10^2 \,|\, 10^1 \,|\, 10^0$$
$$1 \,|\, 9 \,|\, 3$$

The number 193 is really $\{(1*10^2)+(9*10^1)+(3*10^0)\}$.

As you know, the decimal system uses the digits 0—9 to represent numbers. If we wanted to put a larger number in column 10^n (*e.g.*, 10), we would have to multiply $10*10^n$, which would give $10^{(n+1)}$, and be carried a column to the left.

For example, putting ten in the 10^0 column is impossible, so we put a 1 in the 10^1 column, and a 0 in the 10^0 column, thus using two columns.

Twelve would be $12*10^0$, or $10^0(10+2)$, or $10^1 + 2*10^0$, which also uses an additional column to the left (12).The binary system works under the exact same principles as the decimal system, only it operates in base 2 rather than base 10. In other words, instead of columns being they are

$$10^2 \,|\, 10^1 \,|\, 10^0$$
$$2^2 \,|\, 2^1 \,|\, 2^0$$

Instead of using the digits 0-9, we only use 0-1 (again, if we used anything larger it would be like multiplying $2*2^n$ and getting 2^{n+1}, which would not fit in the 2 ^ n column.

Therefore, it would shift you one column to the left. For example, "3" in binary cannot be put into one column.

The first column we fill is the right-most column, which is 2^0, or 1. Since 3>1, we need to use an extra column to the left, and indicate it as "11" in binary $(1*2^1) + (1*2^\wedge 0)$.

SIGNED BINARY NUMBERS

So far we have considered how computers can represent integers as unsigned quantities. It is more realistic and practical to represent them as signed integers having both positive and negative values. Let's consider how this can be done.

First, it should be observed that we will still be restricted to a uniform (finite) precision for representing signed integers. Thus, the same number of bits will now have to serve for for both negative and positive values. A reasonable approach would be to assign half of the binary values to negative decimal values and half to positive decimal values. For example, suppose that we had a precision of only four bits. Figure 2 shows how these might be assigned in an orderly fashion.

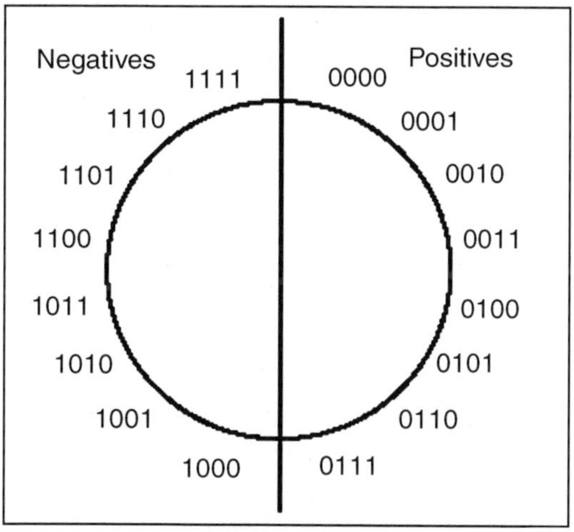

Fig. Dividing 4-bit Precision Binary into Positive and Negative Segments.

The binary values 0-7 could represent positive decimal values and the binary values 8-15 could signify negative decimal values. This divides the range into 8 values for positives and 8 values for negatives.

Notice that the binary values 0–7 representing positives have the most significant bit (MSB) of 0. The binary values 8–15 representing negatives have a MSB of 1. Thus, we could use the MSB to denote the sign and the remaining bits to signify the quantity or magnitude as shown in Figure. This scheme is appropriately called sign and magnitude.

As you can see in Figure 3, each positive has a corresponding negative value that differs only in the sign bit, *i.e.*, the MSB. With a 4-bit precision, our range is small, +7... –7. But, even if we extend the precision to 16 bits or 32 bits, there are other problems that remain. First, notice that there are codes for both +0 and –0; this is not convenient if, for example, we should have to decide whether a number is 0 or not. A more serious problem arises when doing arithmetic.

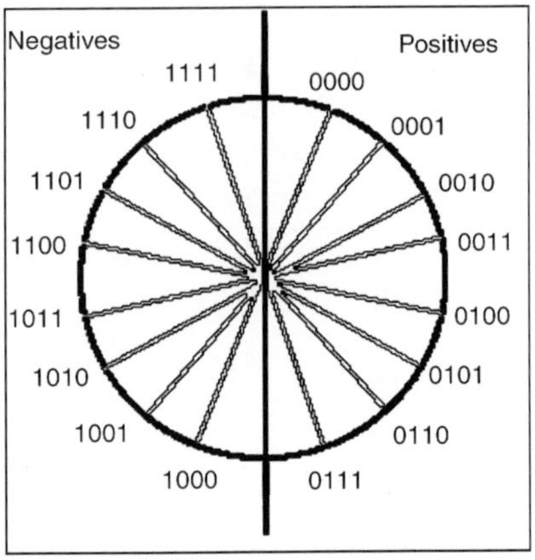

Fig. The Most Significant Bit Denotes the Sign and the Last Three Bits Signify the Magnitude.

Recall that any binary integer can be represented by the sequence

+4 0100

+ -3 +1011

+1 1111

= –7

To do the arithmetic correctly, we have to consider the signs and the relative magnitudes of both numbers. It is much too complicated and therefore potentially expensive to implement as computer circuits.

There is another approach: using Ns-complement. Blaise Pascal, for example, employed this technique for his decimal calculating machine later known as the Pascaline.

Suppose that we had a decimal precision of 4 digits. The range of values would be from 0000 to 9999. In 9s-complement (*i.e.,* N = 9), a negative value is determined by taking its additive inverse. This is the result of subtracting the magnitude from 9999. Here are some examples.

–11 9999

–0011

= 9988

–53 9999

–0053

= 9946

Simple addition operations can be used for both addition and subtraction. For example, adding a positive and negative number works provided that the carry-out value is added to adjust the sum.

+373 0373
+ −53 + 9946
+320 1 0319
(add carry) + 1
0320

Subtraction can also be solved using addition provided that the term subtracted is complemented and then added to the other term. The example illustrates.

2430 2430
−11 + 9988
2419 1 2418
(add carry) + 1
2419

Twos Complement

Most computer systems employ a variation of this method for representing signed integers called twos-complement. The procedures are similar.

Negative values are represented by:

1. Finding the complement (bit-value inverse) of the unsigned quantity.
2. The result is adjusted by adding +1.

Examples

Decimal Value	(Binary) Magnitude	(Binary) Inverse	Adjusted Result
-4	0100	1011	1100
		+1	
		1100	

Decimal Value	(Binary) Magnitude	(Binary) Inverse	Adjusted Result
-7	0111	1000	1001
		+1	
		1001	

One advantage of this scheme is that the MSB still serves to signify the sign of the number. In general, values for which the MSB = 1 are negative; values for which the MSB = 0 are positive. Addition and subtraction can also be handled by performing simple additions.

Examples

+7 0111
+ −4 +1100
+3 1 0011
(ignore carry out) = +3
+3 011
- -4(convert to comp) +0100
+7 0111
= +7

But, in some instances, we can produce a value that it too large for our precision. This causes an overflow error condition. The results are incorrect. For instance,

+7	0111
– –4 (convert to comp) +0100	
+11	1011
(too big)	= –5

In short, adding two numbers with the same sign (MSB) results in overflow when the sign (MSB) changes.

INTEGER

If we are given an *m*-bit computer word B with bits b_{m-1}, b_{m-2},..., b_0, we can interpret it as an unsigned number with value,

$$\text{unsigned(B)} = b_{m-1}*2^{m-1} + b_{m-2}*2^{m-2} + ... + b_0*2^0.$$

This gives us a representation for numbers from 0 to 2^m - 1.

If we want to represent negative values as well, we can just change the sign of the high-order bit b_{m-1}, giving a signed number with value

$$\text{signed(B)} = -b_{m-1}*2^{m-1} + b_{m-2}*2^{m-2} + ... + b_0*2^0.$$

This is called the *m*-bit *2's complement* interpretation of B.

With $b_{m-1} = 0$, we get representations for numbers from 0 to $2^{m-1} - 1$. With $b_{m-1} = 1$, we get representations for numbers from -2^{m-1} to $-2^{m-1} + (2^{m-1} - 1) = -1$. Thus we have several important facts about the *m*-bit 2's complement representation.

- It represents numbers from -2^{m-1} to $2^{m-1} - 1$.
- If b_{m-1} is 1 then the value is negative; otherwise it is non-negative. Thus the high-order bit, b_{m-1} is called the *sign bit*.
- When the sign bit of B is 0, signed(B) = unsigned(B).
- When the sign bit of B is 1, signed(B) = unsigned(B) – 2^m.
- In either case, signed(B) — unsigned(B) is a multiple of 2^m.

The third fact tells you how to convert between a computer word with sign bit 0 and its 2's complement signed value, assuming you know how to convert between unsigned binary numbers and decimal. The fourth fact could be used to do the same when the sign bit is 1, but there is an alternative method described below that does not require that you remember powers of 2. The fifth fact can be used to determine when arithmetic with signed numbers fails.

Negation and the 2's Complement Operation

The one's complement operation is important bitwise operation for computer words. It converts each 0 bit in the word to 1 and each 1 bit in the word to 0. This operation is closely related to the negation operation for numbers. To see this, look at the follwing sum where B' is the one's complement of B.

signed(B) + signed(B') + 1 =

$-b_{m-1}*2^{m-1} + b_{m-2}*2^{m-2} +\ldots + b_0*2^0 + -b'_{m-1}*2^{m-1} + b'_{m-2}*2^{m-2} +\ldots + b'_0*2^0 + 1 =$

$-b_{m-1}*2^{m-1} + -b'_{m\hat{u}1}*2^{m-1} + b_{m-2}*2^{m-2} + b'_{m-2}*2^{m-2} +\ldots + b_0*2^0 + b'_0*2^0 + 1 =$

$-2^{m-1} + 2^{m-2} +\ldots + 2^0 + 1 =$

$-2^{m-1} + 2^{m-1} - 1 + 1 = 0$

The terms in the third line are just the terms in the second line, rearranged to bring together like powers of 2. Then the b_is and the b'_is drop out because one of them is always 0 and the other is always 1. The equality between the fourth and fifth lines follows from the formula for summing a geometric series. The above formulas show that the negative of signed(B) is signed(B') + 1. To make computer arithmetic reflect our arithmetic with numbers, addition on computer words is defined so that,

signed(B + C) = signed(B) + signed(C) Thus we can write

–signed(B) = signed(B' + 1)

The operation on the right-hand side—doing a one's complement and adding one - is called the two's complement operation (not to be confused with the two's complement number representation). It is the computer equivalent of negating a number.

There is one case where the two's complement operation fails to give us the same result as negation. This is to be expected when our computer operations use limited size words. Since the range of integers represented by an *m*-bit two's complement representation is from, -2^{m-1} to $2^{m-1} - 1$, there is number, -2^{m-1}, whose negative is not even representable. In this case, the add one step fails to work correctly.

INTEGERS OF SIGNED BINARY SYSTEM

It was noted previously that we will not be using a minus sign (–) to represent negative numbers. We would like to represent our binary numbers with only two symbols, 0 and 1. There are a few ways to represent negative binary numbers. The simplest of these methods is called ones complement, where the sign of a binary number is changed by simply toggling each bit (0's become 1's and *vice-versa*). This has some difficulties, among them the fact that zero can be represented in two different ways (for an eight bit number these would be 0000 0000 and 1111 1111)., we will use a method called two's complement notation which avoids the pitfalls of one's complement, but which is a bit more complicated.

To represent an n bit signed binary number the leftmost bit, has a special significance. The difference between a signed and an unsigned number is given in the table below for an 8 bit number.

The value of bits in signed and unsigned binary numbers

	Bit 7	Bit 6	Bit 5	Bit 4	Bit 3	Bit 2	Bit 1	Bit 0
Unsigned	$2^7 = 128$	$2^6 = 64$	$2^5 = 32$	$2^4 = 16$	$2^3 = 8$	$2^2 = 4$	$2^1 = 2$	$2^0 = 1$
Signed	$-(2^7) = $ h128	$2^6 = 64$	$2^5 = 32$	$2^4 = 16$	$2^3 = 8$	$2^2 = 4$	$2^1 = 2$	$2^0 = 1$

Let's look at how this changes the value of some binary numbers

Binary	Unsigned	Signed
0010 0011	35	35
1010 0011	163	–93
1111 1111	255	–1
1000 0000	128	–128

If Bit 7 is not set (as in the first example) the representation of signed and unsigned numbers is the same. However, when Bit 7 is set, the number is always negative. For this reason Bit 7 is sometimes called the sign bit. Signed numbers are added in the same way as unsigned numbers, the only difference is in the way they are interpreted. This is important for designers of arithmetic circuitry because it means that numbers can be added by the same circuitry regardless of whether or not they are signed.

To form a two's complement number that is negative you simply take the corresponding positive number, invert all the bits, and add 1. The example below illustrated this by forming the number negative 35 as a two's complement integer:

$35_{10} = 0010\ 0011_2$

invert $\rightarrow 1101\ 1100_2$

add 1 $\rightarrow 1101\ 1101_2$

So 1101 1101 is our two's complement representation of –35. We can check this by adding up the contributions from the individual bits

$$1101\ 1101_2 = -128 + 64 + 0 + 16 + 8 + 4 + 0 + 1 = -35.$$

The same procedure (invert and add 1) is used to convert the negative number to its positive equivalent. If we want to know what what number is represented by 1111 1101, we apply the procedure again

$? = 1111\ 1101_2$

invert $\rightarrow 0000\ 0010_2$

add 1 $\rightarrow 0000\ 0011_2$

Since 0000 0011 represents the number 3, we know that 1111 1101 represents the number –3.

POSITIVE BINARY FRACTIONS

The representation of unsigned binary fractions proceeds in exactly the same way as decimal fractions. For example

$$0.625_{10} = 1*0.5 + 0*0.25 + 1*0.125 =$$
$$1*2^{-1} + 0*2^{-2} + 1*2^{-3} = 0.101_2$$

Each place to the right of the decimal point represents a negative power of 2, just as for decimals they represent a negative power of 10. Likewise, if there are m bits to the right of a decimal, the precision of the number is 2^{-m} (versus 10^{-m} for decimal). Though it is possible to represent numbers greater than one by having digits to the left of the decimal place we will restrict ourselves to numbers less than one. These are commonly used by Digital Signal Processors. The largest number that can be represented by such a representation is $1–2^{-m}$, the smallest number is 2^{-m}. For a fraction with 15 bits of resolution this gives a range of approximately 0.99997 to 3.05E-5.

Note that this representationis easily extended to represent all positive numbers by having the digits to the left of the decimal point represent the integer part, and the digits to the right representing the fractional part. Thus $6.625_{10} = 110.101_2$

SIGNED BINARY FRACTIONS

Signed binary fractions are formed much like signed integers. We will work with a single digit to the left of the decimal point, and this will represent the number –1 (= $-(2^0)$). The rest of the representation of the fraction remains unchanged. Therefore this leftmost bit represents a sign bit just as with two's complement integers. If this bit is set, the number is negative, otherwise the number is positive. The largest positive number that can be represented is still $1–2^{-m}$ but the largest negative number is –1. The resolution is still $1–2^{-m}$. There is a terminology for naming the resolution of signed fractions. If there are m bits to the right of the decimal point, the number is said to be in Q*m*format. For a 16 bit number (15 bits to the right of the decimal point) this results in Q15 notation.

BINARY ARITHMETIC AND NUMBERS VERSUS NUMERATION

It is imperative to understand that the type of numeration system used to represent numbers has no impact upon the outcome of any arithmetical function (addition, subtraction, multiplication, division, roots, powers, or logarithms). A number is a number is a number; one plus one will always equal two (so long as we're dealing with real numbers), no matter how you symbolize one, one, and two.

A prime number in decimal form is still prime if it's shown in binary form, or octal, or hexadecimal? is still the ratio between the circumference and diameter of a circle, no matter what symbol(s) you use to denote its value. The essential functions and interrelations of mathematics are unaffected by the particular system of symbols we might choose to represent quantities. This distinction between numbers and systems of numeration is critical to understand.

The essential distinction between the two is much like that between an object and the spoken word(s) we associate with it. A house is still a house regardless of whether we call it by its English name house or its Spanish name casa. The first is the actual thing, while the second is merely the symbol for the thing.

That being said, performing a simple arithmetic operation such as addition (longhand) in binary form can be confusing to a person accustomed to working with decimal numeration only. In this lesson, we'll explore the techniques used to perform simple arithmetic functions on binary numbers, since these techniques will be employed in the design of electronic circuits to do the same.

You might take longhand addition and subtraction for granted, having used a calculator for so long, but deep inside that calculator's circuitry all those operations are performed"longhand," using binary numeration. To understand how that's accomplished, we need to review to the basics of arithmetic.

BINARY ADDITION

Adding binary numbers is a very simple task, and very similar to the longhand addition of decimal numbers. As with decimal numbers, you start by adding the bits (digits) one column, or place weight, at a time, from right to left. Unlike decimal addition, there is little to memorize in the way of rules for the addition of binary bits:

0 + 0 = 0
1 + 0 = 1
0 + 1 = 1
1 + 1 = 10
1 + 1 + 1 = 11

Just as with decimal addition, when the sum in one column is a two-bit (two-digit) number, the least significant figure is written as part of the total sum and the most significant figure is "carried" to the next left column. Consider the following examples:

- 11 1← Carry bits → 11
- 1001101 1001001 1000111
- \+ 0010010 + 0011001 + 0010110
- ——— ——— ———
- 1011111 1100010 1011101

The addition problem on the left did not require any bits to be carried, since the sum of bits in each column was either 1 or 0, not 10 or 11. In the other two problems, there definitely were bits to be carried, but the process of addition is still quite simple. As we'll see, there are ways that electronic circuits can be built to perform this very task of addition, by representing each bit of each binary number as a voltage signal (either "high," for a 1; or "low" for a 0). This is the very foundation of all the arithmetic which modern digital computers perform.

NEGATIVE BINARY NUMBERS

With addition being easily accomplished, we can perform the operation of subtraction with the same technique simply by making one of the numbers negative.

For example, the subtraction problem of 7 – 5 is essentially the same as the addition problem 7 + (–5). Since we already know how to represent positive numbers in binary, all we need to know now is how to represent their negative counterparts and we'll be able to subtract.

Usually we represent a negative decimal number by placing a minus sign directly to the left of the most significant digit, just as in the example above, with -5.

However, the whole purpose of using binary notation is for constructing on/off circuits that can represent bit values in terms of voltage (2 alternative values: either "high" or "low").

In this context, we don't have the luxury of a third symbol such as a "minus" sign, since these circuits can only be on or off (two possible states). One solution is to reserve a bit (circuit) that does nothing but represent the mathematical sign:

- $101_2 = 5_{10}$ (positive)
- Extra bit, representing sign (0=positive, 1=negative)
- |
- $0101_2 = 5_{10}$ (positive)
- Extra bit, representing sign (0=positive, 1=negative)
- |
- $1101_2 = -5_{10}$ (negative)

As you can see, we have to be careful when we start using bits for any purpose other than standard place-weighted values. Otherwise, 1101_2 could be misinterpreted as the number thirteen when in fact we mean to represent negative five.

To keep things straight here, we must first decide how many bits are going to be needed to represent the largest numbers we'll be dealing with, and then be sure not to exceed that bit field length in our arithmetic operations.

For the above example, I've limited myself to the representation of numbers from negative seven (1111_2) to positive seven (0111_2), and no more, by making the fourth bit the "sign" bit. Only by first establishing these limits can I avoid confusion of a negative number with a larger, positive number.

Representing negative five as 1101_2 is an example of the *sign-magnitude* system of negative binary numeration.

By using the leftmost bit as a sign indicator and not a place-weighted value, I am sacrificing the "pure" form of binary notation for something that gives me a practical advantage: the representation of negative numbers.

The leftmost bit is read as the sign, either positive or negative, and the remaining bits are interpreted according to the standard binary notation: left to right, place weights in multiples of two.

As simple as the sign-magnitude approach is, it is not very practical for arithmetic purposes. For instance, how do I add a negative five (1101_2) to any other number, using the standard technique for binary addition? I'd have to invent a new way of doing addition in order for it to work, and if I do that, I might as well just do the job with longhand subtraction; there's no arithmetical advantage to using negative numbers to perform subtraction through addition if we have to do it with sign-magnitude numeration, and that was our goal!

There's another method for representing negative numbers which works with our familiar technique of longhand addition, and also happens to make more sense from a place-weighted numeration point of view, called *complementation*. With this strategy, we assign the leftmost bit to serve a special purpose, just as we did with the sign-magnitude approach, defining our number limits just as before. However, this time, the leftmost bit is more than just a sign bit; rather, it possesses a negative place-weight value.

For example, a value of negative five would be represented as such:

Extra bit, place weight = negative eight:

- |
- $1011_2 = 5_{10}$ (negative)
- $(1 \times -8_{10}) + (0 \times 4_{10}) + (1 \times 2_{10}) + (1 \times 1_{10}) = -5_{10}$

With the right three bits being able to represent a magnitude from zero through seven, and the leftmost bit representing either zero or negative eight, we can successfully represent any integer number from negative seven,

$(1001_2 = -8_{10} + 7_{10} = -1_{10})$

to positive seven

$(0111_2 = 0_{10} + 7_{10} = 7_{10})$.

Representing positive numbers in this scheme (with the fourth bit designated as the negative weight) is no different from that of ordinary binary notation.

However, representing negative numbers is not quite as straightforward:

Zero	0000	Negative one	1111
Positive one	0001	Negative two	1110
Positive two	0010	Negative three	1101
Positive three	0011	Negative four	1100
Positive four	0100	Negative five	1011
Positive five	0101	Negative six	1010
Positive six	0110	Negative seven	1001
Positive seven	0111	Negative eight	1000

Note that the negative binary numbers in the right column, being the sum of the right three bits' total plus the negative eight of the leftmost bit, don't "count" in the same progression as the positive binary numbers in the left column.

Rather, the right three bits have to be set at the proper value to equal the desired (negative) total when summed with the negative eight place value of the leftmost bit. Those right three bits are referred to as the *two's complement* of the corresponding positive number.

Consider the following comparison:

Positive Number	Two's Complement
001	111
010	110
011	101
100	100
101	011
110	010
111	001

In this case, with the negative weight bit being the fourth bit (place value of negative eight), the two's complement for any positive number will be wthatever value is needed to add to negative eight to make that positive value's negative equivalent.

Thankfully, there's an easy way to figure out the two's complement for any binary number: simply invert all the bits of that number, changing all 1's to 0's and visa-versa (to arrive at what is called the *one's complement*) and then add one!

For example, to obtain the two's complement of five (101_2), we would first invert all the bits to obtain 010_2 (the "one's complement"), then add one to obtain 011_2, or -5_{10} in three-bit, two's complement form.

Interestingly enough, generating the two's complement of a binary number works the same if you manipulate *all* the bits, including the leftmost (sign) bit at the same time as the magnitude bits.

Let's try this with the former example, converting a positive five to a negative five, but performing the complementation process on all four bits.

We must be sure to include the 0 (positive) sign bit on the original number, five (0101_2). First, inverting all bits to obtain the one's complement: 1010_2. Then, adding one, we obtain the final answer: 1011_2, or -5_{10} expressed in four-bit, two's complement form.

It is critically important to remember that the place of the negative-weight bit must be already determined before any two's complement conversions can be done.

If our binary numeration field were such that the eighth bit was designated as the negative-weight bit (10000000_2), we'd have to determine the two's complement based on all seven of the other bits. Here, the two's complement of five (0000101_2) would be 1111011_2. A positive five in this system would be represented as 00000101_2, and a negative five as 11111011_2.

SUBTRACTION

We can subtract one binary number from another by using the standard techniques adapted for decimal numbers (subtraction of each bit pair, right to left, "borrowing" as needed from bits to the left).

However, if we can leverage the already familiar (and easier) technique of binary addition to subtract, that would be better. As we just learned, we can represent negative binary numbers by using the "two's complement" method and a negative place-weight bit. Here, we'll use those negative binary numbers to subtract through addition.

Here's a sample problem:

Subtraction: $7_{10} - 5_{10}$ *Addition equivalent*: $7_{10} + (-5_{10})$

If all we need to do is represent seven and negative five in binary (two's complemented) form, all we need is three bits plus the negative-weight bit:

Positive seven = 0111_2

Negative five = 1011_2

Now, let's add them together:

111 <— Carry bits
0111
\+ 1011
———————
10010
|
Discard extra bit.

Answer = 0010_2

Since we've already defined our number bit field as three bits plus the negative-weight bit, the fifth bit in the answer (1) will be discarded to give us a result of 0010_2, or positive two, which is the correct answer.

Another way to understand why we discard that extra bit is to remember that the leftmost bit of the lower number possesses a negative weight, in this case equal to negative eight. When we add these two binary numbers together, what we're actually doing with the MSBs is subtracting the lower number's MSB from the upper number's MSB. In subtraction, one never "carries" a digit or bit on to the next left place-weight.

Let's try another example, this time with larger numbers. If we want to add -25_{10} to 18_{10}, we must first decide how large our binary bit field must be.

To represent the largest (absolute value) number in our problem, which is twenty-five, we need at least five bits, plus a sixth bit for the negative-weight bit. Let's start by representing positive twenty-five, then finding the two's complement and putting it all together into one numeration:

$+25_{10} = 011001_2$ (showing all six bits)

One's complement of $11001_2 = 100110_2$

One's complement + 1 = two's complement = 100111_2

$-25_{10} = 100111_2$

Essentially, we're representing negative twenty-five by using the negative-weight (sixth) bit with a value of negative thirty-two, plus positive seven (binary 111_2).

Now, let's represent positive eighteen in binary form, showing all six bits:

$18_{10} = 010010_2$.

Now, let's add them together and see what we get:.

11 <— Carry bits
100111
+ 010010

―――――

111001

Since there were no "extra" bits on the left, there are no bits to discard. The leftmost bit on the answer is a 1, which means that the answer is negative, in two's complement form, as it should be. Converting the answer to decimal form by summing all the bits times their respective weight values, we get:

$(1 \times -32_{10}) + (1 \times 16_{10}) + (1 \times 8_{10}) + (1 \times 1_{10}) = -7_{10}$

Indeed -7_{10} is the proper sum of -25_{10} and 18_{10}.

BINARY NOTATION

Converting from decimal to binary notation is slightly more difficult conceptually, but can easily be done once you know how through the use of algorithms. Begin by thinking of a few examples. We can easily see that the number 3= 2+1. and that this is equivalent to $(1*2^1) + (1*2^0)$.

This translates into putting a "1" in the 2^1 column and a "1" in the 2^0 column, to get "11". Almost as intuitive is the number 5: it is obviously 4+1, which is the same as saying [(2*2) +1], or 2^2+1. This can also be written as $[(1*2^2) + (1*2^0)]$. Looking at this in columns,

$2^2 \mid 2^2 \mid 2^0$

1 0 1

or 101.

What we're doing here is finding the largest power of two within the number ($2^2=4$ is the largest power of 2 in 5), subtracting that from the number (5–4=1), and finding the largest power of 2 in the remainder ($2^0=1$ is the largest power of 2 in 1). Then we just put this into columns. This process continues until we have a remainder of 0. Let's take a look at how it works. We know that:

$2^0=1$

$2^1=2$

$2^2=4$

$2^3=8$

$2^4=16$
$2^5=32$
$2^6=64$
$2^7=128$

and so on. To convert the decimal number 75 to binary, we would find the largest power of 2 less than 75, which is 64. Thus, we would put a 1 in the 2^6 column, and subtract 64 from 75, giving us 11. The largest power of 2 in 11 is 8, or 2^3. Put 1 in the 2^3 column, and 0 in 2^4 and 2^5. Subtract 8 from 11 to get 3. Put 1 in the 2^1 column, 0 in 2^2, and subtract 2 from 3. We're left with 1, which goes in 2^0, and we subtract one to get zero. Thus, our number is 1001011.

Making this algorithm a bit more formal gives us:

- Let D=number we wish to convert from decimal to binary
- Repeat until D=0
  - Find the largest power of two in D. Let this equal P.
  - Put a 1 in binary column P.
  - Subtract P from D.
- Put zeros in all columns which don't have ones.

This algorithm is a bit awkward. Particularly step 3, "filling in the zeros." Therefore, we should rewrite it such that we ascertain the value of each column individually, putting in 0's and 1's as we go:

- Let D= the number we wish to convert from decimal to binary
- Find P, such that 2^P is the largest power of two smaller than D.
- Repeat until P<0
  - If $2^P<=D$ then
    - a. put 1 into column P
    - b. subtract 2^P from D
- Else
  - a. put 0 into column P
- End if
- Subtract 1 from P

Now that we have an algorithm, we can use it to convert numbers from decimal to binary relatively painlessly. Let's try the number D=55.

- Our first step is to find P. We know that $2^4=16$, $2^5=32$, and $2^6=64$. Therefore, P=5.
- $2^5<=55$, so we put a 1 in the 2^5 column: 1——-.
- Subtracting 55–32 leaves us with 23. Subtracting 1 from P gives us 4.
- Following step 3 again, $2^4<=23$, so we put a 1 in the 2^4 column: 11——.
- Next, subtract 16 from 23, to get 7. Subtract 1 from P gives us 3.
- $2^3>7$, so we put a 0 in the 2^3 column: 110—-
- Next, subtract 1 from P, which gives us 2.

- $2^2 <= 7$, so we put a 1 in the 2^2 column: 1101—
- Subtract 4 from 7 to get 3. Subtract 1 from P to get 1.
- $2^1 <= 3$, so we put a 1 in the 2^1 column: 11011-
- Subtract 2 from 3 to get 1. Subtract 1 from P to get 0.
- $2^0 <= 1$, so we put a 1 in the 2^0 column: 110111
- Subtract 1 from 1 to get 0. Subtract 1 from P to get –1.
- P is now less than zero, so we stop.

Another algorithm for converting decimal to binary

However, this is not the only approach possible. We can start at the right, rather than the left.

All binary numbers are in the form

$a[n]*2^n + a[n-1]*2^{(n-1)}+...+a[1]*2^1 + a[0]*2^0$

where each a[i] is either a 1 or a 0 (the only possible digits for the binary system). The only way a number can be odd is if it has a 1 in the 2^0 column, because all powers of two greater than 0 are even numbers (2, 4, 8, 16...). This gives us the rightmost digit as a starting point.

Now we need to do the remaining digits. One idea is to "shift" them. It is also easy to see that multiplying and dividing by 2 shifts everything by one column: two in binary is 10, or $(1*2^1)$. Dividing $(1*2^1)$ by 2 gives us $(1*2^0)$, or just a 1 in binary. Similarly, multiplying by 2 shifts in the other direction: $(1*2^1)*2=(1*2^2)$ or 10 in binary. Therefore

$\{a[n]*2^n + a[n-1]*2^{(n-1)} +... + a[1]*2^1 + a[0]*2^0\}/2$

is equal to

$a[n]*2^{(n-1)} + a[n-1]*2^{(n-2)} +... + a[1]2\hat{}0$

Let's look at how this can help us convert from decimal to binary. Take the number 163. We know that since it is odd, there must be a 1 in the 2^0 column (a[0]=1). We also know that it equals 162+1. If we put the 1 in the 2^0 column, we have 162 left, and have to decide how to translate the remaining digits. Two's column: Dividing 162 by 2 gives 81. The number 81 in binary would also have a 1 in the 2^0 column. Since we divided the number by two, we "took out" one power of two. Similarly, the statement $a[n-1]*2^{(n-1)} + a[n-2]*2^{(n-2)} +... + a[1]*2^0$ has a power of two removed. Our "new" 2^0 column now contains a1.

We learned earlier that there is a 1 in the 2^0 column if the number is odd. Since 81 is odd, a[1]=1. Practically, we can simply keep a "running total", which now stands at 11 (a[1]=1 and a[0]=1). Also note that a1 is essentially "remultiplied" by two just by putting it in front of a[0], so it is automatically fit into the correct column.

Four's column: Now we can subtract 1 from 81 to see what remainder we still must place (80). Dividing 80 by 2 gives 40. Therefore, there must be a 0 in the 4's column, (because what we are actually placing is a 2^0 column, and the number is not odd).

Eight's column: We can divide by two again to get 20. This is even, so we put a 0 in the 8's column. Our running total now stands at a[3]=0, a[2]=0, a[1]=1, and a[0]=1.

We can continue in this manner until there is no remainder to place.

Let's formalize this algorithm:

- Let D= the number we wish to convert from decimal to binary.
- Repeat until D=0:
  - If D is odd, put "1" in the leftmost open column, and subtract 1 from D.
  - If D is even, put "0" in the leftmost open column.
  - Divide D by 2.

  End Repeat

For the number 163, this works as follows:

- Let D=163
- D is odd, put a 1 in the 2^0 column.
  Subtract 1 from D to get 162.
- Divide D=162 by 2.
  Temporary Result: 01 New D=81
  D does not equal 0, so we repeat step 2.
- D is odd, put a 1 in the 2^1 column.
  Subtract 1 from D to get 80.
- Divide D=80 by 2.
  Temporary Result: 11 New D=40
  D does not equal 0, so we repeat step 2.
- D is even, put a 0 in the 2^2 column.
- Divide D by 2.
  Temporary Result: 011 New D=20
- D is even, put a 0 in the 2^3 column.
- Divide D by 2.
  Temporary Result: 0011 New D=10
- D is even, put a 0 in the 2^4 column.
- Divide D by 2.
  Temporary Result: 00011 New D=5
- D is odd, put a 1 in the 2^5 column.
  Subtract 1 from D to get 4.
- Divide D by 2.
  Temporary Result: 100011 New D=2
- D is even, put a 0 in the 2^6 column.
- Divide D by 2.
  Temporary Result: 0100011 New D=1
- D is odd, put a 1 in the 27 column.
  Subtract 1 from D to get D=0.

- Divide D by 2.

  Temporary Result: 10100011 New D=0

D=0, so we are done, and the decimal number 163 is equivalent to the binary number 10100011. Since we already knew how to convert from binary to decimal, we can easily verify our result. 10100011 = $(1*2^0)$ + $(1*2^1)+(1*2^5)+(1*2^7)$=1+2+32+128= 163.

SUBTRACTION OF LARGE BIT BINARY NUMBERS

We can subtract one binary number from another by using the standard techniques adapted for decimal numbers (subtraction of each bit pair, right to left, "borrowing" as needed from bits to the left). However, if we can leverage the already familiar (and easier) technique of binary addition to subtract, that would be better. As we just learned, we can represent negative binary numbers by using the "two's complement" method and a negative place-weight bit. Here, we'll use those negative binary numbers to subtract through addition. Here's a sample problem:

CONCEPT

Subtraction: $7_{10} - 5_{10}$ Addition equivalent: $7_{10} + (-5_{10})$

If all we need to do is represent seven and negative five in binary (two's complemented) form, all we need is three bits plus the negative-weight bit:

Positive seven = 0111_2

Negative five = 1011_2

Now, let's add them together:

```
 1111 ←Carry bits
  0111
+1011
------
 10010
 |
Discard extra bit
```

Answer = 0010_2

Since we've already defined our number bit field as three bits plus the negative-weight bit, the fifth bit in the answer (1) will be discarded to give us a result of 0010_2, or positive two, which is the correct answer. Another way to understand why we discard that extra bit is to remember that the leftmost bit of the lower number possesses a negative weight, in this case equal to negative eight. When we add these two binary numbers together, what we're actually doing with the MSBs is subtracting the lower number's MSB from the upper number's MSB. In subtraction, one never "carries" a digit or bit on to the next left place-weight.

Let's try another example, this time with larger numbers. If we want to add -25_{10} to 18_{10}, we must first decide how large our binary bit field must be.

To represent the largest (absolute value) number in our problem, which is twenty-five, we need at least five bits, plus a sixth bit for the negative-weight bit. Let's start by representing positive twenty-five, then finding the two's complement and putting it all together into one numeration:

$+25_{10} = 011001_2$ (showing all six bits)

One's complement of $11001_2 = 100110_2$

One's complement + 1 = two's complement = 100111_2

$-25_{10} = 100111_2$

Essentially, we're representing negative twenty-five by using the negative-weight (sixth) bit with a value of negative thirty-two, plus positive seven (binary 111_2).

Now, let's represent positive eighteen in binary form, showing all six bits:

$$18_{10} = 010010_2$$

Now, let's add them together and see what we get:

```
 11  ← Carry bits
 1001111
+010010
--------
 111001
```

Since there were no "extra" bits on the left, there are no bits to discard. The leftmost bit on the answer is a 1, which means that the answer is negative, in two's complement form, as it should be.

Converting the answer to decimal form by summing all the bits times their respective weight values, we get:

$$(1 \times -32_{10}) + (1 \times 16_{10}) + (1 \times 8_{10}) + (1 \times 1_{10}) = -7_{10}$$

Indeed -7_{10} is the proper sum of -25_{10} and 18_{10}.

Overflow

One caveat with signed binary numbers is that of overflow, where the answer to an addition or subtraction problem exceeds the magnitude which can be represented with the alloted number of bits. Remember that the place of the sign bit is fixed from the beginning of the problem. With the last example problem, we used five binary bits to represent the magnitude of the number, and the left-most (sixth) bit as the negative-weight, or sign, bit. With five bits to represent magnitude, we have a representation range of 2^5, or thirty-two integer steps from 0 to maximum.

This means that we can represent a number as high as $+31_{10}$ (011111_2), or as low as -32_{10} (100000_2). If we set up an addition problem with two binary numbers, the sixth bit used for sign, and the result either exceeds $+31_{10}$ or is less than -32_{10}, our answer will be incorrect. Let's try adding 17_{10} and 19_{10} to see how this overflow condition works for excessive positive numbers:

$17_{10} = 10001_2$ $19_{10} = 10011_2$

```
                 1 11    ← Carry bits
(Showing sign bits)  010001
                 +1010011
                 --------
                  100100
```

The answer (100100_2), interpreted with the sixth bit as the -32_{10} place, is actually equal to -28_{10}, not $+36_{10}$ as we should get with $+17_{10}$ and $+19_{10}$ added together! Obviously, this is not correct. What went wrong? The answer lies in the restrictions of the six-bit number field within which we're working Since the magnitude of the true and proper sum (36_{10}) exceeds the allowable limit for our designated bit field, we have an overflow error.

Simply put, six places doesn't give enough bits to represent the correct sum, so whatever figure we obtain using the strategy of discarding the left-most "carry" bit will be incorrect. A similar error will occur if we add two negative numbers together to produce a sum that is too low for our six-bit binary field. Let's try adding -17_{10} and -19_{10} together to see how this works (or doesn't work, as the case may be!):

```
.       -17₁₀  = 10111₁₂                 -19₁₀  = 101101₂
.
.                                   1 1111  ← Carry bits
.     (Showing sign bits)             101111
.                                   + 101101
.                                   --------
.                                    1011100
.                                    |
.                        Discard extra bit
.
  Final Answer: 011100₂ = +28₁₀
```

The (incorrect) answer is a positive twenty-eight. The fact that the real sum of negative seventeen and negative nineteen was too low to be properly represented with a five bit magnitude field and a sixth sign bit is the root cause of this difficulty. Let's try these two problems again, except this time using the seventh bit for a sign bit, and allowing the use of 6 bits for representing the magnitude:

```
.          17₁₀ + 19₁₀           (-17₁₀) + (-19₁₀)
.
.           1  11                11   1111
.          0010001                 1101111
.        + 0010011              +   1101101
.        ---------              ---------
.          0100100₂             11011100₂
.                                |
```

```
.                    Discard extra bit
.
. ANSWERS:    0100100₂  =  +36₁₀
.               1011100₂  =  -36₁₀
```

By using bit fields sufficiently large to handle the magnitude of the sums, we arrive at the correct answers. In these sample problems we've been able to detect overflow errors by performing the addition problems in decimal form and comparing the results with the binary answers. For example, when adding $+17_{10}$ and $+19_{10}$ together, we knew that the answer was supposed to be $+36_{10}$, so when the binary sum checked out to be -28_{10}, we knew that something had to be wrong. Although this is a valid way of detecting overflow, it is not very efficient.

After all, the whole idea of complementation is to be able to reliably add binary numbers together and not have to double-check the result by adding the same numbers together in decimal form. This is especially true for the purpose of building electronic circuits to add binary quantities together: the circuit has to be able to check itself for overflow without the supervision of a human being who already knows what the correct answer is.

What we need is a simple error-detection method that doesn't require any additional arithmetic. Perhaps the most elegant solution is to check for the sign of the sum and compare it against the signs of the numbers added. Obviously, two positive numbers added together should give a positive result, and two negative numbers added together should give a negative result. Notice that whenever we had a condition of overflow in the example problems, the sign of the sum was always opposite of the two added numbers: $+17_{10}$ plus $+19_{10}$ giving -28_{10}, or -17_{10} plus -19_{10} giving $+28_{10}$.

By checking the signs alone we are able to tell that something is wrong. But what about cases where a positive number is added to a negative number? What sign should the sum be in order to be correct. Or, more precisely, what sign of sum would necessarily indicate an overflow error?

The answer to this is equally elegant: there will never be an overflow error when two numbers of opposite signs are added together! The reason for this is apparent when the nature of overflow is considered. Overflow occurs when the magnitude of a number exceeds the range allowed by the size of the bit field. The sum of two identically-signed numbers may very well exceed the range of the bit field of those two numbers, and so in this case overflow is a possibility. However, if a positive number is added to a negative number, the sum will always be closer to zero than either of the two added numbers: its magnitude must be less than the magnitude of either original number, and so overflow is impossible.Fortunately, this technique of overflow detection is easily implemented in electronic circuitry, and it is a standard feature in digital adder circuits: a subject for a later chapter.

Bit Groupings

The singular reason for learning and using the binary numeration system in electronics is to understand how to design, build, and troubleshoot circuits that represent and process numerical quantities in digital form. Since the bivalent (two-valued) system of binary bit numeration lends itself so easily to representation by "on" and "off" transistor states (saturation and cutoff, respectively), it makes sense to design and build circuits leveraging this principle to perform binary calculations.

If we were to build a circuit to represent a binary number, we would have to allocate enough transistor circuits to represent as many bits as we desire. In other words, in designing a digital circuit, we must first decide how many bits (maximum) we would like to be able to represent, since each bit requires one on/off circuit to represent it. This is analogous to designing an abacus to digitally represent decimal numbers: we must decide how many digits we wish to handle in this primitive "calculator" device, for each digit requires a separate rod with its own beads.

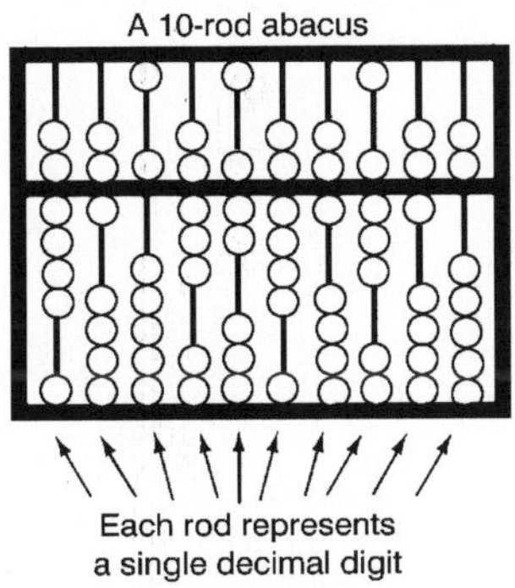

A ten-rod abacus would be able to represent a ten-digit decimal number, or a maximum value of 9,999,999,999. If we wished to represent a larger number on this abacus, we would be unable to, unless additional rods could be added to it. In digital, electronic computer design, it is common to design the system for a common "bit width:" a maximum number of bits allocated to represent numerical quantities. Early digital computers handled bits in groups of four or eight. More modern systems handle numbers in clusters of 32 bits or more.

To more conveniently express the "bit width" of such clusters in a digital computer, specific labels were applied to the more common groupings. Eight bits, grouped together to form a single binary quantity, is known as a byte. Four bits, grouped together as one binary number, is known by the humorous title of nibble, often spelled as nybble. A multitude of terms have followed byte and nibble for labeling specfiic groupings of binary bits. Most of the terms shown here are informal, and have not been made "authoritative" by any standards group or other sanctioning body.

However, their inclusion into this chapter is warranted by their occasional appearance in technical literature, as well as the levity they add to an otherwise dry subject:

- *Bit:* A single, bivalent unit of binary notation. Equivalent to a decimal "digit."
- *Crumb, Tydbit, or Tayste:* Two bits.
- *Nibble, or Nybble:* Four bits.
- *Nickle:* Five bits.
- *Byte:* Eight bits.
- *Deckle:* Ten bits.
- *Playte:* Sixteen bits.
- *Dynner:* Thirty-two bits.
- *Word:* (system dependent).

The most ambiguous term by far is word, referring to the standard bit-grouping within a particular digital system. For a computer system using a 32 bit-wide "data path," a "word" would mean 32 bits. If the system used 16 bits as the standard grouping for binary quantities, a "word" would mean 16 bits. The terms playte and dynner, by contrast, always refer to 16 and 32 bits, respectively, regardless of the system context in which they are used. Context dependence is likewise true for derivative terms of word, such as double word and longword (both meaning twice the standard bit-width), half-word (half the standard bit-width), and quad (meaning four times the standard bit-width). One humorous addition to this somewhat boring collection of word-derivatives is the term chawmp, which means the same as half-word. For example, a chawmp would be 16 bits in the context of a 32-bit digital system, and 18 bits in the context of a 36-bit system. Also, the term gawble is sometimes synonymous with word.

PRIME INTENT OF OCTAL AND HEXADECIMAL NUMERATION SYSTEMS

Although the prime intent of octal and hexadecimal numeration systems is for the "shorthand" representation of binary numbers in digital electronics, we sometimes have the need to convert from either of those systems to decimal form. Of course, we could simply convert the hexadecimal or octal format to binary, then convert from binary to decimal, since we already know how to do both, but we can also convert directly.

Because octal is a base-eight numeration system, each place-weight value differs from either adjacent place by a factor of eight. For example, the octal number 245.37 can be broken down into place values as such:

```
octal
digits =        2   4   5   .   3   7
```

```
.               -    -    -    -    -    -
weight =        6    8    1         1    1
(in deci-       4                   /    /
mal notation)                       8    6
.                                        4
```

The decimal value of each octal place-weight times its respective cipher multiplier can be determined as follows:

$(2 \times 64_{10}) + (4 \times 8_{10}) + (5 \times 1_{10}) + (3 \times 0.125_{10}) + (7 \times 0.015625_{10}) = 165.484375_{10}$

The technique for converting hexadecimal notation to decimal is the same, except that each successive place-weight changes by a factor of sixteen. Simply denote each digit's weight, multiply each hexadecimal digit value by its respective weight (in decimal form), then add up all the decimal values to get a total. For example, the hexadecimal number $30F.A9_{16}$ can be converted like this:

```
hexadecimal
digits =        3    0    F    .    A    9
.               -    -    -    -    -    -
weight =        2    1    1         1    1
(in decimal     5    6              /    /
notation)       6                   1    2
.                                   6    5
.                                        6
```

$(3 \times 256_{10}) + (0 \times 16_{10}) + (15 \times 1_{10}) + (10 \times 0.0625_{10}) + (9 \times 0.00390625_{10}) = 783.66015625_{10}$

These basic techniques may be used to convert a numerical notation of any base into decimal form, if you know the value of that numeration system's base.

CONVERSION FROM DECIMAL NUMERATION

Because octal and hexadecimal numeration systems have bases that are multiples of binary (base 2), conversion back and forth between either hexadecimal or octal and binary is very easy. Also, because we are so familiar with the decimal system, converting binary, octal, or hexadecimal to decimal form is relatively easy (simply add up the products of cipher values and place-weights). However, conversion from decimal to any of these "strange" numeration systems is a different matter.

The method which will probably make the most sense is the "trial-and-fit" method, where you try to "fit" the binary, octal, or hexadecimal notation to the desired value as represented in decimal form.

For example, let's say that I wanted to represent the decimal value of 87 in binary form. Let's start by drawing a binary number field, complete with place-weight values:.

```
.               -   -   -   -   -   -   -   -
weight=         1   6   3   1   8   4   2   1
(in decimal     2   4   2   6
notation)       8
```

Well, we know that we won't have a "1" bit in the 128's place, because that would immediately give us a value greater than 87. However, since the next weight to the right (64) is less than 87, we know that we must have a "1" there.

```
            .   1
            .   -   -   -   -   -   -   -   Decimal  value  so  far
= 64₁₀   weight  =  6   3   1   8   4   2   1
(in deci-   4   2   6
mal notation)
```

If we were to make the next place to the right a "1" as well, our total value would be 64_{10} + 32_{10}, or 96_{10}. This is greater than 87_{10}, so we know that this bit must be a "0". If we make the next (16's) place bit equal to "1," this brings our total value to 64_{10} + 16_{10}, or 80_{10}, which is closer to our desired value (87_{10}) without exceeding it:

```
.               1   0   1
.               -   -   -   -   -   -   -   Decimal  value  so  far
= 80₁₀
weight  =       6   3   1   8   4   2   1
(in decimal     4   2   6
notation)
```

By continuing in this progression, setting each lesser-weight bit as we need to come up to our desired total value without exceeding it, we will eventually arrive at the correct figure:

```
.               1   0   1   0   1   1   1
.               -   -   -   -   -   -   -   Decimal  value  so  far
= 87₁₀
weight  =       6   3   1   8   4   2   1
(in decimal     4   2   6
notation)
```

This trial-and-fit strategy will work with octal and hexadecimal conversions, too. Let's take the same decimal figure, 87_{10}, and convert it to octal numeration:.

```
.               -   -   -
weight  =       6   8   1
(in decimal     4
notation)
```

If we put a cipher of "1" in the 64's place, we would have a total value of 64_{10} (less than 87_{10}). If we put a cipher of "2" in the 64's place, we would have a total value of 128_{10} (greater than 87_{10}). This tells us that our octal numeration must start with a "1" in the 64's place:

```
.               1
.               -   -   - Decimal value so far = 64₁₀
weight =        6   8   1
(in decimal     4
notation)
```

Now, we need to experiment with cipher values in the 8's place to try and get a total (decimal) value as close to 87 as possible without exceeding it. Trying the first few cipher options, we get:

$$"1" = 64_{10} + 8_{10} = 72_{10}$$
$$"2" = 64_{10} + 16_{10} = 80_{10}$$
$$"3" = 64_{10} + 24_{10} = 88_{10}$$

A cipher value of "3" in the 8's place would put us over the desired total of 87_{10}, so "2" it is!

```
.               1   2
.               -   -   - Decimal value so far = 80₁₀
weight =        6   8   1
(in decimal     4
notation)
```

Now, all we need to make a total of 87 is a cipher of "7" in the 1's place:

```
.               1   2   7
.               -   -   - Decimal value so far = 87₁₀
weight =        6   8   1
(in decimal     4
notation)
```

Of course, if you were paying attention during the last section on octal/binary conversions, you will realise that we can take the binary representation of (decimal) 87_{10}, which we previously determined to be 1010111_2, and easily convert from that to octal to check our work:

```
.     Implied zeros
.          ||
.          001 010 111 Binary
.          —   —   —
.          1   2   7   Octal
.
Answer: 1010111₂ = 127₈
```

Can we do decimal-to-hexadecimal conversion the same way? Sure, but who would want to? This method is simple to understand, but laborious to carry out. There is another way to do these conversions, which is essentially the same (mathematically), but easier to accomplish.

This other method uses repeated cycles of division (using decimal notation) to break the decimal numeration down into multiples of binary, octal, or hexadecimal place-weight values. In the first cycle of division, we take the original decimal number and divide it by the base of the numeration system that we're converting to (binary=2 octal=8, hex=16).

Then, we take the whole-number portion of division result (quotient) and divide it by the base value again, and so on, until we end up with a quotient of less than 1. The binary, octal, or hexadecimal digits are determined by the "remainders" left over by each division step. Let's see how this works for binary, with the decimal example of 87_{10}:

```
. 87            Divide 87 by 2, to get a quotient of 43.5
. — = 43.5      Division "remainder" = 1, or the < 1 portion
. 2             of the quotient times the divisor(0.5 × 2)
.
. 43            Take the whole-number portion of 43.5(43)
. — = 21.5      and divide it by 2 to get 21.5, or 21 with
. 2             a remainder of 1
.
. 21            And so on... remainder = 1 (0.5 × 2)
. — = 10.5
. 2
.
. 10            And so on... remainder = 0
. — = 5.0
. 2
.
. 5             And so on... remainder = 1 (0.5 × 2)
. — = 2.5
. 2
.
. 2             And so on... remainder = 0
. — = 1.0
. 2
.
. 1             ... until we get a quotient of less than 1
. — = 0.5       remainder = 1 (0.5 × 2)
. 2
```

The binary bits are assembled from the remainders of the successive division steps, beginning with the LSB and proceeding to the MSB. In this case, we arrive at a binary notation of 1010111_2.

When we divide by 2, we will always get a quotient ending with either ".0" or ".5", *i.e.* a remainder of either 0 or 1. As was said before, this repeat-division technique for conversion will work for numeration systems other than binary. If we were to perform successive divisions using a different number, such as 8 for conversion to octal, we will necessarily get remainders between 0 and 7.

Let's try this with the same decimal number, 87_{10}:

```
.  87           Divide 87 by 8, to get a quotient of 10.875
.  — = 10.875   Division "remainder" = 7, or the < 1 portion
.  8               of the quotient times the divisor(.875 ×
8)
.
.  10
.  — = 1.25     Remainder = 2
.  8
.
.  1
.  — = 0.125    Quotient is less than 1, so we'll stop here.
.  8            Remainder = 1
.
. RESULT: 87₁₀  = 127₈
```

We can use a similar technique for converting numeration systems dealing with quantities less than 1, as well. For converting a decimal number less than 1 into binary, octal, or hexadecimal, we use repeated multiplication, taking the integer portion of the product in each step as the next digit of our converted number.

Let's use the decimal number 0.8125_{10} as an example, converting to binary:

```
. 0.8125 × 2 = 1.625       Integer portion of product = 1
.
. 0.625 × 2 = 1.25         Take < 1 portion of product and
remultiply
.                          Integer portion of   product = 1
.
. 0.25 × 2 = 0.5           Integer portion of   product = 0
.
. 0.5 × 2 = 1.0            Integer portion of   product = 1
.                          Stop when product is a pure
integer
.                          (ends with.0)
.
. RESULT: 0.8125₁₀      =  0.1101₂
```

As with the repeat-division process for integers, each step gives us the next digit (or bit) further away from the "point." With integer (division), we worked from the LSB to the MSB (right-to-left), but with repeated multiplication, we worked from the left to the right.

To convert a decimal number greater than 1, with a < 1 component, we must use both techniques, one at a time. Take the decimal example of 54.40625_{10}, converting to binary:

REPEATED DIVISION FOR THE INTEGER PORTION:

```
.  54
.  —  = 27.0        Remainder = 0
.  2
.
.  27
.  —  = 13.5        Remainder = 1 (0.5 × 2)
.  2
.
.  13
.  —  = 6.5         Remainder = 1 (0.5 × 2)
.  2
.
.  6
.  —  = 3.0         Remainder = 0
.  2
.
.  3
.  —  = 1.5         Remainder = 1 (0.5 × 2)
.  2
.
.  1
.  —  = 0.5         Remainder = 1 (0.5 × 2)
.  2
.
```

PARTIAL ANSWER: $54_{10} = 110110_2$

REPEATED MULTIPLICATION FOR THE < 1 PORTION:

```
.
. 0.40625 × 2 = 0.8125    Integer portion of product  = 0
.
. 0.8125 × 2 = 1.625      Integer portion of product = 1
.
. 0.625 × 2 = 1.25        Integer portion of product = 1
.
. 0.25 × 2 = 0.5          Integer portion of product = 0
.
. 0.5 × 2 = 1.0           Integer portion of product = 1
.
```

. PARTIAL ANSWER: $0.40625_{10} = 0.01101_2$

.

. COMPLETE ANSWER: $54_{10} + 0.40625_{10} = 54.40625_{10}$

.

. $110110_2 + 0.01101_2 = 110110.01101_2$

NUMBER SYSTEMS AND CODES

Computers are information processing tools that give meaning to raw data. Data should be represented in a form that permits efficient storage and ease of manipulation (although these two goals might be conflicting at times). In data communications, we are also concerned with issues such as security, error detection and error correction.

Data are usually represented in some form of code for compactness and uniformity. For instance, a 'yes' response may be represented by '*Y*'', and a 'no' by '*N*'; the four seasons by '*S*' (spring), '*M*' (summer), '*A*' (autumn) and '*W*' (winter). We can see that the choices are rather arbitrary. In real-life, codes are used for identification purposes, such as the Singapore NRIC number and the NUS student's matriculation number.

Computers are also number crunching devices. Numeric values are represented in a form that eases arithmetic computations and preserves accuracy as much as possible.

The atomic unit of data is the *bit* (*b*inary dig*it*), stemming from the fact that the elementary storage units in a computer are electronic switches where each switch holds one of the two states: *on* (usually represented by 1), or *off* (0), as shown below.

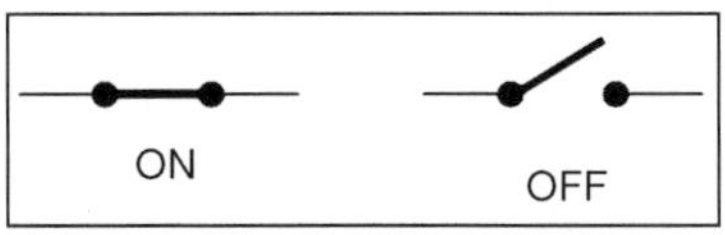

Fig. Two States of a Switch

Ultimately, all data, numeric or non-numeric, are represented in sequence of bits of zeroes and ones. Eight bits constitute a *byte* (a *nibble* is half a byte, or four bits, but this is rarely used these days), and a *word*, which is a unit for data storage and transfer, is usually in multiples of byte, depending on the width of the system bus. Computers nowadays typically use 32-bit or 64-bit words.

A sequence of bits allows for a range of representations. Storage units can be grouped together to provide larger range of representations. For convenience, we shall refer to these representations as values, though they may represent non-numeric data. For example, the four seasons could be represented by two switches, to cover a range of four values 00, 01, 10 and 11, as shown below.

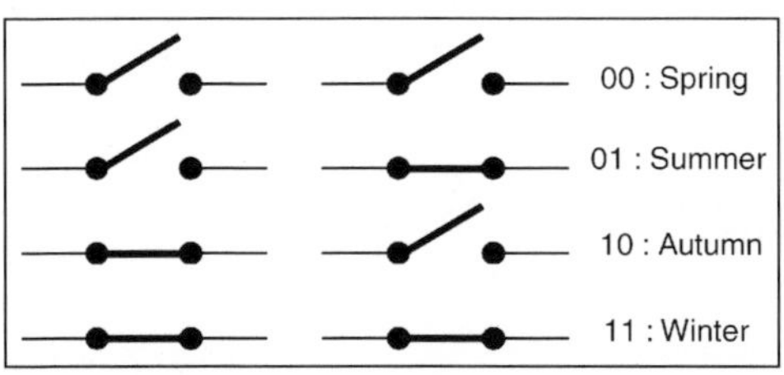

Fig. Representing Four States Using Two Switches (Two Bits)

In general, N bits can represent up to 2^N distinct values. Conversely, to represent a range of M values, the number of bits required is $\lceil \log_2 M \rceil$.

1 bit → represents up to 2 values (0, 1)
2 bits → represents up to 4 values (00, 01, 10, 11)
3 bits → represents up to 8 values (000, 001, 010, 011, 100, 101, 110, 111)
4 bits → represents up to 16 values (0000, 0001, 0010, ..., 1110, 1111)
32 values → requires 5 bits
40 values → requires 6 bits
64 values → requires 6 bits
100 values → requires 7 bits
1024 values → requires 10 bits

POSITIONAL NUMERALS

Different numeral systems have emerged in civilizations over the history. They can be roughly categorised as follows:

- Non-positional numeral systems
- Relative-position numeral systems
- Positional numeral systems

In the decimal numeral system, the positional weights are powers of ten. In general, a decimal number $(a_n\, a_{n-1} \ldots a_0 . f_1 f_2 \ldots f_m)$ has the value,

$$(a_n \times 10^n) + (a_{n-1} \times 10^{n-1}) + \ldots + (a_0 \times 10^0) + (f_1 \times 10^{-1}) + (f_2\ 10^{-2}) + \ldots + (f_m \times 10^{-m})$$

For example, $28.75 = (2 \times 10^1) + (8 \times 10^0) + (7 \times 10^{-1}) + (5 \times 10^{-2})$

BASES OF NUMBER SYSTEMS

The *base* or *radix* of a number system is the number of digits present. The decimal numeral system has a base or radix of 10, where the set of 10 symbols (digits) is {0, 1, 2, 3, 4, 5, 6, 7, 8, 9}. The weights are in powers of ten.

Number systems on other bases are defined likewise. For instance, a base-four number system consists of the set of four symbols {0, 1, 2, 3} with weights in powers of four. Hence, a base-four number 123.1, also written as $(123.1)_4$ for clarity, has the value,

$(1 \times 4^2) + (2 \times 4^1) + (3 \times 4^0) + (1 \times 4^{-1})$ or 16 + 8 + 3 + 0.25 or 27.25 in decimal, or $(27.25)_{10}$. Listing base-four integers in increasing value yields

0, 1, 2, 3, 10, 11, 12, 13, 20, 21, 22, 23, 30, 31, 32, 33, 100, 101, 102, 103, 110, ...

In general, a base-b number $(a_n\, a_{n-1} \ldots a_0 . f_1 f_2 \ldots f_m)_b$ has the value,

$$(a_n\ ´\ b^n) + (a_{n-1}\ ´\ b^{n-1}) + \ldots + (a_0\ ´\ b^0) + (f_1\ ´\ b^{-1}) + (f_2\ ´\ b^{-2}) + \ldots + (f_m\ ´\ b^{-m})$$

The point that separates the integer part and fraction part is known as the *radix point*. The weights are in powers of b. The above forms the basis for conversion of a base-*b* number to decimal. We may also adopt the computationally more efficient *Horner's rule* to convert a base-*b* integer to decimal. This method simply factors out the powers of *b*, by rewriting,

$(a_n$ ´ $b^n) + (a_{n-1}$ ´ $b^{n-1}) + \ldots + (a_0$ ´ $b^0)$ as $(((a_n$ ´ $b) + a_{n-1})$ ´ $b + \ldots)$ ´ $b + a_0$.

For example,

$$(123)_4 = ((1 ´ 4) + 2) ´ 4 + 3 = 27.$$

For fractions, we may adopt the same method by replacing multiplication with division. Special names are given to number systems on certain bases.If you think that the base is a small value, you are wrong. The Babylonians used a sexagesimal system with base 60!

Table. Bases of Positional Numeral Systems

Base	Name	Base	Name
2	Binary	9	Nonary
3	Ternary	10	Decimal (or Denary)
4	Quaternary	12	Duodecimal
5	Quinary	16	Hexadecimal
6	Senary	20	Vigesimal
7	Septimal	60	Sexagesimal
8	Octal (or Octonary)		

For a base *b* that is less than ten, the symbols used are the first *b* Arabic symbols: 0, 1, 2, ..., *b*–1. For bases that exceed ten, more symbols beyond the ten Arabic symbols must be introduced.

The hexadecimal system uses the additional symbols *A*, *B*, *C*, *D*, *E* and *F* to represent 10, 11, 12, 13, 14 and 15 respectively. The binary number system is of particular interest to us as the two symbols 0 and 1 correspond to the two bits that represent the states of a switch. The octal and hexadecimal number systems are also frequently encountered in computing.

Examples: Convert the following numbers into their decimal equivalent.

$(1101.101)_2 = (1 ´ 2^3) + (1 ´ 2^2) + (1 ´ 2^0) + (1 ´ 2^{-1}) + (1 ´ 2^{-3})$

$= 8 + 4 + 1 + 0.5 + 0.125 = 13.625$

$(572.6)_8 = (5 ´ 8^2) + (7 ´ 8^1) + (2 ´ 8^0) + (6 ´ 8^{-1})$

$= 320 + 56 + 2 + 0.75 = 378.75$

$(2A.8)_{16} = (2 ´ 16^1) + (10 ´ 16^0) + (8 ´ 16^{-1})$

$= 32 + 10 + 0.5 = 42.5$

$(341.24)_5 = (3 ´ 5^2) + (4 ´ 5^1) + (1 ´ 5^0) + (2 ´ 5^{-1}) + (4 ´ 5^{-2})$

$= 75 + 20 + 1 + 0.4 + 0.16 = 96.56$

DECIMAL TO BINARY CONVERSION

One way to con vert a decimal number to its binary equivalent is the *sum-of-weights* method. Here, we determine the set of weights (which are in powers of two) whose sum is the number in question.

Examples: Convert decimal numbers to their binary equivalent using sum-of-weights.

$(9)_{10} = 8 + 1 = 2^3 + 2^0 = (1001)_2$
$(18)_{10} = 16 + 2 = 2^4 + 2^1 = (10010)_2$
$(58)_{10} = 32 + 16 + 8 + 2 = 2^5 + 2^4 + 2^3 + 2^1 = (111010)_2$
$(0.625)_{10} = 0.5 + 0.125 = 2^{-1} + 2^{-3} = (0.101)_2$

We observe that multiplying a number x by 2^n is equivalent to shifting left the binary representation of x by n positions, with zeroes appended on the right. For example, the value $(5)_{10}$ is represented as $(101)_2$, and $(40)_{10}$ (which is 5 $\times 2^3$) is $(101000)_2$. Similarly, dividing a number x by 2^n is equivalent to shifting right its binary representation by n positions. This may help in quicker conversion.

The second method deals with the integral portion and the fractional portion separately, as follows:

- Integral part: *Repeated division-by-two*
- Fractional part: *Repeated multiplication-by-two*

CONVERSION BETWEEN BASES

Decimal to Base-*R* Conversion

We may extend the repeated-division and repeated-multiplication techniques to convert a decimal value into its equivalent in base R:

- Integral part: *Repeated division-by-R*
- Fractional part: *Repeated multiplication-by-R*

Truncation and Rounding

In the course of converting values – especially fractional values – between bases, there might be instances when the values are to be corrected within a specific number of places. This may be done by *truncation* or *rounding*.

In truncation, we simply chop a portion off from the fraction. Table shows the truncation of the value $(0.12593)_{10}$ with respect to the number of decimal places desired.

Table. Truncation versus Rounding

Numberof Decimal Places	Truncated Value	Rounded Value
4	$(0.1259)_{10}$	$(0.1259)_{10}$
3	$(0.125)_{10}$	$(0.126)_{10}$
2	$(0.12)_{10}$	$(0.13)_{10}$
1	$(0.1)_{10}$	$(0.1)_{10}$

In rounding, we need to examine the leading digit of the portion we intend to remove. We divide the digits of the numeral system into two equal-sized sets, where the first set contains the 'lighter' digits (Example: 0, 1, 2, 3, 4 in the decimal system), and the second set the 'heavier' digits (5, 6, 7, 8, 9).

If that digit being examined belongs to the former set, we remove the unwanted portion; if it belongs to the latter set, we 'promote' (round up) the last digit in the retained portion to its next higher value (propagating the promotion if necessary). Table 2-2 above shows the rounding of the value $(0.12593)_{10}$ with respect to the specified number of decimal places.

For bases that are odd numbers, there exists a 'middle' digit that may be classified either as a 'light' digit or a 'heavy' digit. For example, $(6.143)_7$ may be rounded to two places either as $(6.14)_7$ or $(6.15)_7$.

Conversion between Binary and Octal/Hexadecimal

There exist simple techniques for conversion between binary numbers and octal (or hexadecimal) numbers.

They are given below:

- *Binary* ® *Octal:* Partition (from the radix point outwards) in groups of 3; each group of 3 bits corresponds to an equivalent octal digit.
- *Octal* ® *Binary:* Expand each octal digit into an equivalent group of 3 bits.
- *Binary* ® *Hexadecimal:* Partition (from the radix point outwards) in groups of 4; each group of 4 bits corresponds to an equivalent hexadecimal digit.
- *Hexadecimal* ® *Binary:* Expand each hexadecimal digit into an equivalent group of 4 bits.

Examples: Conversion between binary and octal/hexadecimal.

$(10\ 111\ 011\ 001.\ 101\ 11)_2 = (2731.56)_8$

$(2731.56)_8 = (010\ 111\ 011\ 001.\ 101\ 110)_2$

$(101\ 1101\ 1001.\ 1011\ 1)_2 = (5D9.B8)_{16}$

$(5D9.B8)_{16} = (0101\ 1101\ 1001.\ 1011\ 1000)_2$

Octal and hexadecimal notations are commonly encountered in computing literature as they provide more compact writing compared to binary notation, and they can be easily converted into binary form.

On closer observation, it is not difficult to uncover the underlying principle behind these techniques. (Hint: 8 is 2^3, and 16 is 2^4.) With this, we can design similar techniques to convert between related bases, such as between binary and quaternary (base 4), and between ternary (base 3) and nonary (base 9).

General Conversion

Apart from the techniques for conversion between related bases such as those discussed above, for general conversion between two bases, the approach is to convert the given value first into decimal, followed by converting the decimal value into the target base.

5

Analysis of Circuits and Transistor Amplifier

COMMON EMITTER AMPLIFIER CIRCUIT

In the Bipolar Transistor tutorial, we saw that the most common circuit configuration for an NPN transistor is that of the Common Emitter Amplifier and that a family of curves known commonly as the Output Characteristics Curves, relates the Collector current (I_c), to the output or Collector voltage (V_{ce}), for different values of Base current (Ib).

All types of transistor amplifiers operate using AC signal inputs which alternate between a positive value and a negative value so some way of "presetting" the amplifier circuit to operate between these two maximum or peak values is required. This is achieved using a process known as Biasing. Biasing is very important in amplifier design as it establishes the correct operating point of the transistor amplifier ready to receive signals, thereby reducing any distortion to the output signal.

We also saw that a static or DC load line can be drawn onto these output characteristics curves to show all the possible operating points of the transistor from fully "ON" to fully "OFF", and to which the quiescent operating point or Q–point of the amplifier can be found. The aim of any small signal amplifier is to amplify all of the input signal with the minimum amount of distortion possible to the output signal, in other words, the output signal must be an exact reproduction of the input signal but only bigger (amplified).

To obtain low distortion when used as an amplifier the operating quiescent point needs to be correctly selected. This is in fact the DC operating point of the amplifier and its position may be established at any point along the load line by a suitable biasing arrangement. The best possible position for this Q–point is as close to the centre position of the load line as reasonably possible, thereby producing a Class A type amplifier operation, ie. Vce = 1/2Vcc. Consider the Common Emitter Amplifier circuit shown below.

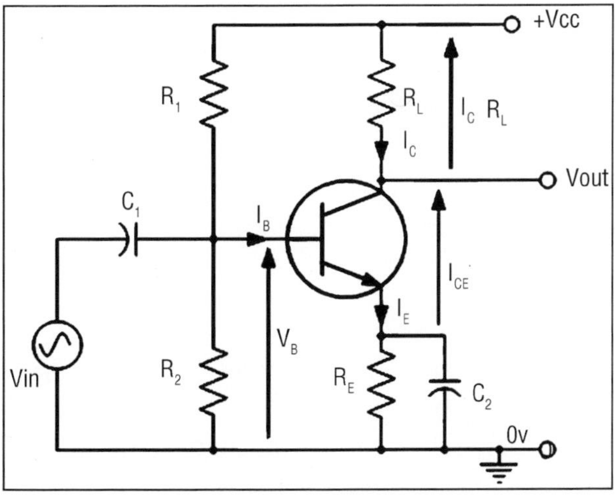

Fig: The Common Emitter Amplifier Circuit

The single stage common emitter amplifier circuit shown above uses what is commonly called "Voltage Divider Biasing". This type of biasing arrangement uses two resistors as a potential divider network and is commonly used in the design of bipolar transistor amplifier circuits. This method of biasing the transistor,

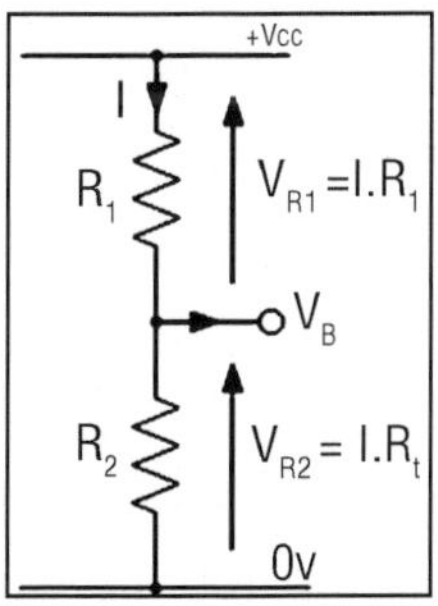

greatly reduces the effects of varying Beta, (β) by holding the Base bias at a constant steady voltage level allowing for best stability. The quiescent Base voltage (V_b) is determined by the potential divider network formed by the two resistors, R_1, R_2 and the power supply voltage Vcc as shown with the current flowing through both resistors. Then the total resistance RT will be equal to $R_1 + R_2$ giving the current as i = Vcc/RT. The voltage level generated at the junction of resistors R_1 and R_2 holds the Base voltage (V_b) constant at a value below the supply voltage. Then the potential divider network used in the common emitter amplifier circuit divides the input signal in proportion to the resistance.

This bias reference voltage can be easily calculated using the simple voltage divider formula below:

$$V_B = \frac{VccR_2}{R_1 + R_2}.$$

The same supply voltage, (Vcc) also determines the maximum Collector current, I_C when the transistor is switched fully "ON" (saturation), $V_{ce} = 0$. The Base current Ib for the transistor is found from the Collector current, Ic and the DC current gain Beta, β of the transistor.

$$\beta = \frac{\Delta I_C}{\Delta I_B}.$$

Beta is sometimes referred to as h_{FE} which is the transistors forward current gain in the common emitter configuration. Beta has no units as it is a fixed ratio of the two currents, Ic and Ib so a small change in the Base current will cause a large change in the Collector current. One final point about Beta. Transistors of the same type and part number will have large variations in their Beta value for example, the BC107 NPN Bipolar transistor has a DC current gain Beta value of between 110 and 450 (data sheet value) this is because Beta is a characteristic of their construction and not their operation.

As the Base/Emitter junction is forward–biased, the Emitter voltage, Ve will be one junction voltage drop different to the Base voltage. If the voltage across the Emitter resistor is known then the Emitter current,Ie can be easily calculated using Ohm's Law. The Collector current, Ic can be approximated, since it is almost the same value as the Emitter current.

Example

A common emitter amplifier circuit has a load resistance, R_L of 1.2kΩs and a supply voltage of 12v. Calculate the maximum Collector current (Ic) flowing through the load resistor when the transistor is switched fully "ON" (saturation), assume Vce = 0. Also find the value of the Emitter resistor, R_E with a voltage drop of 1v across it. Calculate the values of all the other circuit resistors assuming an NPN silicon transistor.

$$I_{C(MAX)} = \frac{Vcc - V_{RE}}{R_L} \frac{12-1}{1200} = 9.2mA$$

$$V_{CE} = 0(Saturation)$$

This then establishes point "A" on the Collector current vertical axis of the characteristics curves and occurs when Vce = 0. When the transistor is switched fully "OFF", their is no voltage drop across either resistor R_E or R_L as no current is flowing through them. Then the voltage drop across the transistor, Vceis equal to the supply voltage, Vcc. This then establishes point

"B" on the horizontal axis of the characteristics curves. Generally, the quiescent Q–point of the amplifier is with zero input signal applied to the Base, so the Collector sits half–way along the load line between zero volts and the supply voltage, ($V_{cc}/2$). Therefore, the Collector current at the Q–point of the amplifier will be given as:

$$I_{C(Q)} = \frac{\frac{12-1}{2}}{1200} = \frac{5.5}{1200} = 4.58mA$$

This static DC load line produces a straight line equation whose slope is given as:$-1/(R_L + R_E)$ and that it crosses the vertical Ic axis at a point equal to $Vcc/(R_L + R_E)$. The actual position of the Q–point on the DC load line is determined by the mean value of Ib.

As the Collector current, Ic of the transistor is also equal to the DC gain of the transistor (Beta), times the Base current ($\beta \times I_b$), if we assume a Beta (β) value for the transistor of say 100, (one hundred is a reasonable average value for low power signal transistors) the Base current Ib flowing into the transistor will be given as:

$$\beta = \frac{I_C}{I_B}$$

$$\therefore I_B = \frac{I_C}{\beta} = \frac{4.58mA}{100} = 45.8\mu A$$

Instead of using a separate Base bias supply, it is usual to provide the Base Bias Voltage from the main supply rail (V_{cc}) through a dropping resistor, R_1. Resistors, R_1 and R_2 can now be chosen to give a suitable quiescent Base current of 45.8μA or 46μA rounded off. The current flowing through the potential divider circuit has to be large compared to the actual Base current, Ib, so that the voltage divider network is not loaded by the Base current flow. A general rule of thumb is a value of at least 10 times Ib flowing through the resistor R_2. Transistor Base/Emitter voltage, Vbe is fixed at 0.7V (silicon transistor) then this gives the value of R_2 as:

$$R_2 = \frac{V_{(RE)} + V_{(BE)}}{10 \times I_B} = \frac{1+0.7}{458 \times 10^{-6}} = 3.71k\Omega$$

If the current flowing through resistor R_2 is 10 times the value of the Base current, then the current flowing through resistor R_1 in the divider network must be 11 times the value of the Base current. The voltage across resistor R_1 is equal to Vcc–1.7v (V_{RE} + 0.7 for silicon transistor) which is equal to 10.3V, therefore R_1 can be calculated as:

$$R_1 = \frac{V_{CC} - \left(V_{(RE)} + V_{(BE)}\right)}{11 \times I_B}$$

$$= \frac{12 - 1.7}{504 \times 10^{-6}}$$

$$= 20.45 k\Omega$$

The value of the Emitter resistor, R_E can be easily calculated using Ohm's Law. The current flowing through R_E is a combination of the Base current, Ib and the Collector current Ic and is given as:

$$I_E = I_C + I_B = 4.58mA + 45.8\mu A = 4.63mA .$$

Resistor, R_E is connected between the Emitter and ground and we said previously that it has a voltage of 1 volt across it. Then the value of R_E is given as:

$$R_E = \frac{V_{RE}}{I_E} = \frac{Iv}{4.63mA} = 216\Omega .$$

So, for our example above, the preferred values of the resistors chosen to give a tolerance of 5 per cent (E24) are:

$$R_1 = 20k\Omega, R_2 = 3.6k\Omega, R_L = 1.2k\Omega, R_E = 220\Omega$$

Then, our original Common Emitter Amplifier circuit above can be rewritten to include the values of the components that we have just calculated above.

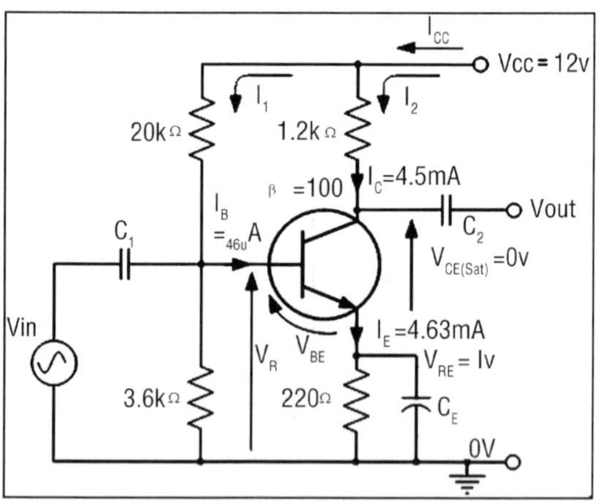

Fig. Completed Common Emitter Circuit

AMPLIFIERS

The basic action of most of the electronic circuits used in behavioural research is one of amplification. That is, an electrical signal of low voltage and/ or amperage is put in and the output is a more powerful signal, correlated in some way with the input. Such amplification is performed not by a process analogous to magnification, but by allowing a low power input to control the larger power of the output, just as the small effort required to open a valve may produce large changes in the amount of power coming out of a water pipe. (The electron tube is called a valve in England.)

Generally, an amplifier or simply amp, is any device that changes, usually increases, the amplitude of a signal. The "signal" is usually voltage or current. The relationship of the input to the output of an amplifier — usually expressed as a function of the input frequency — is called the transfer function of the amplifier, and the magnitude of the transfer function is termed the gain.

In popular use, the term usually refers to an electronic amplifier, often as in audio applications to operate a loudspeaker that is being used in a PA system to make the human voice louder or play recorded music. Amplifiers may be classified by the input (source) they are designed to amplify (such as a guitar amplifier to perform with an electric guitar), or named for the device they are intended to drive (such as a headphone amplifier), or by the frequency range of the signals (Audio, IF, RF and VHF amplifiers for example), or grouped by whether they invert the signal (inverting amplifiers and non-inverting amplifiers, or by the types of device used in the amplification (valve or tube amplifiers, FET amplifiers, etc.).

A related device that emphasizes conversion of signals of one type to another (for example, a light signal in photons to a DC signal in amperes) is a transducer, a transformer, or a sensor. However, none of these amplify power. The quality of an amplifier can be characterized by a number of specifications, listed below.

Gain

The gain of an amplifier is the ratio of output to input power or amplitude, and is usually measured in decibels. (When measured in decibels it is logarithmically related to the power ratio: $G(\text{dB})=10\ \log(P_{out}/(P_{in}))$. RF amplifiers are often specified in terms of the maximum power gain obtainable, while the voltage gain of audio amplifiers and instrumentation amplifiers will be more often specified (since the amplifier's input impedance will often be much higher than the source impedance, and the load impedance higher than the amplifier's output impedance).

Example: An audio amplifier with a gain given as 20dB will have a *voltage gain* of ten (but a power gain of 100 would only occur in the unlikely event the input and output impedances were identical).

BANDWIDTH

The bandwidth (BW) of an amplifier is the range of frequencies for which the amplifier gives "satisfactory performance". The "satisfactory performance" may be different for different applications. However, a common and well-accepted metric are the half power points (*i.e.* frequency where the power goes down by half its peak value) on the power vs. frequency curve. Therefore bandwidth can be defined as the difference between the lower and upper half power points. This is therefore also known as the "3 dB bandwidth. Bandwidths (otherwise called "frequency responses") for other response tolerances are sometimes quoted ("1 dB, "6 dB etc.) or "plus of minus 1dB" (roughly the sound level difference people usually can detect).

A full-range audio amplifier will be essentially flat between 20 Hz to about 20 kHz (the range of normal human hearing). In minimalist amplifier design, the amp's usable frequency response needs to extend considerably beyond this (one or more octaves either side) and typically a good minimalist amplifier will have "3 dB points < 10 and > 65 kHz.

Professional touring amplifiers often have input and/or output filtering to sharply limit frequency response beyond 20 Hz-20 kHz; too much of the amplifier's potential output power would otherwise be wasted on infrasonic and ultrasonic frequencies, and the danger of AM radio interference would increase. Modern switching amplifiers need steep low pass filtering at the output to get rid of high frequency switching noise and harmonics.

Efficiency

Efficiency is a measure of how much of the input power is usefully applied to the amplifier's output. Class A amplifiers are very inefficient, in the range of 10–20 per cent with a max efficiency of 25 per cent. Class B amplifiers have a very high efficiency but are impractical because of high levels of distortion. In practical design, the result of a tradeoff is the class AB design. Modern Class AB amps are commonly between 35–55 per cent efficient with a theoretical maximum of 78.5 per cent.

Commercially available Class D switching amplifiers have reported efficiencies as high as 90 per cent. Amplifiers of Class C-F are usually known to be very high efficiency amplifiers. The efficiency of the amplifier limits the amount of total power output that is usefully available. Note that more efficient amplifiers run much cooler, and often do not need any cooling fans even in multi-kilowatt designs. The reason for this is that the loss of efficiency produces heat as a by-product of the energy lost during the conversion of power. In more efficient amplifiers there is less loss of energy so in turn less heat.

Linearity

An ideal amplifier would be a totally linear device, but real amplifiers are only linear within certain practical limits. When the signal drive to the amplifier is increased, the output also increases until a point is reached where some part of the amplifier becomes saturated and cannot produce any more output; this is called clipping, and results in distortion.

Some amplifiers are designed to handle this in a controlled way which causes a reduction in gain to take place instead of excessive distortion; the result is a *compression* effect, which (if the amplifier is an audio amplifier) will sound much less unpleasant to the ear. For these amplifiers, the 1 dB compression point is defined as the input power (or output power) where the gain is 1 dB less than the small signal gain.

Linearization is an emergent field, and there are many techniques, such as feedforward, predistortion, postdistortion, EER, LINC, CALLUM, cartesian feedback, etc., in order to avoid the undesired effects of the non-linearities.

Noise

This is a measure of how much noise is introduced in the amplification process. Noise is an undesirable but inevitable product of the electronic devices and components. The metric for noise performance of a circuit is Noise Factor. Noise Factor is the ratio of input signal to that of the output signal.

OUTPUT DYNAMIC RANGE

Output dynamic range is the range, usually given in dB, between the smallest and largest useful output levels. The lowest useful level is limited by output noise, while the largest is limited most often by distortion. The ratio of these two is quoted as the amplifier dynamic range. More precisely, if S = maximal allowed signal power and N = noise power, the dynamic range DR is $DR = (S + N)/N$.

Slew Rate

Slew rate is the maximum rate of change of output variable, usually quoted in volts per second (or microsecond). Many amplifiers are ultimately slew rate limited (typically by the impedance of a drive current having to overcome capacitive effects at some point in the circuit), which may limit the full power bandwidth to frequencies well below the amplifier's small-signal frequency response.

Rise Time

The rise time, t_r, of an amplifier is the time taken for the output to change from 10 per cent to 90 per cent of its final level when driven by a step input. For a Gaussian response system (or a simple RC roll off), the rise time is approximated by:

$t_r * BW = 0.35$, where t_r is rise time in seconds and BW is bandwidth in Hz.

Settling Time And Ringing

Time taken for output to settle to within a certain percentage of the final value (say 0.1 per cent). This is called the settle time, and is usually specified for oscilloscope vertical amplifiers and high accuracy measurement systems. Ringing refers to an output that cycles above and below its final value, leading to a delay in reaching final value quantified by the settling time above.

Overshoot

In response to a step input, the overshoot is the amount the output exceeds its final, steady-state value.

Stability Factor

Stability is a major concern in RF and microwave amplifiers. The degree of an amplifiers stability can be quantified by a so-called stability factor. There are several different stability factors, such as the Stern stability factor and the Linvil stability factor, which specify a condition that must be met for the absolute stability of an amplifier in terms of its two-port parameters.

ELECTRONIC AMPLIFIERS

There are many types of electronic amplifiers, commonly used in radio and television transmitters and receivers, high-fidelity ("hi-fi") stereo equipment, microcomputers and other electronic digital equipment, and guitar and other instrument amplifiers. Critical components include active devices, such as vacuum tubes or transistors. A brief introduction to the many types of electronic amplifier follows.

The term amplifier as used in this article can mean either a circuit (or stage) using a single active device or a complete system such as a packaged audio hi-fi amplifier.

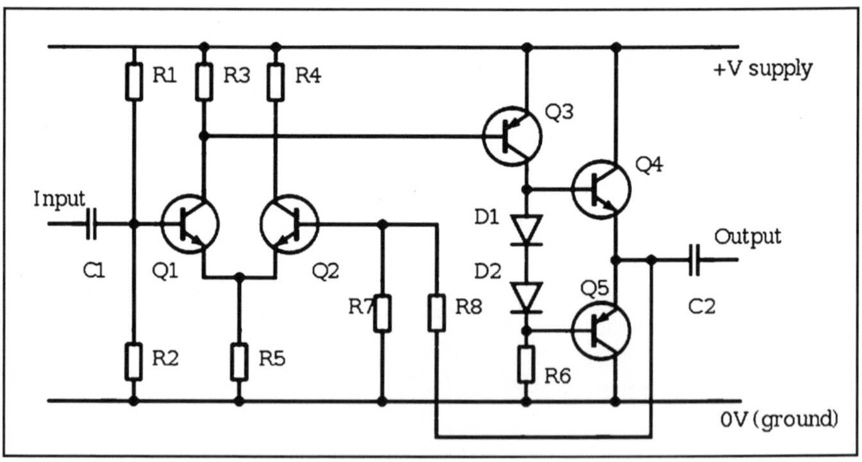

Fig. Types of Amplifier

An electronic amplifier is a device for increasing the power and/or amplitude of a signal. It does this by taking energy from a power supply and controlling the output to match the input signal shape but with a larger amplitude. In this sense, an amplifier may be considered as modulating the output of the power supply.

Amplifiers can be specified according to their input and output properties. They have some kind of gain, or multiplication factor relating the magnitude of the output signal to the input signal. The gain may be specified as "output voltage/input voltage", "output power/input power" or any other combination of current, voltage and power. In many cases, with input and output in the same units, gain will be unitless; for others this is not necessarily so – for example, a transconductance amplifier has a gain with units of conductance (output current per input voltage).

In most cases an amplifier should be linear, that is the gain should be constant for any combination of input and output signal. If the gain is not linear, *e.g.* by clipping the output signal at the limits of its capabilities, the output signal is distorted.

CLASSIFICATION OF AMPLIFIER STAGES AND SYSTEMS

There are many alternative classifications that address different aspects of amplifier designs, and they all express some particular perspective relating the design parameters to the objectives of the circuit. Amplifier design is always a compromise of numerous factors, such as cost, power consumption, real-world device imperfections, and a multitude of performance specifications. Below are several different approaches to classification:

Input and output variables

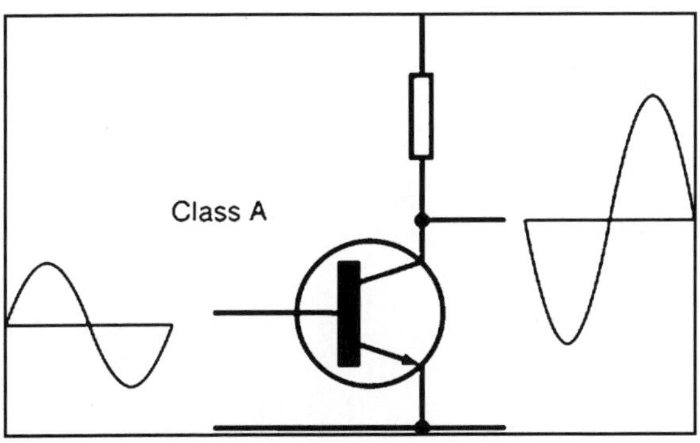

Fig. The Four Types of Dependent Source; control Variable on left, output Variable on Right

Electronic amplifiers use two variables: current and voltage. Either can be used as input, and either as output leading to four types of amplifiers. In practice the ideal impedances are only approximated. For any particular circuit, a small-signal analysis often is used to find the impedance actually achieved. A small-signal AC test current I_x is applied to the input or output node, all external sources are set to zero, and the corresponding alternating voltage V_x across the test current source determines the impedance seen at that node as $R = V_x/I_x$.

Amplifiers designed to attach to a transmission line at input and/or output, especially RF amplifiers, do not fit into this classification approach. Rather than dealing with voltage or current individually, they ideally couple with an input and/or output impedance matched to the transmission line impedance, that is, match *ratios* of voltage to current. Many real RF amplifiers come close to this ideal. Although, for a given appropriate source and load impedance, RF amplifiers can be characterized as amplifying voltage or current, they fundamentally are amplifying power.

Common Terminal

One set of classifications for amplifiers is based on which device terminal is common to both the input and the output circuit. In the case of bipolar junction transistors, the three classes are common emitter, common base, and common collector. For field-effect transistors, the corresponding configurations are common source, common gate, and common drain; for triode vacuum devices, common cathode, common grid, and common plate.

Unilateral or Bilateral

When an amplifier has an output that exhibits no feedback to its input side, it is called unilateral. One consequence is the amplifier has an input impedance that is independent of the load attached to the amplifier, and an output impedance that is independent of the signal source driving the amplifier.

The opposite case is the bilateral amplifier, where feedback connects the output to the input side of the amplifier. Such feedback often is deliberate, for example negative feedback often is used to tailor amplifier behaviour. However, at least as often, feedback is both undesirable and unavoidable; introduced, for example, by parasitic elements like inherent, undesirable capacitances in transistors that couple input to output. In any case, a bilateral amplifier has an input impedance that depends upon the load attached to the amplifier, and an output impedance that depends on the source driving the amplifier.

Linear unilateral and bilateral amplifiers can be represented by two-port networks. Most amplifiers are bilateral to some degree, however they may often be modeled as unilateral under certain operating conditions to simplify the analysis.

Inverting or Non-Inverting

Another way to classify amps is the phase relationship of the input signal to the output signal. An inverting amplifier produces an output 180 degrees out of phase with the input signal (that is, an inversion or mirror image of the input as seen on an oscilloscope). A non-inverting amplifier maintains the phase of the input signal waveforms. An emitter follower is a type of non-inverting amplifier, indicating that the signal at the emitter of a transistor is following (that is, matching with unity gain but perhaps an offset) the input signal. This description can apply to a single stage of an amplifier, or to a complete amplifier system.

Function

Other amplifiers may be classified by their function or output characteristics. These functional descriptions usually apply to complete amplifier systems or sub-systems and rarely to individual stages.

- A servo amplifier indicates an integrated feedback loop to actively control the output at some desired level. A DC servo indicates use at frequencies down to DC levels, where the rapid fluctuations of an audio or RF signal do not occur. These are often used in mechanical actuators, or devices such as DC motors that must maintain a constant speed or torque. An AC servo amp can do this for some ac motors.
- A linear amplifier responds to different frequency components independently, and does not generate harmonic distortion or intermodulation distortion. A nonlinear amplifier does generate distortion.
- A wideband amplifier has a precise amplification factor over a wide range of frequencies, and is often used to boost signals for relay in communications systems. A narrowband amp is made to amplify only a specific narrow range of frequencies, to the exclusion of other frequencies.
- An RF amplifier refers to an amplifier designed for use in the radio frequency range of the electromagnetic spectrum, and is often used to increase the sensitivity of a receiver or the output power of a transmitter.
- An audio amplifier is designed for use in reproducing audio frequencies. This category subdivides into small signal amplification, and power amps which are optimised for driving speakers, sometimes with multiple amps grouped together as separate or bridgeable channels to accommodate different audio reproduction requirements.
- A special type of amplifier is widely used in instruments and for signal processing, among many other varied uses. These are known as operational amplifiers, (or op-amps). This is because this type of amplifier is used in circuits that perform mathematical algorithmic functions, or "operations" on input signals to obtain specific types of output signals. A typical op-amp has differential inputs (one "inverting", one, "non-inverting" relative to the output) and one

output. An idealised op-amp has the following characteristics:

- Infinite input impedance (so as to not load circuitry it is sampling as a control input)
- Zero output impedance
- Infinite gain
- Zero propagation delay

The performance of an op-amp with these characteristics would be entirely defined by the (usually passive) components forming a negative feedback loop around it, that is, *the amplifier itself has no effect on the output*.

Today, op-amps are usually provided as integrated circuits, rather than constructed from discrete components. All real-world op-amps fall short of the idealised specification above – but some modern components have remarkable performance and come close in some respects.

Interstage Coupling Method

Amplifiers are sometimes classified by the coupling method of the signal at the input, output, or between stages. Different types of these include:

Resistive-capacitive (RC) coupled amplifier, using a network of resistors and capacitors

By design these amplifiers cannot amplify DC signals as the capacitors block the DC component of the input signal. RC-coupled amplifiers were used very often in circuits with vacuum tubes or discrete transistors. In the days of the integrated circuit a few more transistors on a chip are much cheaper and smaller than a capacitor.

Inductive-capacitive (LC) coupled amplifier, using a network of inductors and capacitors

This kind of amplifier is most often used in selective radio-frequency circuits.

Transformer coupled amplifier, using a transformer to match impedances or to decouple parts of the circuits

Quite often LC-coupled and transformer-coupled amplifiers cannot be distinguished as a transformer is some kind of inductor.

Direct coupled amplifier, using no impedance and bias matching components

This class of amplifier was very uncommon in the vacuum tube days when the anode (output) voltage was at greater than several 100 V and the grid (input) voltage at a few volts minus. So they were only used if the gain was specified down to DC (*e.g.*, in an oscilloscope). In the context of modern electronics developers are encouraged to use direcly coupled amplifiers whenever possible.

Frequency Range

- Depending on the frequency range and other properties amplifiers are designed according to different principles.

- Frequency ranges down to DC are only used when this property is needed. DC amplification leads to specific complications that are avoided if possible.
- Depending on the frequency range specified different design principles must be used. Up to the MHz range only "discrete" properties need be considered; *e.g.*, a terminal has an input impedance.
- As soon as any connection within the circuit gets longer than perhaps 1 per cent of the wavelength of the highest specified frequency (*e.g.*, at 100 MHz the wavelength is 3 m, so the critical connection length is approx. 3 cm) design properties radically change. For example, a specified length and width of a PCB trace can be used as a selective or impedance-matching entity.
- Above a few 100 MHz, it gets difficult to use discrete elements, especially inductors. In most cases PCB traces of very closely defined shapes are used instead.

Type of Load

- Untuned
  - Audio
  - Video

Tuned (RF amps) - used for amplifying a single radio frequency or band of frequencies

Implementation

Amplifiers are implemented using active elements of different kinds:

- The first active elements were relays. They were for example used in trans-continental telegraph lines: A weak current was used to switch the voltage of a battery to the outgoing line.
- For transmitting audio, carbon microphones were used as the active element. This was used to modulate a radio-frequency source in one of the first AM audio transmissions, by Reginald Fessenden on Dec. 24, 1906.
- Up to the early 1970s, most amplifiers used vacuum tubes (*valves* in the UK). Today, tubes are only generally used for very high-power, high-frequency amplifiers and for specialist audio applications, in which field they have recently achieved a new popularity. Many broadcast transmitters still use vacuum tubes. Additionally, their imperviousness to electromagnetic flash damage may have led to their retention in certain defence contexts.
- In the 1960s, the transistor started to take over. These days, discrete transistors are still used in high-power amplifiers and in specialist audio devices.

- Beginning in the 1970s, more and more transistors were connected on a single chip therefore creating the integrated circuit. Nearly all amplifiers commercially available today are based on integrated circuits.

For exotic purposes, other active elements have been used. For example, in the early days of the communication satellite parametric amplifiers were used. The core circuit was a diode whose capacity was changed by an RF signal created locally. Under certain conditions, this RF signal provided energy that was modulated by the extremely weak satellite signal received at the earth station. The operating principle of a parametric amplifier is somewhat similar to the principle by which children keep their swings in motion: as long as the swing moves you only need to change a parameter of the swinging entity; *e.g.*, you must move your centre of gravity up and down. In our case, the capacity of the diode is changed periodically.

Power amplifier classes

Angle of flow or conduction angle

Power amplifier circuits (output stages) are classified as A, B, AB and C for analog designs, and class D and E for switching designs based upon the conduction angle or *angle of flow*, È, of the input signal through the amplifying device, that is, the portion of the input signal cycle during which the amplifying device conducts. The image of the conduction angle is derived from amplifying a sinusoidal signal. (If the device is always on, È = 360°.) The angle of flow is closely related to the amplifier power efficiency. The various classes are introduced below, followed by more detailed discussion under individual headings later on.

Class A

100 per cent of the input signal is used (conduction angle È = 360° or 2ð; *i.e.*, the active element works in its linear range all of the time). Where efficiency is not a consideration, most small signal linear amplifiers are designed as Class A, which means that the output devices are always in the conduction region. Class A amplifiers are typically more linear and less complex than other types, but are very inefficient. This type of amplifier is most commonly used in small-signal stages or for low-power applications (such as driving headphones).

Class B

50 per cent of the input signal is used (È = 180° or ð; *i.e.*, the active element works in its linear range half of the time and is more or less turned off for the other half). In most Class B, there are two output devices (or sets of output devices), each of which conducts alternately (push–pull) for exactly 180° (or half cycle) of the input signal; selective RF amplifiers can also be implemented using a single active element.

These amplifiers are subject to *crossover distortion* if the transition from one active element to the other is not perfect, as when two complementary transistors (*i.e.*, one PNP, one NPN) are connected as two emitter followers with their base and emitter terminals in common, requiring the base voltage to slew across the region where both devices are turned off.

Class AB

Here the two active elements conduct more than half of the time as a means to reduce the cross-over distortions of Class B amplifiers. In the example of the complementary emitter followers a bias network allows for more or less quiescent current thus providing an operating point somewhere between Class A and Class B. Sometimes a figure is added (*e.g.*, AB_1 or AB_2) with higher figures implying a higher quiescent current and therefore more of the properties of Class A.

Class D

Main article: Switching amplifier

These use switching to achieve a very high power efficiency (more than 90 per cent in modern designs). By allowing each output device to be either fully on or off, losses are minimized. The analog output is created by pulse-width modulation; *i.e.*, the active element is switched on for shorter or longer intervals instead of modifying its resistor. There are more complicated switching schemes like sigma-delta modulation, to improve some performance aspects like lower distortions or better efficiency.

Other classes

There are several other amplifier classes, although they are mainly variations of the previous classes. For example, Class G and Class H amplifiers are marked by variation of the supply rails (in discrete steps or in a continuous fashion, respectively) following the input signal. Wasted heat on the output devices can be reduced as excess voltage is kept to a minimum. The amplifier that is fed with these rails itself can be of any class. These kinds of amplifiers are more complex, and are mainly used for specialized applications, such as very high-power units. Also, Class E and Class F amplifiers are commonly described in literature for radio frequencies applications where efficiency of the traditional classes deviate substantially from their ideal values. These classes use harmonic tuning of their output networks to achieve higher efficiency and can be considered a subset of Class C due to their conduction angle characteristics.

More detail on the various classes is provided below.

Class A

Class A amplifying devices operate over the whole of the input cycle such that the output signal is an exact scaled-up replica of the input with no clipping. Class A amplifiers are the usual means of implementing small-signal amplifiers. They are not very efficient; a theoretical maximum of 50 per cent is obtainable with inductive output coupling and only 25 per cent with capacitive coupling.

In a Class A circuit, the amplifying element is biased so the device is always conducting to some extent, and is operated over the most linear portion of its characteristic curve (known as its transfer characteristic or transconductance curve). Because the device is always conducting, even if there is no input at all, power is drawn from the power supply. This is the chief reason for its inefficiency.

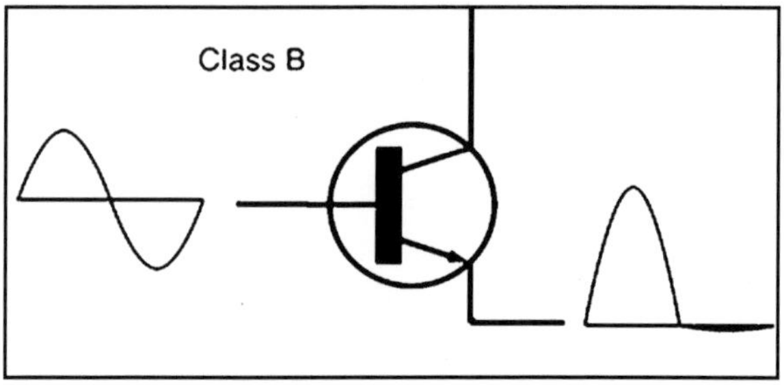

Fig.. Class A Amplifier

If high output powers are needed from a Class A circuit, the power waste (and the accompanying heat) will become significant. For every watt delivered to the load, the amplifier itself will, at best, dissipate another watt. For large powers this means very large and expensive power supplies and heat sinking. Class A designs have largely been superseded for audio amplifiers, though some audiophiles believe that Class A gives the best sound quality, due to it being operated in as linear a manner as possible which provides a small market for expensive high fidelity Class A amps. In addition, some aficionados prefer thermionic valve (or "tube") designs instead of transistors, for several claimed reasons:

- Tubes are more commonly used in class A designs, which have an asymmetrical transfer function. This means that distortion of a sine wave creates both odd- and even-numbered harmonics. The claim is that this sounds more "musical" than the higher level of odd harmonics produced by a symmetrical push–pull amplifier.
- Though good amplifier design can reduce harmonic distortion patterns to almost nothing, distortion is essential to the sound of electric guitar amplifiers, for example, and is held by recording engineers to offer more flattering microphones and to enhance "clinical-sounding" digital technology.
- Valves use many more electrons at once than a transistor, and so statistical effects lead to a "smoother" approximation of the true waveform. Junction field-effect transistors (JFETs) have similar

characteristics to valves, so these are found more often in high quality amplifiers than bipolar transistors. Historically, valve amplifiers often used a Class A power amplifier simply because valves are large and expensive; many Class A designs use only a single device.

Transistors are much cheaper, and so more elaborate designs that give greater efficiency but use more parts are still cost-effective. A classic application for a pair of class A devices is the long-tailed pair, which is exceptionally linear, and forms the basis of many more complex circuits, including many audio amplifiers and almost all op-amps. Class A amplifiers are often used in output stages of op-amps; they are sometimes used as medium-power, low-efficiency, and high-cost audio amplifiers. The power consumption is unrelated to the output power. At idle (no input), the power consumption is essentially the same as at high output volume. The result is low efficiency and high heat dissipation.

Class B and AB

Class B amplifiers only amplify half of the input wave cycle. As such they create a large amount of distortion, but their efficiency is greatly improved and is much better than Class A. Class B has a maximum theoretical efficiency of 78.5 per cent (*i.e.*, ð/4). This is because the amplifying element is switched off altogether half of the time, and so cannot dissipate power. A single Class B element is rarely found in practice, though it can be used in RF power amplifier where the distortion levels are less important. However Class C is more commonly used for this.

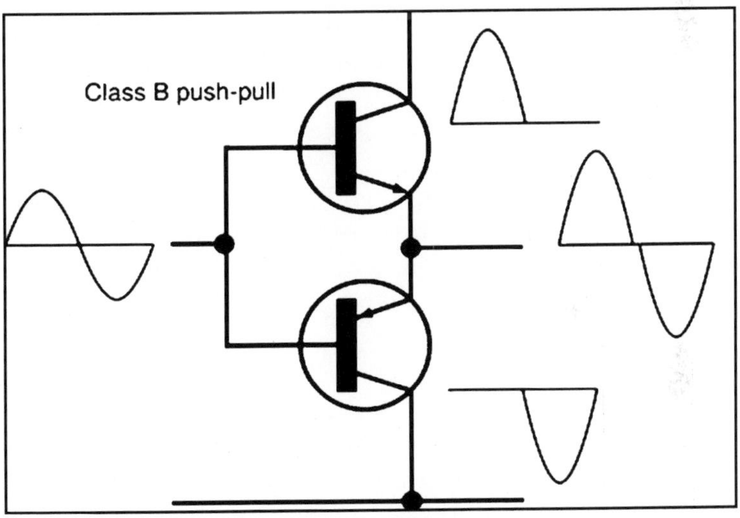

Fig. Class B Amplifier

A practical circuit using Class B elements is the complementary pair or "push–pull" arrangement. Here, complementary or quasi-complementary devices are used to each amplify the opposite halves of the input signal, which is then recombined at the output. This arrangement gives excellent efficiency, but can suffer from the drawback that there is a small mismatch at the "joins" between the two halves of the signal. This is called crossover distortion. An improvement is to bias the devices so they are not completely off when they're not in use. This approach is called Class AB operation.

In Class AB operation, each device operates the same way as in Class B over half the waveform, but also conducts a small amount on the other half. As a result, the region where both devices simultaneously are nearly off (the "dead zone") is reduced. The result is that when the waveforms from the two devices are combined, the crossover is greatly minimised or eliminated altogether.

Class AB sacrifices some efficiency over class B in favour of linearity, so will always be less efficient (below 78.5 per cent). It is typically much more efficient than class A.

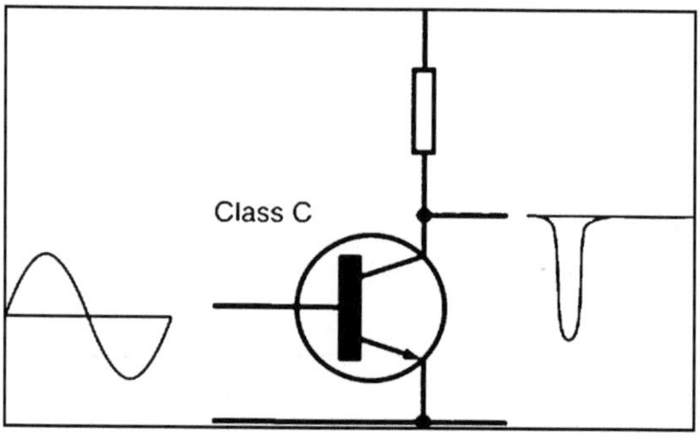

Fig. Class B push–pull Amplifier

Class B or AB push–pull circuits are the most common design type found in audio power amplifiers. Class AB is widely considered a good compromise for audio amplifiers, since much of the time the music is quiet enough that the signal stays in the "class A" region, where it is amplified with good fidelity, and by definition if passing out of this region, is large enough that the distortion products typical of class B are relatively small. The crossover distortion can be reduced further by using negative feedback. Class B and AB amplifiers are sometimes used for RF linear amplifiers as well. Class B amplifiers are also favoured in battery-operated devices, such as transistor radios.

Digital Class B

A limited power output Class-B amplifier with a single-ended supply rail of 5±0.5 V.

Class C

Class C amplifiers conduct less than 50 per cent of the input signal and the distortion at the output is high, but high efficiencies (up to 90 per cent) are possible. Some applications (for example, megaphones) can tolerate the distortion. A much more common application for Class C amplifiers is in RF transmitters, where the distortion can be vastly reduced by using tuned loads on the amplifier stage. The input signal is used to roughly switch the amplifying device on and off, which causes pulses of current to flow through a tuned circuit.

The Class C amp. has two modes of operation: tuned, and untuned. The diagram below shows a waveform from a simple class C circuit without the tuned load. This is called untuned operation, and the analysis of the waveforms shows the massive distortion that appears in the signal. When the proper load (*e.g.*, a pure inductive-capacitive filter) is used, two things happen. The first is that the output's bias level is clamped, so that the output variation is centered at one-half of the supply voltage. This is why tuned operation is sometimes called a *clamper*. This action of elevating bias level allows the waveform to be restored to its proper shape, allowing a complete waveform to be re-established despite having only a one-polarity supply. This is directly related to the second phenomenon: the waveform on the centre frequency becomes much less distorted. The distortion that is present is dependent upon the bandwidth of the tuned load, with the centre frequency seeing very little distortion, but greater attenuation the farther from the tuned frequency that the signal gets.

The tuned circuit will only resonate at particular frequencies, and so the unwanted frequencies are dramatically suppressed, and the wanted full signal (sine wave) will be extracted by the tuned load (*e.g.*, a high-quality bell will ring at a particular frequency when it is hit periodically with a hammer). Provided the transmitter is not required to operate over a very wide band of frequencies, this arrangement works extremely well. Other residual harmonics can be removed using a filter.

Class C Amplifier

Class D

Block diagram of a basic switching or PWM (Class-D) amplifier.

Class D amplifiers are much more efficient than Class AB power amplifiers. As such, Class D amplifiers do not need large transformers and heavy heatsinks, which means that they are smaller and lighter in weight than an equivalent Class AB amplifier. All power devices in a Class D amplifier are operated in on/off mode. Output stages such as those used in pulse generators are examples of class D amplifiers. The term usually applies to devices intended to reproduce signals with a bandwidth well below the switching frequency.

These amplifiers use pulse width modulation, pulse density modulation (sometimes referred to as pulse frequency modulation) or more advanced form of modulation such as Delta-sigma modulation (for example, in the Analog Devices AD1990 Class-D audio power amplifier).

The input signal is converted to a sequence of pulses whose averaged value is directly proportional to the instantaneous amplitude of the signal. The frequency of the pulses is typically ten or more times the highest frequency of interest in the input signal. The output of such an amplifier contains unwanted spectral components (that is, the pulse frequency and its harmonics) which must be removed by a passive filter. The resulting filtered signal is then an amplified replica of the input.

The main advantage of a class D amplifier is power efficiency. Because the output pulses have a fixed amplitude, the switching elements (usually MOSFETs, but valves and bipolar transistors were once used) are switched either on or off, rather than operated in linear mode. This means that very little power is dissipated by the transistors, except during the very short interval between the on and off states. The wasted power is low because the instantaneous power dissipated in the transistor is the product of voltage and current, and one or the other is almost always close to zero. The lower losses permit the use of a smaller heat sink while the power supply requirements are lessened too.

Class D amplifiers can be controlled by either analog or digital circuits. The digital control introduces additional distortion called *quantization error* caused by its conversion of the input signal to a digital value.

Class D amplifiers have been widely used to control motors, and almost exclusively for small DC motors, but they are now also used as audio amplifiers, with some extra circuitry to allow analogue to be converted to a much higher frequency pulse width modulated signal. The relative difficulty of achieving good audio quality means that nearly all are used in applications where quality is not a factor, such as modestly-priced bookshelf audio systems and "DVD-receivers" in mid-price home theater systems.

High quality Class D audio amplifiers are now, however, starting to appear in the market:

- Tripath have called their revised Class D designs Class T.
- Bang and Olufsen's ICEPower Class D system has been used in the Alpine PDX range and some Pioneer's PRS range and for other manufacturers' equipment.

These revised designs have been said to rival good traditional AB amplifiers in terms of quality.

Before these higher quality designs existed an earlier use of Class D amplifiers and prolific area of application was high-powered, subwoofer amplifiers in cars. Because subwoofers are generally limited to a bandwidth of no higher than 150 Hz, the switching speed for the amplifier does not have to be as high as for a full range amplifier. The drawback with Class D designs being used to power subwoofers is that their output filters (typically inductors that convert the pulse width signal back into an analogue waveform) lower the damping factor of the amplifier.

This means that the amplifier cannot prevent the subwoofer's reactive nature from lessening the impact of low bass sounds (as explained in the feedback part of the Class AB section). Class D amplifiers for driving subwoofers are relatively inexpensive, in comparison to Class AB amplifiers. A 1000 watt Class D subwoofer amplifier that can operate at about 80 per cent to 95 per cent efficiency costs about $250 USD, much less than a Class AB amplifier of this power, which would cost several thousand dollars.

The letter *D* used to designate this amplifier class is simply the next letter after *C*, and does not stand for *digital*. Class D and Class E amplifiers are sometimes mistakenly described as "digital" because the output waveform superficially resembles a pulse-train of digital symbols, but a Class D amplifier merely converts an input waveform into a continuously pulse-width modulated (square wave) analog signal. (A digital waveform would be pulse-code modulated.)

Special classes

Class E

The class E/F amplifier is a highly efficient switching power amplifier, typically used at such high frequencies that the switching time becomes comparable to the duty time. As said in the class-D amplifier the transistor is connected via a serial-LC-circuit to the load, and connected via a large L (inductance) to the supply voltage. The supply voltage is connected to ground via a large capacitor to prevent any RF-signals leaking into the supply. The class-E amplifier adds a *C* between the transistor and ground and uses a defined L_1 to connect to the supply voltage.

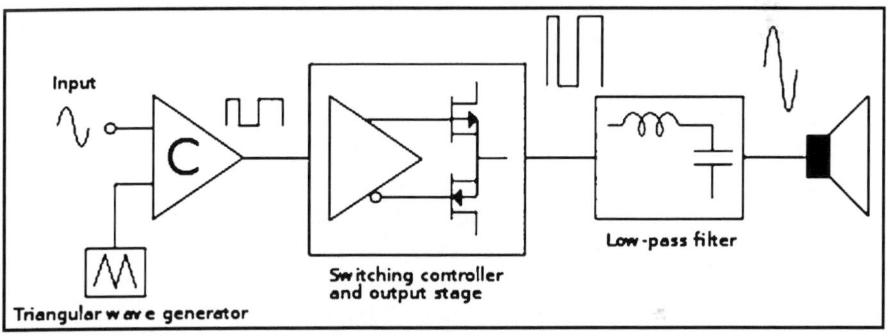

Class E Amplifier

The following description ignores DC, which can be added afterwards easily. The above mentioned *C* and L are in effect a parallel LC-circuit to ground. When the transistor is on, it pushes through the serial LC-circuit into the load and some current begins to flow to the parallel LC-circuit to ground. Then the serial LC-circuit swings back and compensates the current into the parallel LC-circuit. At this point the current through the transistor is zero and it is switched off. Both LC-circuits are now filled with energy in the C and the L_0. The whole circuit performs a damped oscillation. The damping by the load has been adjusted so that some time later the energy from the Ls is gone into the load, but the energy in both C_0 peaks at the original value, to in turn restore the original voltage, so that the voltage across the transistor is zero again and it can be switched on.

With load, frequency, and duty cycle (0.5) as given parameters and the constraint that the voltage is not only restored, but peaks at the original voltage, the four parameters (L, L_0, C and C_0) are determined. The class F-amplifier takes the finite on resistance into account and tries to make the current touch the bottom at zero. This means the voltage and the current at the transistor are symmetric with respect to time. The Fourier transform allows an elegant formulation to generate the complicated LC-networks. It says that the first harmonic is passed into the load, all even harmonics are shorted and all higher odd harmonics are open.

Class F and the even harmonics

In push–pull amplifiers and in CMOS, the even harmonics of both transistors just cancel. Experiment shows that a square wave can be generated by those amplifiers and theory shows that square waves do consist of odd harmonics only. In a class D amplifier, the output filter blocks all harmonics; *i.e.*, the harmonics see an open load. So even small currents in the harmonics suffice to generate a voltage square wave. The current is in phase with the voltage applied to filter, but the voltage across the transistors is out of phase. Therefore, there is a minimal overlap between current through the transistors and voltage across the transistors. The sharper the edges the lower the overlap.

While class D sees the transistors and the load as two separate modules, the class F admits imperfections like the parasitics of the transistor and tries to optimise the global system to have a high impedance at the harmonics. Of course there has to be a finite voltage across the transistor to push the current across the on state resistance. Because the combined current through both transistors is mostly in the first harmonic it looks like a sine. That means that in the middle of the square the maximum of current has to flow, so it may make sense to have a dip in the square or in other words to allow some over swing of the voltage square wave. A class F load network by definition has to transmit below a cut off frequency and to reflect above.

Any frequency lying below the cut off and having its second harmonic above the cut off can be amplified, that is an octave bandwidth. On the other hand, an inductive-capacitive series circuit with a large inductance and a tunable capacitance may be simpler to implement. By reducing the duty cycle below 0.5, the output amplitude can be modulated. The voltage square waveform will degrade, but any overheating is compensated by the lower overall power flowing. Any load mismatch behind the filter can only act on the first harmonic current waveform, clearly only a purely resistive load makes sense, then the lower the resistance the higher the current.

Class F can be driven by sine or by a square wave, for a sine the input can be tuned by an inductor to increase gain. If class F is implemented with a single transistor, the filter is complicated to short the even harmonics. All previous designs use sharp edges to minimise the overlap.

Class E uses a significant amount of second harmonic voltage. The second harmonic can be used to reduce the overlap with edges with finite sharpness. For this to work energy on the second harmonic has to flow from the load into the transistor, and no source for this is visible in the circuit diagram. In reality, the impedance is mostly reactive and the only reason for it is that class E is a class F amplifier with a much simplified load network and thus has to deal with imperfections.

In many amateur simulations of class E amplifiers, sharp current edges are assumed nullifying the very motivation for class E and measurements near the transit frequency of the transistors show very symmetric curves, which look much similar to class F simulations. The class E amplifier was invented in 1972 by Nathan O. Sokal and Alan D. Sokal, and details were first published in 1975. Some earlier reports on this operating class have been published in Russian.

Class G and H

There is a variety of amplifier designs that couple a class AB output stage with other more efficient techniques to achieve a higher efficiency with low distortion. These designs are common in large audio amplifiers since the heatsinks and power transformers would be prohibitively large (and costly) without the increase in efficiency. The terms "class G" and "class H" are used interchangeably to refer to different designs, varying in definition from one manufacturer or paper to another.

Class G amplifiers (which use "rail switching" to decrease power consumption and increase efficiency) are more efficient than class AB amplifiers. The class G amplifier has several power rails at different voltages, and switches between rails as the signal output approaches each. Thus the amplifier increases efficiency by reducing the wasted power at the output transistors.

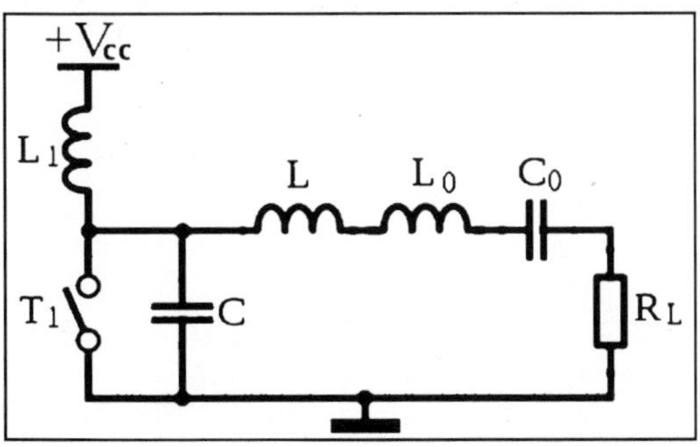

A Class H amplifier takes the idea of Class G one step further creating an infinitely variable supply rail. This is done by modulating the supply rails so that the rails are only a few volts larger than the output signal at any given time.

The output stage operates at its maximum efficiency all the time. Switched mode power supplies can be used to create the tracking rails. Significant efficiency gains can be achieved but with the drawback of more complicated supply design and reduced THD performance.

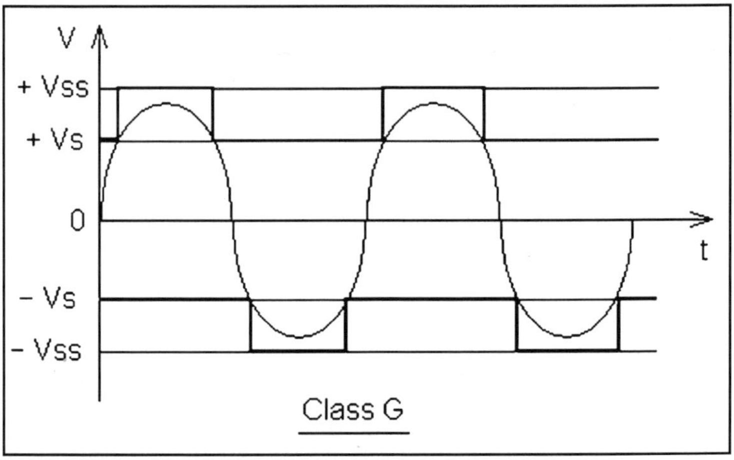

Efficiency Class H

The classes can be most easily understood using the diagrams in each section below. For the sake of illustration, a bipolar junction transistor is shown as the amplifying device, but in practice this could be a MOSFET or vacuum tube device. In an analog amplifier (the most common kind), the signal is applied to the input terminal of the device (base, gate or grid), and this causes a proportional output drive current to flow out of the output terminal. The output drive current comes from the power supply.

The voltage signal shown is thus a larger version of the input, but has been changed in sign (inverted) by the amplification. Other arrangements of amplifying device are possible, but that given (that is, common emitter, common source or common cathode) is the easiest to understand and employ in practice. If the amplifying element is linear, then the output will be faithful copy of the input, only larger and inverted. In actual practice, transistors are not linear, and the output will only approximate the input. Non-linearity from any of several sources is the origin of distortion within an amplifier. Which class of amplifier (A, B, AB or C) depends on how the amplifying device is biased — in the diagrams the bias circuits are omitted for clarity.

Any real amplifier is an imperfect realization of an ideal amplifier. One important limitation of a real amplifier is that the output it can generate is ultimately limited by the power available from the power supply. An amplifier will saturate and clip the output if the input signal becomes too large for the amplifier to reproduce or if operational limits for a device are exceeded.

DOHERTY AMPLIFIERS

A hybrid configuration receiving new attention is the Doherty amplifier, invented in 1934 by William H. Doherty for Bell Laboratories (whose sister company, Western Electric, was then an important manufacturer of radio transmitters). The Doherty amplifier consists of a class-B main (or *carrier*) stage in parallel with a class-C auxiliary (or *peaking*) stage. The input signal is split evenly to drive the two amplifiers, and a combining network sums the two output signals and corrects for phase differences between the two amplifiers.

During periods of low signal level, the class-B amplifier efficiently operates on the signal and the class-C amplifier is inactive and consumes no power. During high signal peaks the class-B amplifier saturates and the class-C amplifier kicks in. The efficiency of previous AM transmitter designs was proportional to modulation, but with average modulation typically 20 per cent, transmitters were limited to less than 50 per cent efficiency. In Doherty's design, even with zero modulation a transmitter could achieve at least 60 per cent efficiency.

As a successor to Western Electric for broadcast transmitters, the Doherty concept was considerably refined by Continental Electronics Manufacturing Company of Dallas, Texas. Perhaps the ultimate refinement was the screen grid modulation scheme invented by Joseph B. Sainton. The Sainton amplifier consists of a Class C main (or *carrier*) stage in parallel with a Class C auxiliary (or *peak*) stage. The stages are split and combined through 90 degree phase shifting networks as in the Doherty amplifier. The unmodulated radio frequency carrier is applied to the control grids of both tubes.

Carrier modulation is applied to the screen grids of both tubes. The bias of the carrier and peak tubes are different, and are established so that the peak tube is quiescent when modulation is absent (and the amplifier is producing rated unmodulated carrier power) whereas both tubes contribute twice the rated carrier power during 100 per cent modulation (as four times the carrier power is required to achieve 100 per cent modulation). As both tubes operate in Class C, a significant improvement in efficiency is thereby achieved in the final stage. And, as the tetrode carrier and peak tubes require very little drive power, a significant improvement in efficiency within the driver stage is achieved as well (317C, et al.). The released version of the Sainton amplifier employs a cathode follower modulator, not a push-pull modulator. Previous Continental Electronics designs, by James O. Weldon and others, retained most of the characteristics of the Doherty amplifier but added screen grid modulation of the driver (317B, et al.).

The Doherty amplifier remains in use in very-high-power AM transmitters, but for lower-power AM transmitters, vacuum-tube amplifiers in general were eclipsed in the 1980s by arrays of solid-state amplifiers, which could be switched on and off with much finer granularity in response to the requirements of the input audio. However, interest in the Doherty configuration has been revived by cellular-telephone and wireless-Internet applications where the sum of several constant-envelope users creates an aggregate AM result. The main challenge of the Doherty amplifier for digital transmission modes is in aligning the two stages and getting the class-C amplifier to turn on and off very quickly.

Recently, Doherty amplifiers have found widespread use in cellular base station transmitters for GHz frequencies. Implementations for transmitters in mobile devices have also been demonstrated.

OTHER CLASSES

Several audio amplifier manufacturers have started "inventing" new classes as a way to differentiate themselves. These class names usually do not reflect any revolutionary amplification technique, and are used mostly for marketing purposes. This can easily be determined by the fact that the class name is trademarked or copyrighted. For example, Crown's K and I-Tech Series as well as several other models utilise Crown's patented Class-I (or BCA) technology. Lab.gruppen use a form of class D amplifier called class TD or Tracked Class D which tracks the waveform to more accurately amplify it without the drawbacks of traditional class D amplifiers.

"Class T" is a trademark of TriPath company, which manufactures audio amplifier ICs. This new class "T" is a revision of the common class D amplifier, but with changes to ensure fidelity over the full audio spectrum, unlike traditional class D designs. It operates at different frequencies depending on the power output, with values ranging from as low as 200 kHz to 1.2 MHz, using a proprietary modulator.

"Class Z" is a trademark of Zetex semiconductor and is a direct-digital-feedback technology.

AMPLIFIER CIRCUIT

The practical amplifier circuit below could be the basis for a moderate-power audio amplifier. It features a typical (though substantially simplified) design as found in modern amplifiers, with a class AB push–pull output stage, and uses some overall negative feedback. Bipolar transistors are shown, but this design would also be realizable with FETs or valves.

A Practical Amplifier Circuit

The input signal is coupled through capacitor C1 to the base of transistor Q1. The capacitor allows the AC signal to pass, but blocks the DC bias voltage established by resistors R1 and R2 so that any preceding circuit is not affected by it. Q1 and Q2 form a differential amplifier (an amplifier that multiplies the difference between two inputs by some constant), in an arrangement known as a long-tailed pair. This arrangement is used to conveniently allow the use of negative feedback, which is fed from the output to Q2 via R7 and R8.

The negative feedback into the difference amplifier allows the amplifier to compare the input to the actual output. The amplified signal from Q1 is directly fed to the second stage, Q3, which is a common emitter stage that provides further amplification of the signal and the DC bias for the output stages, Q4 and Q5. R6 provides the load for Q3 (A better design would probably use some form of active load here, such as a constant-current sink).

So far, all of the amplifier is operating in Class A. The output pair are arranged in Class AB push–pull, also called a complementary pair. They provide the majority of the current amplification and directly drive the load, connected via DC-blocking capacitor C2. The diodes D1 and D2 provide a small amount of constant voltage bias for the output pair, just biasing them into the conducting state so that crossover distortion is minimized. That is, the diodes push the output stage firmly into class-AB mode (assuming that the base-emitter drop of the output transistors is reduced by heat dissipation).

This design is simple, but a good basis for a practical design because it automatically stabilises its operating point, since feedback internally operates from DC up through the audio range and beyond. Further circuit elements would probably be found in a real design that would roll off the frequency response above the needed range to prevent the possibility of unwanted oscillation. Also, the use of fixed diode bias as shown here can cause problems if the diodes are not both electrically and thermally matched to the output transistors — if the output transistors turn on too much, they can easily overheat and destroy themselves, as the full current from the power supply is not limited at this stage.

A common solution to help stabilise the output devices is to include some emitter resistors, typically an ohm or so. Calculating the values of the circuit's resistors and capacitors is done based on the components employed and the intended use of the amp.

Notes on implementation

Real world amplifiers are imperfect.

- One consequence is that the power supply itself may influence the output, and must itself be considered when designing the amplifier
- The amplifier circuit has an "open loop" performance, that can be described by various parameters (gain, slew rate, output impedance, distortion, bandwidth, signal to noise ratio...)
- Many modern amplifiers use negative feedback techniques to hold the gain at the desired value.

Different methods of supplying power result in many different methods of bias. Bias is a technique by which the active devices are set up to operate in a particular regime, or by which the DC component of the output signal is set to the midpoint between the maximum voltages available from the power supply. Most amplifiers use several devices at each stage; there are typically matched in specifications except for polarity. Matched inverted polarity devices are called complementary pairs. Class A amplifiers generally use only one device, unless the power supply is set to provide both positive and negative voltages, in which case a dual device symmetrical design may be used. Class C amps, by definition, use a single polarity supply.

Amplifiers often have multiple stages in cascade to increase gain. Each stage of these designs may be a different type of amp to suit the needs of that stage. For instance, the first stage might be a Class A stage, feeding a class AB push–pull second stage, which then drives a class G final output stage, taking advantage of the strengths of each type, while minimizing their weaknesses.

POWER AMPLIFIER

The term "power amplifier" is a relative term with respect to the amount of power delivered to the load and/or sourced by the supply circuit. In general a power amplifier is designated as the last amplifier in a transmission chain (the *output stage*) and is the amplifier stage that typically requires most attention to power efficiency.

VACUUM TUBE (VALVE) AMPLIFIERS

According to Symons, while semiconductor amplifiers have largely displaced valve amplifiers for low power applications, valve amplifiers are much more cost effective in high power applications such as "radar, countermeasures equipment, or communications equipment". Many microwave amplifiers are specially designed valves, such as the klystron, gyrotron, traveling wave tube, and crossed-field amplifier, and these microwave valves provide much greater single-device power output at microwave frequencies than solid-state devices.

A valve amplifier or tube amplifier is a type of electronic amplifier that makes use of vacuum tubes to increase the power and/or amplitude of a signal. They are typically (but not exclusively) used for sound amplification, either in home stereo hi-fi amplifiers or for electric guitar amps, for radio frequency signals.

Low to medium power valve amplifiers for frequencies below the microwaves were largely replaced by solid state amplifiers during the 1960s and 1970s. Valve amplifiers are used for applications such as guitar amplifiers, satellite transponders such as DirecTV and GPS systems, audiophile stereo amplifiers, military applications (such as target acquisition and radar) and very high power radio and UHF television transmitters.

Origins

Until the invention of the transistor in 1947, all practical amplifiers were made using thermionic valves. The simplest valve was invented by John

Ambrose Fleming while working for the Marconi Company in London in 1904 and named the Diode, as it had two electrodes. The diode conducted electricity in one direction only and was used as a radio detector and a rectifier.

Although he may not have at first realised the significance of his invention it was Lee De Forest who added a third electrode and invented the first electronic amplifying device, the Triode which he named the 'Audion'. This additional 'control grid' modulates the current that flows between cathode and anode for a given voltage between the cathode and anode. The relationship between current flow and plate and grid voltage is often represented as a series of "characteristic curves" on a diagram. Depending on the other components in the circuit this modulated current flow can be used to provide current or voltage gain.

The first application of valve amplification was in the regeneration of long distance telephony signals. Later, valve amplification was applied to the 'wireless' market that began in the early thirties. In due course amplifiers for music and later television were also built exclusively using valves until the 1950s. During this period power levels were usually very low (a few watts) and radios often used headphones only, with no loudspeaker at all.

The overwhelmingly dominant circuit topology during this period was the single-ended triode gain stage, operating in class A, which gave very good sound (and reasonable measured distortion performance) despite extremely simple circuitry with very few components: important at a time when components were hand made and extremely expensive. Prior to World War II almost all valve amplifiers were of low gain and with linearity dependent entirely on the inherent linearity of the tube itself, typically 5 per cent distortion at full power.

Post-War Developments

Increasing post-war affluence and industrialised production economies, dramatic technical progress stimulated by the war, and for the first time a substantial and expanding consumer market brought more advanced valve designs to market at affordable prices, with the result that the 1960's saw the increasing spread of Gramophone players, and ultimately the beginnings of "HiFi" able to drive full frequency range loudspeakers (for the first time often with multiple drivers for different frequencies) to significant volume levels, combined with the spread of B&W TV, to produce a 'golden age' in Valve development and also in the development of Valve amplifier circuits.

Negative feedback (NFB) was invented by Harold Stephen Black in 1927, but initially little used since at that time gain was at a premium. This technique allows amplifiers to trade gain for reduced distortion levels (and also gave other benefits such as reduced output impedance). The introduction of the Williamson amplifier in 1947, which was extremely advanced in many respects including very successful use of NFB, was a turning point in audio power amplifier design, operating a push pull output circuit in class AB1 to give performance far ahead of its time.

Similar topologies with only minor variations (notably different phase splitter arrangements and the "ultra-Linear" transformer connection for tetrodes) rapidly became widespread and this family of designs remains the dominant high power amplifier topology to this day for music application.This period also saw continued growth in civilian radio, with valves being used for both transmitters and receivers.

Decline

From the 70s the silicon transistor became increasingly pervasive and valve production was sharply ramped down, with the notable exception of Cathode Ray Tubes, and a dramatically rationalized range of valves for amplifier applications, low power tubes being mostly dual triodes (ECCnn, 12Ax7 series) plus the EF86 pentode, power tubes mostly being Beam Tetrode and Pentodes (EL84, EL34, KT88/6550, 6L6), in both cases with indirect heating and this rationalized set of types remains the core of all subsequent production and remain in production (in the east) today.

The Soviets retained valves to a much greater extent than the west during the cold war, for the majority of their communications and military amplification requirements, in part due to tubes' ability to withstand instantaneous overloads (notably due to a nuclear detonation) that would destroy a transistor.

The dramatic reduction in size, power consumption, reduced distortion levels and above all cost of electronics products based on Transistors has made valves obsolete for mainstream products since ~ the 1970s, although valves remained in niche (mainly high power RF transmitters) applications for somewhat longer. However, the difficulty in producing transistors with good gain and efficiency at very high frequencies, combined with the fragility of transistors (problems such as thermal runaway), resulted in tubes being retained for longer in high power and high frequency applications, notably large radio (and TV) transmitters and guitar amplifier for many more years. However, developments in semiconductor production have now mostly closed that market also.

Resurgence

In audio applications, valves continue to be highly desired by some users, both in the higher-end home audio market and in the guitar amplifier market. Amongst stereo enthusiasts, there is a subgroup of audio buffs who advocate the use of tube amplifiers for home listening; they argue that tube amplifiers produce a "warmer" or more "natural" sound. Companies in Russia, China and Eastern Europe continue to produce valves to cater to this market.

In the guitar amplifier market, some performers continue to use tube amps in the 2000s, including in folk, blues, roots rock, and in harder genres such as metal, where Marshall tube amps are used to create heavy distortion. Audio engineers suggest that the subjectively-pleasing aspects of tube amplification may be due to the non-linear overdrive that is produced with tubes. Tube amps have the following advantages over solid-state amps.

Compared to semiconductors, tubes have a very low "drift" (of specs) over a wide range of operating conditions, specifically high heat/high power. Semiconductors are very heat-sensitive by comparison and this fact usually leads to compromises in solid-state amplifier designs. When a tube fails, it is replaceable. While solid state devices are also replaceable, it is usually a much more involved process (*i.e.*, having the amplifier tested by a professional, removing the faulty component, and replacing it).

For working musicians this is usually a huge problem by comparison to looking in the back of a tube amp at the tubes and simply replacing the faulty tube. In addition, tubes can easily be removed and tested, while transistors cannot. Tube amplifiers respond differently from transistor amplifiers when signal levels approach and reach the point of clipping. In a tube-powered amplifier, the transition from linear amplification to limiting is less abrupt than in a solid state unit, resulting in a less grating form of distortion at the onset of clipping. For this reason, some guitarists prefer the sound of an all-tube amplifier; the aesthetic properties of tube versus solid state amps, though, are a topic of debate in the guitarist community.

Characteristics

Valves are high voltage/low current devices in comparison with transistors (and especially MOSFETs). The high working voltage makes them well suited for radio transmitters, for example, and valves remain in use today for very high power radio transmitters, where there is still no other technology available. However, for most applications requiring an appreciable output current, a matching transformer is required. The transformer is a critical component and heavily influences the performance (and cost) of the amplifier.

Many power valves have good linearity but modest gain or transconductance. Signal amplifiers using tubes are capable of very high frequency response ranges - up to radio frequency. Indeed, many of the Directly Heated Single Ended Triode (DH-SET) audio amplifiers are in fact radio transmitting tubes designed to operate in the megahertz range. In practice, however, tube amplifier designs typically "couple" stages either capacitively, limiting bandwidth at the low end, or inductively with transformers, limiting the bandwidth at high end.

Advantages

- Very linear (especially triodes) making it viable to use them in low distortion linear circuits with little or no negative feedback.

- Extremely high input impedance (cf bipolar transistors but a characteristic shared by FETs).
- High voltage devices and thus inherently suitable for very high voltage circuits.
- Can be constructed on a scale that can dissipate large amounts of heat (some extreme devices even being water cooled). For this reason valves remained the only viable technology for very high power, and especially high power/high voltage applications such as Radio & TV transmitters long into the age when transistors had displaced valves in most other applications. However, today these also are becoming obsolete.
- Electrically very robust, they can tolerate overloads for minutes which would destroy bipolar transistor systems in milliseconds. In the worst case a failed tube can simply be unplugged and replaced by the user.

Disadvantages

- Heater supplies are usually required for the cathodes, and dangerously high voltages are usually required for the anodes.
- They are significantly larger than equivalent solid-state transistors
- High impedance/low current output is unsuitable for direct drive of many real world loads, notably various forms of electric motors.
- Valved audio equipment is normally heavy because of the weight of transformers.
- Valves may have a shorter working life than solid state parts due to various failure mechanisms (cathode poisoning, breakages (*i.e.*, open circuit) or shorts internally, notably of the heater or grid structures, or in the case of glass valves, physical breakage, although this should not be overstated; many valve types typically have operation lives in the tens of thousands of hours and an indefinite shelf life (many 60 year old tubes are still in regular use).
- Available in a single polarity only whereas transistors are available in complementary polarities (*e.g.*, NPN/PNP), making possible many circuit configurations that cannot be realized directly with valves.
- In comparison to the lower impedance environment of transistors special consideration must be made to the physical layout of valve circuits in order to avoid instability and the introduction of noise from radio frequency interference and ac heater supplies.

Operation

All amplifier circuits are classified by "class of operation" as A, B, AB and C etc. However, the nature of valves results in significantly different circuit topologies and characteristics than transistor designs, and as a sweeping generalisation, valve amplifiers tend to operate in class A (or class AB1 with a heavy class A overlap), whereas transistor amplifiers tend to operate more in class B (there are however significant exceptions in both directions).

- The grid (where the input signal is presented) needs to be biased substantially negative with respect to the cathode (typically ~ -75 V). This makes it extremely difficult to direct-couple the output of one valve (typically sitting at about + 100 V) to the input of a following valve as is normally done in modern transistor designs. (cf transistors which are usually biased just a few volts positive), at a voltage between that of the collector and emitter, facilitating direct coupling)
- Valve stages thus need to be coupled using some component able to totally block and withstand several hundred volts, typically a capacitor, occasionally a coupling transformer, adding phase shifts and possibly coloration to the signal. These introduced phase shifts can become problematic in circuits that have feedback
- There is no valve analog of the complementary devices widely used in "totem pole" output stages of silicon circuits. This is because valves work based on the flow of electrons from the cathode to the anode and it is not possible to construct a hypothetical valve in which an "electron hole" migrated the other way. Push-pull valve topologies therefore typically require a transformer.
- The very high output impedance of valves (compared with transistors) usually demands the use of matching transformers if low impedance loads (notably loudspeakers or various forms of motor, such as cutting lathe heads, etc.) are to be driven. The transformer is used as the load, in place of the resistor usually used in small-signal and driver stages. NB the impedance of the transformer primary at the frequencies in use is much higher than the DC resistance of the windings, often kOhms. High performance transformers are however severe engineering compromises, are expensive, and in operation are far from ideal. Transformers dramatically increase the cost of a valve amplifier circuit compared to a direct-coupled transistor alternative.
- The (typically) much lower open loop gain but enhanced open loop linearity of valves, especially triodes, makes it possible to use little or no negative feedback in circuits whilst retaining acceptable or even excellent distortion performance (especially for small-signal circuits).

Topologies

- Linear small signal circuits almost invariably use a triode in the single ended gain stage topology (in class A), including the output stage (cf silicon circuits, notably the very widely use "op-amp" configuration, which normally have a totem pole output stage).
- Broadband valve amplifiers typically use class A1 or AB1,

- Modern high power output stages are usually push pull, often necessitating some form of phase splitter to derive a differential/ balanced drive signal from a single ended input, typically followed by a further gain stage (the "driver") prior to the output tubes.
- Single ended" power stages using very large valves exist and dominate in radio transmitter applications. A sidebar is the observation that the niche "DH-SET" (directly heated single-ended triode) topology favoured by some audiophiles are extremely simple and typically constructed using valve types originally designed for use in radio transmitters
- More complex topologies (notably the use of active loads) can improve linearity and frequency response (by removing miller capacitance effects).

Output Transformerless

Julius Futterman pioneered a type of amplifier known as "output transformerless" (OTL). These forego the typical output transformer by paralleling (electrically connecting and operating side-by-side) perhaps one dozen or more output tubes in an attempt to reduce effective plate resistance and more closely match it with speaker impedances (typically 8 ohms). This design and its various incarnations tend to require numerous tubes, run hot, and because they attempt to match impedances in a way fundamentally different from a transformer, they often have a unique sound quality.

Output Impedance

The high voltage/low current/High output impedance (Z out) of the output (anode circuit in the overwhelming majority of valve circuits) is suitable to drive another following valve stage, and can drive an antenna system that has been arranged to resonate at the required drive frequency, but usually is not suitable to drive low impedance loads, such as (notably) loudspeakers and motors. The main techniques are used to resolve this: the use of a matching transformer, and the use of negative feedback to reduce that active output impedance (in proportion to the amount of feedback applied). In combination these techniques can reduce the Z out from hundreds of ohms to a fraction of an ohm.

Applications

Audio Frequency (AF)/Broadband amplifiers

Valves remain in widespread use in guitar and high-end audio amplifiers due to the sound quality they produce, which is subjectively preferred by some users. They are largely obsolete for most other applications, mainly due to the cost effectiveness advantages of the transistor, and due to the lower weight and heat production of transistor amps.

Telephony

(Medium voltage/low to medium power) Telephony was the original, and for many years was a driving application for audio amplification. Although a telecoms voice connection is very undemanding compared with modern data communication (it has a very narrow frequency range, typically ~ 300 - 3000 Hz, and also has very poor signal to noise ratio), a specific issue for the telecomms industry was the technique of multiplexing many (up to a thousand) voice lines onto a single cable, at different frequencies.

The advantage of this is a that a single valve "repeater" amplifier can amplify many calls at once, this being very cost effective. The problem is that the amplifiers need to be extremely linear, otherwise "Intermodulation Distortion" (IMD) will result in "crosstalk" between the multiplexed channels. This stimulated development emphasis towards low distortion far beyond the nominal needs of a single voice channel.

Audio

Today the main application for valves is audio amplifiers for high-end Hi-Fi and musical performance use with electric guitars, electric basses, and Hammond organs, although these applications have different requirements regarding distortion which result in different design compromises, although the same basic design techniques are generic and widely applicable to all broadband amplification applications, not only audio.

Post WW-II, the majority of valve power amplifiers are of the Class AB-1 "Push Pull" ultralinear topology, but niche products using the DH-SET and even OTL topologies still exist in small numbers.

INSTRUMENTATION AMPLIFIERS

The basic moving coil volt/ammeter itself takes a small current and thus loads the circuit to which it is attached. This can significantly alter the operating conditions in the circuit being measured, clearly an undesirable feature. The Vacuum Tube Volt Meter (VTVM) was developed by taking advantage of the near infinite input impedance of a valve to buffer the circuit being measured from the load of the ammeter. VTVMs have become obsolete since the introduction of the modern Digital Volt Meter (DVM) which typically also has an extremely high input impedance (FET) input.

Valve oscilloscopes share this very high input impedance and thus can be used to measure voltages even in very high impedance circuits. There may typically be 3 or 4 sets of amplification per display channel. In later oscilloscopes, a type of amplifier using a series of tubes connected at equal distances along transmission lines, known as a distributed amplifier was employed to amplify very high frequency vertical signals before application to the display tube. Valve oscilloscopes are now obsolete.

In the closing years of the valve era, valves were even used to make simple "operational amplifiers" - the building blocks of much modern linear electronics. An Op-amp typically has a differential input stage and a totem pole output, the circuit usually having a minimum of five active devices. A number of "packages" were produced that integrated such circuits (typically using two or more glass envelopes) into a single module that could be plugged into a larger circuit (such as an analog computer). Such Valve op-amps were very far from ideal and quickly became obsolete, being replaced with "integrated" (planar silicon) types.

Narrow band/Radio Frequency (RF)/tuned amplifiers

Historically (pre WWII) "transmitting tubes" were among the most powerful tubes available, were usually direct heated by fragile thoriated filaments that glowed like light bulbs.

Some tubes were capable of being driven so hard that the anode would itself glow cherry red, the anodes being machined from solid material (rather than fabricated from thin sheet) to be able to withstand this without distorting when heated. Notable tubes of this type are the 845 and 211. Later tetrode/pentodes such as 817 and (direct heated) 813 were also used in large numbers in (especially military) radio transmitters

RF circuits (in particular transmitters, which is something more than simply an amplification gain stage) are significantly different from broadband amplifier circuits. The antenna or following circuit stage typically contains one or more adjustable capacitive or inductive component allowing the resonance of the stage to be accurately matched with carrier frequency in use, to optimize power transfer from and loading on the valve, a so called "tuned circuit".

Broadband circuits often go down to near DC (10 Hz or below) and up to tens or hundreds of kilohertz/low megahertz, and are required to have essentially flat frequency response over this entire range (4 or more orders of magnitude). RF circuits by contrast are typically required to operate over higher frequencies (which makes capacitive and inductive parasitic effects much more of a design challenge) but often a very narrow frequency range. For example, an RF device might be required to operate over the range 144 to 146 MHz (just 1.4 per cent of an octave)

Historically, distortion and out of band emission was less of an issue and Class C could be used. In the 2000s, radio transmitters are overwhelmingly silicon based, even at microwave frequencies (notably consider cellular radio base stations). However an ever decreasing minority of especially high power radio frequency amplifiers (notably for TV) continue to have valve construction. Nevertheless, since the development of radio is inseparable from valve technology, the field of valve amplification remains of historic interest, notably to radio amateurs.

Transistor Amplifiers

The essential role of this active element is to magnify an input signal to yield a significantly larger output signal. The amount of magnification (the "forward gain") is determined by the external circuit design as well as the active device.

Many common active devices in transistor amplifiers are bipolar junction transistors (BJTs) and metal oxide semiconductor field-effect transistors (MOSFETs). Applications are numerous, some common examples are audio amplifiers in a home stereo or PA system, RF high power generation for semiconductor equipment, to RF and Microwave applications such as radio transmitters.

Transistor-based amplifier can be realized using various configurations: for example with a bipolar junction transistor we can realize common base, common collector or common emitter amplifier; using a MOSFET we can realize common gate, common source or common drain amplifier. Each configuration has different characteristic (gain, impedance...).

Operational Amplifiers (op-amps)

An operational amplifier is an amplifier circuit with very high open loop gain and differential inputs which employs external feedback for control of its transfer function or gain. Although the term is today commonly applied to integrated circuits, the original operational amplifier design was implemented with valves.

Fully Differential Amplifiers (FDA)

A fully differential amplifier is a solid state integrated circuit amplifier which employs external feedback for control of its transfer function or gain. It is similar to the operational amplifier but it also has differential output pins.

A fully differential amplifier, usually referred to as an 'FDA' for brevity, is a DC-coupled high-gain electronic voltage amplifier with differential inputs and differential outputs. In its ordinary usage, the output of the FDA is controlled by two feedback paths which, because of the amplifier's high gain, almost completely determines the output voltage for any given input.

The Ideal FDA

For any input voltages the *ideal* FDA has infinite open-loop gain, infinite bandwidth, infinite input impedances resulting in zero input currents, infinite slew rate, zero output impedance and zero noise.

A Real FDA can only approximate this ideal, and the actual parameters are subject to drift over time and with changes in temperature, input conditions, etc. Modern integrated FET or MOSFET FDAs approximate more closely to these ideals than bipolar ICs where large signals must be handled at room temperature over a limited bandwidth; input impedance, in particular, is much higher, although the bipolar FDA usually exhibit superior (*i.e.*, lower) input offset drift and noise characteristics.

Where the limitations of real devices can be ignored, an FDA can be viewed as a Black Box with gain; circuit function and parameters are determined by feedback, usually negative. An FDA as implemented in practice is moderately complex integrated circuit

Limitations of real FDAs

DC imperfections

- Finite gain — the effect is most pronounced when the overall design attempts to achieve gain close to the inherent gain of the FDA.
- Finite input resistance — this puts an upper bound on the resistances in the feedback circuit.
- Non-zero output resistance — important for low resistance loads. Except for very small voltage output, power considerations usually come into play first. (Output impedance is inversely proportional to the idle current in the output stage — very low idle current results in very high output impedance.)
- Input bias current — a small amount of current (typically ~10 nA for bipolar FDAs, or picoamperes for CMOS designs) flows into the inputs. This current is mismatched slightly between the inverting and non-inverting inputs (there is an input offset current). This effect is usually important only for very low power circuits.
- Input offset voltage — the FDA will produce an output even when the input pins are at exactly the same voltage. For circuits which require precise DC operation, this effect must be compensated for.
- Common mode gain — A perfect operational amplifier amplifies only the voltage difference between its two inputs, completely rejecting all voltages that are common to both. However, the differential input stage of an FDA is never perfect, leading to the amplification of these identical voltages to some degree. The standard measure of this defect is called the common-mode rejection ratio (denoted, CMRR). Minimization of common mode gain is usually important in non-inverting amplifiers (described below) that operate at high amplification.
- Temperature effects — all parameters change with temperature. Temperature drift of the input offset voltage is especially important.

AC Imperfections

- Finite bandwidth — all amplifiers have a finite bandwidth. This is because FDAs use internal frequency compensation to increase the phase margin.

- Input capacitance — most important for high frequency operation because it further reduces the open loop bandwidth of the amplifier.
- Common mode gain
- Noise - all real electronic components (except superconductor) generate noise. You can find devices with 0.8 to several hundreds nv/rtHz noise performance.

Non-linear Imperfections

- Saturation — output voltage is limited to a peak value, usually slightly less than the power supply voltage. Saturation occurs when the differential input voltage is too high for the op-amp's gain, driving the output level to that peak value.
- Slewing — the amplifier's output voltage reaches its maximum rate of change. Measured as the slew rate, it is usually specified in volts per microsecond. When slewing occurs, further increases in the input signal have no effect on the rate of change of the output. Slewing is usually caused by internal capacitances in the amplifier, especially those used to implement its frequency compensation, particularly using pole splitting.
- Non-linear transfer function — The output voltage may not be accurately proportional to the difference between the input voltages. It is commonly called distortion when the input signal is a waveform. This effect will be very small in a practical circuit if substantial negative feedback is used.

Power Considerations

- Limited output power — if high power output is desired, an op-amp specifically designed for that purpose must be used. Most op-amps are designed for low power operation and are typically only able to drive output resistances down to 2 kÙ.
- Limited output current — the output current must obviously be finite. In practice, most op-amps are designed to limit the output current so as not to exceed a specified level thus protecting the FDA and associated circuitry from damage.

DC Behaviour

Open-loop gain is defined as the amplification from input to output without any feedback applied. For most practical calculations, the open-loop gain is assumed to be infinite; in reality it is obviously not. Typical devices exhibit open-loop DC gain ranging from 100,000 to over 1 million; this is sufficiently large for circuit gain to be determined almost entirely by the amount of negative feedback used. Op-amps have performance limits that the designer must keep in mind and sometimes work around. In particular, instability is possible in a DC amplifier if AC aspects are neglected.

AC Behaviour

The FDA gain calculated at DC does not apply at higher frequencies. To a first approximation, the gain of a typical FDA is inversely proportional to frequency. This means that an FDA is characterized by its gain-bandwidth product. For example, an FDA with a gain bandwidth product of 1 MHz would have a gain of 5 at 200 kHz, and a gain of 1 at 1 MHz. This low-pass characteristic is introduced deliberately, because it tends to stabilize the circuit by introducing a dominant pole. This is known as frequency compensation. Typical low cost, a general purpose FDA exhibits a gain bandwidth product of a few megahertz. Specialty and high speed FDAs can achieve gain bandwidth products of hundreds of megahertz. Some FDAs are even capable of gain bandwidth products greater than a gigahertz.

Video Amplifiers

These deal with video signals and have varying bandwidths depending on whether the video signal is for SDTV, EDTV, HDTV 720p or 1080i/p etc.. The specification of the bandwidth itself depends on what kind of filter is used and which point (-1 dB or -3 dB for example) the bandwidth is measured. Certain requirements for step response and overshoot are necessary in order for acceptable TV images to be presented.

Oscilloscope Vertical Amplifiers

These are used to deal with video signals to drive an oscilloscope display tube and can have bandwidths of about 500 MHz. The specifications on step response, rise time, overshoot and aberrations can make the design of these amplifiers extremely difficult. One of the pioneers in high bandwidth vertical amplifiers was the Tektronix company.

Distributed Amplifiers

These use transmission lines to temporally split the signal and amplify each portion separately in order to achieve higher bandwidth than can be obtained from a single amplifying device. The outputs of each stage are combined in the output transmission line. This type of amplifier was commonly used on oscilloscopes as the final vertical amplifier. The transmission lines were often housed inside the display tube glass envelope.

Distributed amplifiers are circuit designs that incorporate transmission line theory into traditional amplifier design to obtain a larger gain-bandwidth product than is realizable by conventional circuits.

The design of the distributed amplifiers was first formulated by William S. Percival in 1936. In that year Percival proposed a design by which the transconductances of individual vacuum tubes could be added linearly without lumping their element capacitances at the input and output, thus arriving at a circuit that achieved a gain-bandwidth product greater than that of an individual tube. Percival's design did not gain widespread awareness however, until a publication on the subject was authored by Ginzton, Hewlett, Jasberg, and Noe in 1948. It is to this later paper that the term *distributed amplifier* can actually be traced. Traditionally, DA design architectures were realized using valve technology.

Current Rechnology

More recently, III-V semiconductor technologies, such as GaAs and InP have been used. These have superior performance resulting from higher bandgaps (higher electron mobility), higher saturated electron velocity, higher breakdown voltages and higher-resistivity substrates. The latter contributes much to the availability of higher quality-factor (Q-factor or simply Q) integrated passive devices in the III-V semiconductor technologies.

To meet the marketplace demands on cost, size, and power consumption of monolithic microwave integrated circuits (MMICs), research continues in the development of mainstream digital bulk-CMOS processes for such purposes. The continuous scaling of feature sizes in current IC technologies has enabled microwave and mm-wave CMOS circuits to directly benefit from the resulting increased unity-gain frequencies of the scaled technology. This device scaling, along with the advanced process control available in today's technologies, has recently made it possible to reach an fT of 170 GHz and a maximum oscillation frequency (fmax) of 240 GHz in a 90nm CMOS process.

Theory of Operation

The operation of the DA can perhaps be most easily understood when explained in terms of the traveling-wave tube amplifier (TWTA). The DA consists of a pair of transmission lines with characteristic impedances of Z_0 independently connecting the inputs and outputs of several active devices. An RF signal is thus supplied to the section of transmission line connected to the input of the first device. As the input signal propagates down the input line, the individual devices respond to the forward traveling input step by inducing an amplified complementary forward traveling wave on the output line. This assumes the delays of the input and output lines are made equal through selection of propagation constants and lengths of the two lines and as such the output signals from each individual device sum in phase. Terminating resistors Z_g and Z_d are placed to minimize destructive reflections.

The transconductive gain of each device is g_m and the output impedance seen by each transistor is half the characteristic impedance of the transmission line. So that the overall voltage gain of the DA is: where *n* is the number of stages.

Neglecting losses, the gain demonstrates a linear dependence on the number of devices (stages). Unlike the multiplicative nature of a cascade of conventional amplifiers, the DA demonstrates an additive quality. It is this synergistic property of the DA architecture that makes it possible for it to provide gain at frequencies beyond that of the unity-gain frequency of the individual stages. In practice, the number of stages is limited by the diminishing input signal resulting from attenuation on the input line.

Means of determining the optimal number of stages are discussed below. Bandwidth is typically limited by impedance mismatches brought about by frequency dependent device parasitics. The DA architecture introduces delay in order to achieve its broadband gain characteristics. This delay is a desired feature in the design of another distributive system called the distributed oscillator.

Lumped Elements

Delay lines are made of lumped elements of L and C. The parasitic L and the C from the transistors are used for this and usually some L is added to raise the line impedance. Due to the Miller effect in the common source amplifier the input and the output transmission line are coupled. For example for voltage inverting and current amplifying the input and the output form a shielded balanced line.

Due to the current increasing in the output transmission line with every subsequent transistor, less and less L is added to keep the voltage constant and more and more extra C is added to keep the velocity constant. This C can come from parasitics of a second stage. These delay lines do not have a flat dispersion near their cut off, so it is important to use the same L-C periodicity in the input and the output. If inserting transmission lines, input and output will disperse away from each other.

For a distributed amplifier the input is feed in series into the amplifiers and parallel out of them. To avoid losses in the input, no input signal is allowed to leak through. This avoided by using a balanced input and output also known as push-pull amplifier. Then all signals which leak through the parasitic capacitances cancel. The output is combined in a delay line with decreasing impedance. For narrow band operation other methods of phase-matching are possible, which avoid feeding the signal through multiple coils and capacitors. This may be useful for power-amplifiers.

The single amplifiers can be of any class. There maybe some synergy between distributed class E/F amplifiers and some phase-matching methods. Only the fundamental frequency is used in the end, so this is the only frequency, which travels through the delay line version.

Due to this Miller effect a common source transistor acts as a capacitor (non-inverting) at high frequencies and has an inverting transconductance at low frequencies. The channel of the transistor has three dimensions. One dimension, the width, is chosen depending on the current needed. The trouble is for a single transistor parasitic capacitance and gain both scale linearly with the width. For the distributed amplifier the capacitance - that is the width - of the single transistor is chosen based on the highest frequency and the width needed for the current is split across all transistors.

Applications

Note that those termination resistors are usually not used in CMOS, but the losses due to these are small in typical applications. In solid state power amplifiers often multiple discrete transistors are used for power reasons anyway. If all transistors are driven in a synchronized fashion a very high gate drive power is needed. For frequencies at which small and efficient coils are available distributed amplifiers are more efficient.

Voltage can be amplified by a common gate transistor, which shows no miller effect and no unit gain frequency cut off. Adding this yields the cascode configuration. The common gate configuration is incompatible with CMOS; it adds a resistor, that means loss, and is more suited for broadband than for high efficiency applications.

- Radio
- Acousto-optic modulator
- Time to digital converter

MICROWAVE AMPLIFIERS

Travelling wave tube (TWT) amplifiers

Used for high power amplification at low microwave frequencies. They typically can amplify across a broad spectrum of frequencies; however, they are usually not as tunable as klystrons.

A traveling-wave tube (TWT) is an electronic device used to amplify radio frequency signals to high power, usually in an electronic assembly known as a traveling-wave tube amplifier (TWTA).

The bandwidth of a broadband TWT can be as high as three octaves, although tuned (narrowband) versions exist, and operating frequencies range from 300 MHz to 50 GHz. The voltage gain of the tube can be of the order of 70 decibels.

The device is an elongated vacuum tube with an electron gun (a heated cathode that emits electrons) at one end. A magnetic containment field around the tube focuses the electrons into a beam, which then passes down the middle of a wire helix that stretches from the RF input to the RF output, the electron beam finally striking a collector at the other end. A directional coupler, which can be either a waveguide or an electromagnetic coil, fed with the low-powered radio signal that is to be amplified, is positioned near the emitter, and induces a current into the helix.

The helix acts as a delay line, in which the RF signal travels at near the same speed along the tube as the electron beam. The electromagnetic field due to the current in the helix interacts with the electron beam, causing bunching of the electrons (an effect called *velocity modulation*), and the electromagnetic field due to the beam current then induces more current back into the helix (*i.e.* the current builds up and thus is amplified as it passes down). A second directional coupler, positioned near the collector, receives an amplified version of the input signal from the far end of the helix. An attenuator placed on the helix, usually between the input and output helicies, prevents reflected wave from travelling back to the cathode.

Invention, Development and early use

The invention of the TWT is widely attributed to Rudolf Kompfner in 1942–1943, though Nils Lindenblad did patent a device in May 1940 which was remarkably similar to Kompfner's TWT. Kompfner invented the TWT in a British radar lab during World War II. His first sketch of a TWT is dated November 12, 1942, and he built the first TWT in early 1943. The TWT was refined by Kompfner and John Pierce at Bell Labs.

By the sixties TWTs were produced by such companies as the English Electric Valve Company, followed by Ferranti in the seventies.

On July 10, 1962, the first communications satellite, Telstar 1, was launched with a 2 W, 4 GHz RCA-designed TWT transponder used for transmitting RF signals back to the earth. Syncom 2, the first synchronous satellite (Syncom 1 did not reach its final orbit), launched on July 26, 1963 with two 2 W, 1850 MHz Hughes-designed TWT transponders (one active and one spare).

COUPLED-CAVITY TWT

Helix TWTs are limited in peak RF power by the current handling (and therefore thickness) of the helix wire. As power level increases, the wire can overheat and cause the helix geometry to warp. Wire thickness can be increased to improve matters, but if the wire is too thick it becomes impossible to obtain the required helix pitch for proper operation. Typically helix TWTs achieve less than 2.5 kW output power.

The coupled-cavity TWT overcomes this limit by replacing the helix with a series of coupled cavities arranged axially along the beam. Conceptually, this structure provides a helical waveguide and hence amplification can occur via velocity modulation. Helical waveguides have very non-linear dispersion and thus are only narrowband (but wider than klystron). A coupled-cavity TWT can achieve 60 kW output power.

Operation is similar to that of a klystron, except that coupled-cavity TWTs are designed with attenuation between the slow-wave structure instead of a drift tube. The slow-wave structure gives the TWT its wide bandwidth. A free electron laser allows higher frequencies.

Traveling-wave tube Amplifier

A TWT integrated with a regulated power supply and protection circuits is referred to as a traveling-wave tube amplifier (abbreviated TWTA and often pronounced "TWEET-uh"). It is used to produce high-power radio frequency signals. The bandwidth of a broadband TWTA can be as high as one octave, although tuned (narrowband) versions exist; operating frequencies range from 300 MHz to 50 GHz.

A TWTA consists of a traveling-wave tube coupled with its protection circuits (as in klystron) and regulated power supply (EPC, electronic power conditioner), which may be supplied and integrated by a different manufacturer. The main difference between most power supplies and those for vacuum tubes is that efficient vacuum tubes have depressed collectors to recycle kinetic energy of the electrons and therefore the secondary winding of the power supply needs up to 6 taps of which the helix voltage needs precise regulation.

The subsequent addition of a linearizer (as for inductive output tube) can, by complementary compensation, improve the gain compression and other characteristics of the TWTA; this combination is called a linearized TWTA (LTWTA, "EL-tweet-uh"). Broadband TWTAs generally use a helix TWT, and achieve less than 2.5 kW output power. TWTAs using a coupled cavity TWT can achieve 15 kW output power, but at the expense of bandwidth.

Uses

TWTAs are commonly used as amplifiers in satellite transponders, where the input signal is very weak and the output needs to be high power. A TWTA whose output drives an antenna is a type of transmitter. TWTA transmitters are used extensively in radar, particularly in airborne fire-control radar systems, and in electronic warfare and self-protection systems. In these types of applications, a control grid is typically introduced between the TWT's electron gun and slow-wave structure to allow pulsed operation. The circuit that drives the control grid is usually referred to as a grid modulator.

Another major use of TWTAs is for the electromagnetic compatibility (EMC) testing industry for immunity testing of electronic devices. A TWT has sometimes been referred to as a "traveling-wave amplifier tube" (TWAT), although this term was never widely adopted. "TWT" has been pronounced by engineers as "twit", and "TWTA" as "tweeta".

Klystrons

Very similar to TWT amplifiers, but more powerful and with a specific frequency "sweet spot". They generally are also much heavier than TWT amplifiers, and are therefore ill-suited for light-weight mobile applications. Klystrons are tunable, offering selective output within their specified frequency range. A klystron is a specialized linear-beam vacuum tube (evacuated electron tube). Klystrons are used as amplifiers at microwave and radio frequencies to produce both low-power reference signals for superheterodyne radar receivers and to produce high-power carrier waves for communications and the driving force for modern particle accelerators.

Klystron amplifiers have the advantage (over the magnetron) of coherently amplifying a reference signal so its output may be precisely controlled in amplitude, frequency and phase. Many klystrons have a waveguide for coupling microwave energy into and out of the device, although it is also quite common for lower power and lower frequency klystrons to use coaxial couplings instead. In some cases a coupling probe is used to couple the microwave energy from a klystron into a separate external waveguide.

All modern klystrons are amplifiers, since reflex klystrons, which were used as oscillators in the past, have been surpassed by alternative technologies. The pseudo-Greek word *klystron* comes from the stem form *klys* of a Greek verb referring to the action of waves breaking against a shore, and the end of the word *electron*.

The brothers Russell and Sigurd Varian of Stanford University are generally considered to be the inventors of the klystron. Their prototype was completed in August 1937. Upon publication in 1939, news of the klystron immediately influenced the work of US and UK researchers working on radar equipment. The Varians went on to found Varian Associates to commercialize the technology (for example to make small linear accelerators to generate photons for external beam radiation therapy). In their 1939 paper, they acknowledged the contribution of A. Arsenjewa-Heil and O. Heil (wife and husband) for their velocity modulation theory in 1935.

During the second World War, the Axis powers relied mostly on (then low-powered) klystron technology for their radar system microwave generation, while the Allies used the far more powerful but frequency-drifting technology of the cavity magnetron for microwave generation. Klystron tube technologies for very high-power applications, such as synchrotrons and radar systems, have since been developed.

Explanation

Klystrons amplify RF signals by converting the kinetic energy in a DC electron beam into radio frequency power. A beam of electrons is produced by a thermionic cathode (a heated pellet of low work function material), and accelerated by high voltage electrodes (typically in the tens of kilovolts). This beam is then passed through an input cavity. RF energy is fed into the input cavity at, or near, its natural frequency to produce a voltage which acts on the electron beam.

The electric field causes the electrons to bunch: electrons that pass through during an opposing electric field are accelerated and later electrons are slowed, causing the previously continuous electron beam to form bunches at the input frequency. To reinforce the bunching, a klystron may contain additional "buncher" cavities. The RF current carried by the beam will produce an RF magnetic field, and this will in turn excite a voltage across the gap of subsequent resonant cavities. In the output cavity, the developed RF energy is coupled out. The spent electron beam, with reduced energy, is captured in a collector.

Two-cavity klystron Amplifier

In the two-chamber klystron, the electron beam is injected into a resonant cavity. The electron beam, accelerated by a positive potential, is constrained to travel through a cylindrical *drift tube* in a straight path by an axial magnetic field. While passing through the first cavity, the electron beam is velocity modulated by the weak RF signal. In the moving frame of the electron beam, the velocity modulation is equivalent to a plasma oscillation.Plasma oscillations are rapid oscillations of the electron density in conducting media such as plasmas or metals.(The frequency only depends weakly on the wavelength). So in a quarter of one period of the plasma frequency, the velocity modulation is converted to density modulation, *i.e.* bunches of electrons. As the bunched electrons enter the second chamber they induce standing waves at the same frequency as the input signal. The signal induced in the second chamber is much stronger than that in the first.

Two-cavity klystron Oscillator

The two-cavity amplifier klystron is readily turned into an oscillator klystron by providing a feedback loop between the input and output cavities. Two-cavity oscillator klystrons have the advantage of being among the lowest-noise microwave sources available, and for that reason have often been used in the illuminator systems of missile targeting radars. The two-cavity oscillator klystron normally generates more power than the reflex klystron—typically watts of output rather than milliwatts.

Since there is no reflector, only one high-voltage supply is necessary to cause the tube to oscillate, the voltage must be adjusted to a particular value. This is because the electron beam must produce the bunched electrons in the second cavity in order to generate output power. Voltage must be adjusted to vary the velocity of the electron beam (and thus the frequency) to a suitable level due to the fixed physical separation between the two cavities. Often several "modes" of oscillation can be observed in a given klystron.

Reflex Klystron

In the reflex klystron (also known as a 'Sutton' klystron after its inventor), the electron beam passes through a single resonant cavity. The electrons are fired into one end of the tube by an electron gun. After passing through the resonant cavity they are reflected by a negatively charged reflector electrode for another pass through the cavity, where they are then collected. The electron beam is velocity modulated when it first passes through the cavity. The formation of electron bunches takes place in the drift space between the reflector and the cavity.

The voltage on the reflector must be adjusted so that the bunching is at a maximum as the electron beam re-enters the resonant cavity, thus ensuring a maximum of energy is transferred from the electron beam to the RF oscillations in the cavity.The voltage should always be switched on before providing the input to the reflex klystron as the whole function of the reflex klystron would be destroyed if the supply is provided after the input. The reflector voltage may be varied slightly from the optimum value, which results in some loss of output power, but also in a variation in frequency. This effect is used to good advantage for automatic frequency control in receivers, and in frequency modulation for transmitters. The level of modulation applied for transmission is small enough that the power output essentially remains constant. At regions far from the optimum voltage, no oscillations are obtained at all. This tube is called a reflex klystron because it repels the input supply or performs the opposite function of a [Klystron].

There are often several regions of reflector voltage where the reflex klystron will oscillate; these are referred to as modes. The electronic tuning range of the reflex klystron is usually referred to as the variation in frequency between half power points—the points in the oscillating mode where the power output is half the maximum output in the mode. It should be noted that the frequency of oscillation is dependent on the reflector voltage, and varying this provides a crude method of frequency modulating the oscillation frequency, albeit with accompanying amplitude modulation as well. Modern semiconductor technology has effectively replaced the reflex klystron in most applications.

Multicavity Klystron

In all modern klystrons, the number of cavities exceeds two. A larger number of cavities may be used to increase the gain of the klystron, or to increase the bandwidth.

Tuning a Klystron

Some klystrons have cavities that are tunable. Tuning a klystron is delicate work which, if not done properly, can cause damage to equipment or injury to the technician. By adjusting the frequency of individual cavities, the technician can change the operating frequency, gain, output power, or bandwidth of the amplifier. The technician must be careful not to exceed the limits of the graduations, or damage to the klystron can result. Manufacturers generally send a card with the unique calibrations for a klystron's performance characteristics, that lists the graduations to be set to attain any of a set of listed frequencies. No two klystrons are exactly identical (even when comparing like part/model number klystrons), and so every card is specific to the individual unit. Klystrons have serial numbers on each of them to uniquely identify each unit, and for which manufacturers may (hopefully) have the performance characteristics in a database. If not, loss of the calibration card may be an insoluble problem, making the klystron unusable or perform marginally un-tuned.

Other precautions taken when tuning a klystron include using non-ferrous tools. Some klystrons employ permanent magnets. If a technician uses ferrous tools, (which are ferromagnetic), and comes too close to the intense magnetic fields that contain the electron beam, such a tool can be pulled into the unit by the intense magnetic force, smashing fingers, injuring the technician, or damaging the unit. Special lightweight non-magnetic (aka diamagnetic) tools made of beryllium alloy have been used for tuning U.S. Air Force klystrons.

Precautions are routinely taken when transporting klystron devices in aircraft, as the intense magnetic field can interfere with magnetic navigation equipment. Special overpacks are designed to help limit this field "in the field," and thus allow such devices to be transported safely.

Optical Klystron

In an optical klystron the cavities are replaced with undulators. Very high voltages are needed. The electron gun, the drift tube and the collector are still used.

Floating drift tube Klystron

The floating drift tube klystron has a single cylindrical chamber containing an electrically isolated central tube. Electrically, this is similar to the two cavity oscillator klystron with a lot of feedback between the two cavities. Electrons exiting the source cavity are velocity modulated by the electric field as they travel through the drift tube and emerge at the destination chamber in bunches, delivering power to the oscillation in the cavity. This type of oscillator klystron has an advantage over the two-cavity klystron on which it is based. It only needs one tuning element to effect changes in frequency. The drift tube is electrically insulated from the cavity walls, and DC bias is applied separately. The DC bias on the drift tube may be adjusted to alter the transit time through it, thus allowing some electronic tuning of the oscillating frequency. The amount of tuning in this manner is not large and is normally used for frequency modulation when transmitting.

Collector

After the RF energy has been extracted from the electron beam, the beam is destroyed in a collector. Some klystrons include depressed collectors, which recover energy from the beam before collecting the electrons, increasing efficiency. Multistage depressed collectors enhance the energy recovery by "sorting" the electrons in energy bins.

Applications

Klystrons produce microwave power far in excess of that developed by solid state. In modern systems, they are used from UHF (100's of MHz) up through hundreds of gigahertz (as in the Extended Interaction Klystrons in the CloudSat satellite). Klystrons can be found at work in radar, satellite and wideband high-power communication (very common in television broadcasting and EHF satellite terminals), and high-energy physics (particle accelerators and experimental reactors). At SLAC, for example, klystrons are routinely employed which have outputs in the range of 50 megawatts (pulse) and 50 kilowatts (time-averaged) at frequencies nearing 3 GHz Popular Science's "Best of What's New 2007" included a company using a klystron to convert the hydrocarbons in everyday materials, automotive waste, coal, oil shale, and oil sands into natural gas and diesel fuel.

In St. Petersburg, FL, local cable news station Bay News 9 uses a Klystron Tube in its Klystron 9 Weather Radar to increase its range and resolution. The radar site is actually located behind the Bright House Sports Network building in Pinellas Park, FL.

Confusion with Krytron

A misleadingly similarly named tube, the krytron, is used in simple switching applications. It has recently gained fame as a rapid switch which can be used in nuclear weapons to precisely detonate explosives at high speeds, in order to start the fission process. Krytrons have also been used in photocopiers, raising issues of war technology transfer to countries for items such as this, which have a "dual use."

Musical Instrument (audio) Amplifiers

An audio amplifier is usually used to amplify signals such as music or speech. An instrument amplifier is an electronic amplifier that converts the often barely audible or purely electronic signal from musical instruments such as an electric guitar, an electric bass, or an electric keyboard into an electronic signal capable of driving a loudspeaker that can be heard by the performers and audience. Combination ("combo") amplifiers include a preamplifier, a power amplifier, tone controls, and one or more speakers in a cabinet, a housing usually made of plywood, particleboard, or, less commonly, moulded plastic. Instrument amplifiers for some instruments are also available without an integral speaker; these amplifiers have to be plugged into an external speaker cabinet.

Instrument amplifiers are available for specific instruments, including the electric guitar, electric bass, electric keyboards, and acoustic instruments such as the mandolin and banjo. Some amplifiers are designed for specific styles of music, such as the "traditional"-style "tweed" guitar amplifiers used by blues and country musicians, and the Marshall amplifiers used by hard rock and heavy metal bands.

Unlike home "hi-fi" amplifiers or public address systems, which are designed to reproduce accurately the source sound signals with as little harmonic distortion as possible, instrument amplifiers are often designed to add additional tonal coloration to the original signal or emphasize (or de-emphasize) certain frequencies. The two exceptions are keyboard amplifiers and "acoustic" instrument amplifiers, which typically aim for a relatively flat frequency response.

Types

"Traditional" amplifiers

Standard amps

Standard amplifiers, such as the Fender "tweed"-style amps and Gibson amps, are often used by traditional rock, blues, and country musicians who wish to create a "vintage" 1950s-style sound. They are used by electric guitarists, pedal steel guitar players, and blues harmonica ("harp") players. Combo amplifiers such as the Fender Bassman have tube amplifiers, four 10" speakers, and built-in reverb and "vibrato" effects units.

These amps are designed to produce a variety of sounds ranging from a clean, warm sound (when used in country and soft rock) to a growling, natural overdrive, when the volume is set near its maximum, (when used for blues, rockabilly, and roots rock). These amplifiers usually have a sharp treble roll-off at 5 kHz to reduce the extreme high frequencies, and a bass roll-off at 60–100 Hz to reduce boominess. The nickname "tweed" refers to the lacquered beige-light brown fabric covering used on these amplifiers.

The smallest "combo" amplifiers, which are mainly used for rehearsal and warm-up purposes, may have only a single 8" or 10" speaker. Some harmonica players use these small combo amplifiers for concert performances, though, because it is easier to create natural overdrive with these lower-powered amplifiers. Larger combo amplifiers, with one 12 inch speaker or two or four 10 or 12 inch speakers are used for club performances. For large concert venues, performers may also use an amplifier "head" with several separate speaker cabinets (which usually contain two or four 12" speakers).

Hard Rock and Heavy Metal

These electric guitar amplifiers add an aggressive "drive", intensity, and "edge" to the guitar sound with distortion effects, preamplification boost controls, and tone filters. While many of the most expensive, high-end models use tube amplifiers, there also many models that use transistor amplifiers, or a mixture of the two technologies (*i.e.*, a tube preamplifier with a transistor power amplifier). Amplifiers of this type, such as Marshall amplifiers, are used in a range of the louder, heavier genres of rock, including hard rock, heavy metal, and hardcore punk. This type of amplifier is available in a range of formats, ranging from smaller combo amplifiers for rehearsal and warm-up purposes to heavy "heads" which are designed to be used with separate speaker cabinets, which is colloquially referred to as a "stack."

In the late 1960s and early 1970s, public address systems at rock concerts were used mainly for the vocals. As a result, to get a loud electric guitar sound, early heavy metal and rock-blues bands often used "stacks" of 4x12" Marshall speaker cabinets on the stage. In 1969 Jimi Hendrix used four stacks to create a powerful lead sound, and in the early 1970s by the band Blue Öyster Cult used an entire wall of Marshall Amplifiers to create a roaring wall of sound. In the 1980s, metal bands such as Slayer and Yngwie Malmsteen also used "walls" of over 20 Marshall cabinets. However, by the 1980s and 1990s, most of the sound at live concerts was produced by the sound reinforcement system rather than the onstage guitar amplifiers, so most of these cabinets were not connected to an amplifier. Instead, these walls of speaker cabinets were used for aesthetic reasons.

Amplifiers for harder, heavier genres often use valve amplifiers (known as "tube amplifiers" in North America). Valve amplifiers have a "warmer" tone than those of transistor amps, particularly when overdriven. Instead of abruptly clipping off the signal at cut-off and saturation levels, the signal is rounded off more smoothly. Vacuum tubes also exhibit different harmonic effects than transistors. In contrast to the "tweed"-style amplifiers, which use 10" speakers in an open-backed cabinet, companies such as Marshall tend to use 12" speakers in a closed-back cabinet. These amplifiers usually allow users to switch between "clean" and distorted tones (or a rhythm guitar-style "crunch" tone and a sustained "lead" tone) with a foot-operated switch.

Bass

Bass amplifiers are designed for bass guitars or more rarely, for upright bass. They differ from amplifiers for the regular electric guitar in several respects. They have extended bass response and tone controls optimised for bass instruments, which produce pitches of 40 Hz, in the case of a standard four-string electric bass, or even lower for five- or six-string electric basses.

Higher-end bass amplifiers sometimes include compressor or limiter features, which help to keep the amplifier from distorting at high volume levels, and an XLR DI output for patching the bass signal directly into a mixing board or PA systems. Larger, more powerful bass amplifiers (300 or more watts) are often provided with external metal heat sinks or fans to help keep the amplifier cool.

Speaker cabinets designed for bass instrument amplification usually use larger loudspeakers (or more loudspeakers, in the case of the popular 4 X 10" cabinets, which contain four 10" speakers) than the cabinets used for other instruments, so that they can move the larger amounts of air needed to reproduce low frequencies. While the largest speakers commonly used for regular electric guitar are 12" speakers, electric bass speaker cabinets often use 15" speakers. Bass players who play styles of music that require an extended low-range response, such as death metal, sometimes use speaker cabinets with 18" speakers. The speakers used for bass instrument amplification tend to be more heavy-duty than speakers used for regular electric guitar, and the speaker cabinets are typically more rigidly constructed and heavily braced, to prevent unwanted buzzes and rattles. Bass cabinets often include bass reflex ports or openings in the cabinet, which improve the bass response, especially at high volumes.

Keyboard

This type of amplifier is used to amplify a range of electric and electronic keyboards, such as synthesizers, Hammond organ-style keyboards, stage pianos and electric pianos. Since keyboard instruments contain a wide frequency range, from very low bass notes to extremely high pitches, keyboard amplifiers are often provided with a large woofer speaker to handle the low notes and a horn (or tweeter) for the high notes.

Keyboard amplifiers intended for general use for a range of keyboard applications usually have very low distortion and extended, flat frequency response in both directions. The exception to this rule is keyboard amplifiers designed for the Hammond organ, such as the vintage Leslie speaker cabinet and modern recreations, which have a tube amplifier which is often turned up to add a warm, "growling" overdrive to the organ sound.

Unlike bass amplifiers and electric guitar amplifiers, keyboard amplifers are rarely used in the "amplifier head" and separate speaker cabinets configuration. Instead, most keyboard amplifiers are "combo" amplifiers that integrate the amplifier, tone controls, and speaker into a single wooden cabinet. Another unusual aspect of keyboard amplifiers is that they are often designed with a "wedge" shape, as used with monitor speakers. This allows the cabinet to be rocked back so that it will project sound upwards at a roughly 45' angle, which is more suitable for a seated keyboardist. Keyboard amplifiers often have a simple onboard mixer, so that keyboardists can control the tone and level of several keyboards. In some genres, such as progressive rock, for example, keyboardists may perform with several synthesizers, electric pianos, and electro-mechanical keyboards.

Acoustic

These amplifiers are designed to be used with acoustic instruments such as violin ("fiddle"), mandolin, and acoustic guitar, especially for the way these instruments are used in relatively quiet genres such as folk and bluegrass. They are similar in many ways to keyboard amplifiers, in that they have a relatively flat frequency response, and they are usually designed so that neither the power amplifier nor the speakers will introduce additional coloration.

To produce this relatively "clean" sound, these amplifiers often have very powerful amplifiers (providing from 400 to 800 watts RMS), to provide additional "headroom" and prevent unwanted distortion. Since an 800 watt amplifier built with standard Class AB technology would be very heavy, some acoustic amplifier manufacturers use lightweight Class D amplifiers, which are also called "switching amplifiers."

Acoustic amplifiers are designed to produce a "clean", transparent, "acoustic" sound when used with acoustic instruments with built-in transducer pickups and/or microphones. The amplifiers often come with a simple mixer, so that the signals from a pickup and microphone can be blended. Since the early 2000s, it has become increasingly common for acoustic amplifiers to be provided with a range of digital effects, such as reverb and compression. As well, these amplifiers often contain feedback-suppressing devices, such as notch filters or parametric equalizers.

An audio amplifier is an electronic amplifier that amplifies low-power audio signals (signals composed primarily of frequencies between 20 hertz to 20,000 hertz, the human range of hearing) to a level suitable for driving loudspeakers and is the final stage in a typical audio playback chain. The preceding stages in such a chain are low power audio amplifiers which perform tasks like pre-amplification, equalization, tone control, mixing/effects, or audio sources like record players, CD players, and cassette players. Most audio amplifiers require these low-level inputs to adhere to line levels. While the input signal to an audio amplifier may measure only a few hundred microwatts, its output may be tens, hundreds, or thousands of watts.

The audio amplifier was invented in 1906 by Lee De Forest when he invented the triode vacuum tube. The triode was a three terminal device with a control grid that can modulate the flow of electrons from the filament to the plate. The triode vacuum amplifier was used to make the first AM radio.

Early audio amplifiers were based on vacuum tubes (also known as *valves*), and some of these achieved notably high quality (*e.g.*, the Williamson amplifier of 1947-9). Most modern audio amplifiers are based on solid state devices (transistors such as BJTs, FETs and MOSFETs), but there are still some who prefer tube-based amplifiers, due to a perceived 'warmer' valve sound. Audio amplifiers based on transistors became practical with the wide availability of inexpensive transistors in the late 1960s.

Design Parameters

Key design parameters for audio amplifiers are frequency response, gain, noise, and distortion. These are interdependent; increasing gain often leads to undesirable increases in noise and distortion. While negative feedback actually reduces the gain, it also reduces distortion. Most audio amplifiers are linear amplifiers operating in class AB.

Filters and Preamplifiers

Historically, the majority of commercial audio preamplifiers made had complex filter circuits for equalization and tone adjustment, due to the far from ideal quality of recordings, playback technology, and speakers of the day.

Using today's high quality (often digital) source material, speakers, and etc., such filter circuits are usually not needed. Audiophiles generally agree that filter circuits are to be avoided wherever possible. Today's audiophile amplifiers do not have tone controls or filters.

Since modern digital devices, including CD and DVD players, radio receivers and tape decks already provide a "flat" signal at line level, the preamp. is not needed other than as volume control. One alternative to a separate preamp. is to simply use passive volume and switching controls, sometimes integrated into a power amp. to form an "integrated" amplifier.

Further Developments in Amplifier Design

For some years following the introduction of solid state amplifiers, their perceived sound did not have the excellent audio quality of the best valve amplifiers. This led audiophiles to believe that valve sound had an intrinsic quality due to the vacuum tube technology itself. In 1972, Matti Otala demonstrated the origin of a previously unobserved form of distortion: transitory intermodulation distortion (TIM), also called slew rate distortion. TIM distortion was found to occur during very rapid increases in amplifier output voltage.

TIM did not appear at steady state sine tone measurements, helping to hide it from design engineers prior to 1972. Problems with TIM distortion stem from reduced open loop frequency response of solid state amplifiers. Further works of Otala and other authors found the solution for TIM distortion, including increasing slew rate, decreasing preamp frequency bandwidth, and the insertion of a lag compensation circuit in the input stage of the amplifier. In high quality modern amplifiers the open loop response is at least 20 kHz, canceling TIM distortion. However, TIM distortion is still present in most low price home quality amplifiers.

The next step in advanced design was the Baxandall Theorem, created by Peter Baxandall in England. This theorem introduced the concept of comparing the ratio between the input distortion and the output distortion of an audio amplifier. This new idea helped audio design engineers to better evaluate the distortion processes within an audio amplifier.

Phonograph (vinyl record) Equalization

Since the mid-1950s, LP phonograph records have been mastered using

RIAA equalization, in which the dynamics of the recording have been altered so that the amplitude of the signal that has been cut into the record increases with increasing frequency. Equalization helps to mask the high frequency noise ("hiss") that is generated as the pickup's stylus rubs against the groove walls. The RIAA curve also attenuates the bass, which reduces the maximum excursions of the stylus to a practical level during loud passages. This has the desirable effect of reducing distortion, as well as making the grooves narrower and increasing the potential maximum recording time per record side. Also, with less excursion, less stress is applied to the stylus, which helps to reduce record wear.

During playback, the RIAA curve is reversed by preamplification, resulting in nearly flat frequency response. It should also be noted that the preamplifier is employed to boost the weak signal emitted by a magnetic pickup. Piezoelectric pickups generally produce much higher output voltages and seldom require preamplification.

Prior to the adoption of the RIAA curve, a number of competing and partially incompatible equalization schemes were utilized during record mastering. Early high fidelity systems often had an equalization selector switch to match playback characteristics to the recording curve of the particular label being played. The development and acceptance of the RIAA curve eliminated this requirement.

Applications

Important applications include public address systems, theatrical and concert sound reinforcement, and domestic sound systems. The sound card in a personal computer contains several audio amplifiers (depending on number of channels), as does every stereo or home-theatre system.

Roles

Instrument amplifiers are designed for a different purpose than 'Hi-Fi' (high fidelity) stereo amplifiers used for radios and home stereo systems. Hi-fi home stereo amplifiers are designed to accurately reproduce the source sound signals from pre-recorded music, with as little harmonic distortion as possible.

In contrast, instrument amplifiers are often designed to add additional tonal coloration to the original signal or emphasize certain frequencies. For electric instruments such as electric guitar, the amplifier helps to create the instrument's tone by boosting the input signal gain and distorting the signal, and by emphasizing frequencies deemed to be desirable (*e.g.*, low frequencies) and de-emphasizing frequencies deemed to be undesirable (*e.g.*, very high frequencies).

The two exceptions are keyboard amplifiers and acoustic amplifiers which are used by folk and bluegrass musicians for amplifying acoustic instruments such as acoustic guitar, violin, and mandolin. Acoustic amplifiers typically aim for a relatively flat response across the entire frequency range, much like a Public Address system.

Size and Power Rating

In the 1960s and 1970s, large, heavy, high output power amplifiers were preferred for instrument amplifiers, especially for large concerts, because public address systems were generally only used to amplify the vocals. Moreover, in the 1960s, PA systems typically did not use monitor speaker systems to amplify the music for the onstage musicians. Instead, the musicians were expected to have instrument amplifiers that were powerful enough to provide amplification for the stage and audience. In late 1960s and early 1970s rock concerts, bands often used large stacks of speaker cabinets powered by heavy tube amplifiers such as the Super Valve Technology (SVT) amplifier, which was often used with eight 10" speakers.

However, over subsequent decades, PA systems were substantially improved, and different approaches such as horn-loaded "bass bins" (in the 1980s) and subwoofers (1990s and 2000s) were used to amplify bass frequencies. As well, in the 1980s and 1990s, monitor systems were substantially improved, which allowed sound engineers to provide onstage musicians with a loud, clear, and full-range reproduction of their instruments' sound.

As a result of the improvements to PA systems and monitor systems, musicians in the 2000s no longer need to have huge, powerful amplifier systems; a small combo amplifier patched into the PA will suffice. In the 2000s, virtually all of the sound reaching the audience in large venues comes from the PA system. As well, in the 2000s onstage instrument amplifiers are more likely to be kept at a low volume, because high volume levels onstage makes it harder to control the sound mix and produce a clean sound.

As a result, in many large venues much of the onstage sound reaching the musicians now comes from the monitor speakers, not from the instrument amplifiers. While stacks of huge speaker cabinets and amplifiers are still used in concerts (especially in heavy metal), this is often mainly for aesthetics or to create a more authentic tone. The switch to smaller instrument amplifiers makes it easier for musicians to transport their equipment to performances. As well, it makes concert stage management easier at large clubs and festivals where several bands are performing in sequence, because the bands can be moved on and off the stage more quickly.

Amplifier Technology

Instrument amplifiers may be based on thermionic ("tube" or "valve") or solid state (transistor) technology.

Tube Amplifiers

Vacuum tubes were the dominant active electronic components in amplifiers manufactured from the 1930s through the early 1970s, and tube amplifiers continue to be preferred by some professional musicians. Some musicians believe that tube amplifiers produce a "warmer" or more "natural" sound than solid state units. However, these subjective assessments of the attributes of tube amplifiers' sound qualities are the subject of ongoing debate.

Although tube amplifiers produce more heat than solid state amplifiers, few manufacturers of these units include cooling fans in the chassis. While tube amplifiers do need to attain a proper operating temperature, if the temperature goes above this operating temperature, it may shorten the tubes' lifespan and lead to tonal inconsistencies.

Solid State Amplifiers

By the 1960s and 1970s, semiconductor transistor-based amplifiers began to become more popular because they are less expensive, lighter-weight, and require less maintenance. In some cases, tube and solid state technologies are used together in amplifiers. A common setup is the use of a tube preamplifier with a solid state power amplifier. There are also an increasing range of products that use digital signal processing and digital modeling technology to simulate many different combinations of amp and cabinets.

The output transistors of solid state amplifiers can be passively cooled by using metal fins called heatsinks to radiate away the heat. For high-wattage amplifiers (over 800 watts), a fan is often used to move air across internal heatsinks.

Other amplifier types

Carbon microphone

One of the first devices used to amplify signals was the carbon microphone (effectively a sound-controlled variable resistor). By channeling a large electric current through the compressed carbon granules in the microphone, a small sound signal could produce a much larger electric signal. The carbon microphone was extremely important in early telecommunications; analog telephones in fact work without the use of any other amplifier. Before the invention of electronic amplifiers, mechanically coupled carbon microphones were also used as amplifiers in telephone repeaters for long distance service.

Magnetic Amplifier

A magnetic amplifier is a transformer-like device that makes use of the saturation of magnetic materials to produce amplification. It is a non-electronic electrical amplifier with no moving parts. The bandwidth of magnetic amplifiers extends to the hundreds of kilohertz. The magnetic amplifier (colloquially known as the "mag amp") is an electromagnetic device for amplifying electrical signals.

The magnetic amplifier was invented early in the 20th century, and was used as an alternative to vacuum tube amplifiers where robustness and high current capacity were required. World War II Germany perfected this type of amplifier, and it was used for instance in the V-2 rocket. The magnetic amplifier has now been largely superseded by the transistor-based amplifier, except in a few safety critical, high reliability or extremely demanding applications.

Principle of Operation

Visually a mag amp device may resemble a transformer but the operating principle is quite different from a transformer - essentially the mag amp is a saturable reactor. It makes use of magnetic saturation of the core, a non-linear property of a certain class of transformer cores. For controlled saturation characteristics the magnetic amplifier employs core materials that have been designed to have a specific B-H curve shape that is highly rectangular, in contrast to the slowly tapering B-H curve of softly saturating core materials that are often used in normal transformers.

The typical magnetic amplifier consists of two physically separate but similar transformer magnetic cores, each of which has two windings - a control winding and an AC winding. A small DC current from a low impedance source is fed into the series-connected control windings. The AC windings may be connected either in series or in parallel, the configurations resulting in different types of mag amps. The amount of control current fed into the control winding sets the point in the AC winding waveform at which either core will saturate. In saturation, the AC winding on the saturated core will go from a high impedance state ("off") into a very low impedance state ("on") - that is, the control current controls at which voltage the mag amp switches "on".

A relatively small DC current on the control winding is able to control or switch large AC currents on the AC windings. This results in current amplification.

Applications

Magnetic amplifiers were used extensively as the switching element in early switched-mode (SMPS) power supplies, as well as in lighting control. They have been largely superseded by semiconductor based solid-state switches, though recently there has been some regained interest in using mag amps in compact and reliable switching power supplies. PC ATX power supplies often use mag amps for secondary side voltage regulation.

Magnetic Amplifiers are still used in some arc welders.

Magnetic amplifier transformer cores designed specifically for switch mode power supplies are currently manufactured by several large electromagnetics companies, including Metglas and Mag-Inc. Magnetic amplifiers can be used for measuring high DC-voltages without direct connection to the high voltage and are therefore still used in the HVDC-technique.

Misnomer uses

Late in the 20th century, Robert Carver designed and produced several high quality high powered audio amplifiers, calling them magnetic amplifiers. In fact, they were in most respects conventional audio amplifier designs with an unusual power supply circuit. They were not magnetic amplfiers in the sense of this.

Rotating Electrical Machinery Amplifier

A Ward Leonard control is a rotating machine like an electrical generator that provides amplification of electrical signals by the conversion of mechanical energy to electrical energy. Changes in generator field current result in larger changes in the output current of the generator, providing gain. This class of device was used for smooth control of large motors, primarily for elevators and naval guns. Field modulation of a very high speed AC generator was also used for some early AM radio transmissions.

Johnsen-Rahbek Effect Amplifier

The earliest form of audio power amplifier was Edison's "electromotograph" loud-speaking telephone, which used a wetted rotating chalk cylinder in contact with a stationary contact. The friction between cylinder and contact varied with the current, providing gain. Edison discovered this effect in 1874, but the theory behind the Johnsen-Rahbek effect was not understood until the semiconductor era.

Mechanical Amplifiers

Mechanical amplifiers were used in the pre-electronic era in specialized applications. Early autopilot units designed by Elmer Ambrose Sperry incorporated a mechanical amplifier using belts wrapped around rotating drums; a slight increase in the tension of the belt caused the drum to move the belt. A paired, opposing set of such drives made up a single amplifier. This amplified small gyro errors into signals large enough to move aircraft control surfaces. A similar mechanism was used in the Vannevar Bush differential analyzer.

Optical Amplifiers

Optical amplifiers amplify light through the process of stimulated emission. An optical amplifier is a device that amplifies an optical signal directly, without the need to first convert it to an electrical signal. An optical amplifier may be thought of as a laser without an optical cavity, or one in which feedback from the cavity is suppressed. Stimulated emission in the amplifier's gain medium causes amplification of incoming light. Optical amplifiers are important in optical communication and laser physics.

Laser Amplifiers

Almost any laser active gain medium can be pumped to produce gain for light at the wavelength of a laser made with the same material as its gain medium. Such amplifiers are commonly used to produce high power laser systems. Special types such as regenerative amplifiers and chirped-pulse amplifiers are used to amplify ultrashort pulses.

Doped fibre Amplifiers

Doped fibre amplifiers (DFAs) are optical amplifiers that use a doped optical fibre as a gain medium to amplify an optical signal. They are related to fibre lasers. The signal to be amplified and a pump laser are multiplexed into the doped fibre, and the signal is amplified through interaction with the doping ions. The most common example is the Erbium Doped Fibre Amplifier (EDFA), where the core of a silica fibre is doped with trivalent Erbium ions and can be efficiently pumped with a laser at a wavelength of 980 nm or 1,480 nm, and exhibits gain in the 1,550 nm region.

Amplification is achieved by stimulated emission of photons from dopant ions in the doped fibre. The pump laser excites ions into a higher energy from where they can decay via stimulated emission of a photon at the signal wavelength back to a lower energy level. The excited ions can also decay spontaneously (spontaneous emission) or even through non-radiative processes involving interactions with phonons of the glass matrix. These last two decay mechanisms compete with stimulated emission reducing the efficiency of light amplification.

The *amplification window* of an optical amplifier is the range of optical wavelengths for which the amplifier yields a usable gain. The amplification window is determined by the spectroscopic properties of the dopant ions, the glass structure of the optical fibre, and the wavelength and power of the pump laser.

Although the electronic transitions of an isolated ion are very well defined, broadening of the energy levels occurs when the ions are incorporated into the glass of the optical fibre and thus the amplification window is also broadened. This broadening is both homogeneous (all ions exhibit the same broadened spectrum) and inhomogeneous (different ions in different glass locations exhibit different spectra). Homogeneous broadening arises from the interactions with phonons of the glass, while inhomogeneous broadening is caused by differences in the glass sites where different ions are hosted. Different sites expose ions to different local electric fields, which shifts the energy levels via the Stark effect.

In addition, the Stark effect also removes the degeneracy of energy states having the same total angular momentum (specified by the quantum number J). Thus, for example, the trivalent Erbium ion (Er^{+3}) has a ground state with

J = 15/2, and in the presence of an electric field splits into J + 1/2 = 8 sublevels with slightly different energies. The first excited state has J = 13/2 and therefore a Stark manifold with 7 sublevels. Transitions from the J = 13/2 excited state to the J= 15/2 ground state are responsible for the gain at 1.5 μm wavelength. The gain spectrum of the EDFA has several peaks that are smeared by the above broadening mechanisms. The net result is a very broad spectrum (30 nm in silica, typically). The broad gain-bandwidth of fibre amplifiers make them particularly useful in wavelength-division multiplexed communications systems as a single amplifier can be utilized to amplify all signals being carried on a fibre and whose wavelengths fall within the gain window.

Noise

The principal source of noise in DFAs is Amplified Spontaneous Emission (ASE), which has a spectrum approximately the same as the gain spectrum of the amplifier. Noise figure in an ideal DFA is 3 dB, while practical amplifiers can have noise figure as large as 6-8 dB.

As well as decaying via stimulated emission, electrons in the upper energy level can also decay by spontaneous emission, which occurs at random, depending upon the glass structure and inversion level. Photons are emitted spontaneously in all directions, but a proportion of those will be emitted in a direction that falls within the numerical aperture of the fibre and are thus captured and guided by the fibre. Those photons captured may then interact with other dopant ions, and are thus amplified by stimulated emission.

The initial spontaneous emission is therefore amplified in the same manner as the signals, hence the term *Amplified Spontaneous Emission*. ASE is emitted by the amplifier in both the forward and reverse directions, but only the forward ASE is a direct concern to system performance since that noise will co-propagate with the signal to the receiver where it degrades system performance. Counter-propagating ASE can, however, lead to degradation of the amplifier's performance since the ASE can deplete the inversion level and thereby reduce the gain of the amplifier.

Gain Saturation

Gain is achieved in a DFA due to population inversion of the dopant ions. The inversion level of a DFA is set, primarily, by the power of the pump wavelength and the power at the amplified wavelengths. As the signal power increases, or the pump power decreases, the inversion level will reduce and thereby the gain of the amplifier will be reduced. This effect is known as gain saturation - as the signal level increases, the amplifier saturates and cannot produce any more output power, and therefore the gain reduces. Saturation is also commonly known as gain compression.

To achieve optimum noise performance DFAs are operated under a significant amount of gain compression (10 dB typically), since that reduces the rate of spontaneous emission, thereby reducing ASE. Another advantage of operating the DFA in the gain saturation region is that small fluctuations in the input signal power are reduced in the output amplified signal: smaller input signal powers experience larger (less saturated) gain, while larger input powers see less gain.

Inhomogeneous Broadening Effects

Due to the inhomogeneous portion of the linewidth broadening of the dopant ions, the gain spectrum has an inhomogeneous component and gain saturation occurs, to a small extent, in an inhomogeneous manner. This effect is known as *Spectral hole burning* because a high power signal at one wavelength can 'burn' a hole in the gain for wavelengths close to that signal by saturation of the inhomogeneously broadened ions. Spectral holes vary in width depending on the characteristics of the optical fibre in question and the power of the burning signal, but are typically less than 1 nm at the short wavelength end of the C-band, and a few nm at the long wavelength end of the C-band. The depth of the holes is very small, though, making it difficult to observe in practice.

Polarization Effects

Although the DFA is essentially a polarization independent amplifier, a small proportion of the dopant ions interact preferentially with certain polarizations and a small dependence on the polarization of the input signal may occur (typically < 0.5 dB). This is called Polarization Dependent Gain (PDG). The absorption and emission crossections of the ions can be modeled as ellipsoids with the major axes aligned at random in all directions in different glass sites. The random distribution of the orientation of the ellipsoids in a glass produces a macroscopically isotropic medium, but a strong pump laser induces an anisotropic distribution by selectively exciting those ions that are more aligned with the optical field vector of the pump. Also, those excited ions aligned with the signal field produce more stimulated emission.

The change in gain is thus dependent on the alignment of the polarizations of the pump and signal lasers - ie. whether the two lasers are interacting with the same sub-set of dopant ions or not. In an ideal doped fibre without birefringence, the PDG would be inconveniently large. Fortunately, in optical fibers small amounts of birefringence are always present and, furthermore, the fast and slow axes vary randomly along the fibre length. A typical DFA has several tens of meters, long enough to already show this randomness of the birefringence axes. These two combined effects (which in transmission fibers give rise to Polarization Mode Dispersion) produce a misalignment of the relative polarizations of the signal and pump lasers along the fibre, thus tending to average out the PDG. The result is that PDG is very difficult to observe in a single amplifier (but is noticeable in links with several cascaded amplifiers).

Erbium-doped Fibre Amplifiers

The erbium-doped fibre amplifier (EDFA) is the most deployed fibre amplifier as its amplification window coincides with the third transmission window of silica-based optical fibre.

Two bands have developed in the third transmission window - the *Conventional*, or C-band, from approximately 1525 nm - 1565 nm, and the *Long*, or L-band, from approximately 1570 nm to 1610 nm. Both of these bands can be amplified by EDFAs, but it is normal to use two different amplifiers, each optimized for one of the bands.

The principal difference between C- and L-band amplifiers is that a longer length of doped fibre is used in L-band amplifiers. The longer length of fibre allows a lower inversion level to be used, thereby giving at longer wavelengths (due to the band-structure of Erbium in silica) while still providing a useful amount of gain.

EDFAs have two commonly-used pumping bands - 980 nm and 1480 nm. The 980 nm band has a higher absorption cross-section and is generally used where low-noise performance is required. The absorption band is relatively narrow and so wavelength stabilised laser sources are typically needed. The 1480 nm band has a lower, but broader, absorption cross-section and is generally used for higher power amplifiers. A combination of 980 nm and 1480 nm pumping is generally utilised in amplifiers.

The EDFA was invented by a group including David N. Payne, R. Mears, and L. Reekie, from the University of Southampton and a group from AT&T Bell Laboratories, E. Desurvire, P. Becker, and J. Simpson.

Doped Fibre Amplifiers for other Wavelength Ranges

Thulium doped fibre amplifiers have been used in the S-band (1450-1490 nm) and Praseodymium doped amplifiers in the 1300 nm region. However, those regions have not seen any significant commercial use so far and so those amplifiers have not been the subject of as much development as the EDFA. However, Ytterbium doped fibre lasers and amplifiers, operating near 1 micrometre wavelength, have many applications in industrial processing of materials, as these devices can be made with extremely high output power (tens of kilowatts).

Semiconductor Optical Amplifier (SOA)

Semiconductor optical amplifiers are amplifiers which use a semiconductor to provide the gain medium. These amplifiers have a similar structure to Fabry-Perot laser diodes but with anti-reflection design elements at the endfaces.

Recent designs include anti-reflective coatings and tilted waveguide and window regions which can reduce endface reflection to less than 0.001 per cent. Since this creates a loss of power from the cavity which is greater than the gain it prevents the amplifier from acting as a laser.

Semiconductor optical amplifiers are typically made from group III-V compound semiconductors such as GaAs/AlGaAs, InP/InGaAs, InP/InGaAsP and InP/InAlGaAs, though any direct band gap semiconductors such as II-VI could conceivably be used. Such amplifiers are often used in telecommunication systems in the form of fibre-pigtailed components, operating at signal wavelengths between 0.85 μm and 1.6 μm and generating gains of up to 30 dB.

The semiconductor optical amplifier is of small size and electrically pumped. It can be potentially less expensive than the EDFA and can be integrated with semiconductor lasers, modulators, etc. However, the performance is still not comparable with the EDFA. The SOA has higher noise, lower gain, moderate polarization dependence and high non-linearity with fast transient time. This originates from the short nanosecond or less upper state lifetime, so that the gain reacts rapidly to changes of pump or signal power and the changes of gain also cause phase changes which can distort the signals. This non-linearity presents the most severe problem for optical communication applications. However it provides the possibility for gain in different wavelength regions from the EDFA. "Linear optical amplifiers" using gain-clamping techniques have been developed.

High optical non-linearity makes semiconductor amplifiers attractive for all optical signal processing like all-optical switching and wavelength conversion. There has been much research on semiconductor optical amplifiers as elements for optical signal processing, wavelength conversion, clock recovery, signal demultiplexing, and pattern recognition.

Vertical-Cavity SOA

A recent addition to the SOA family is the vertical-cavity SOA (VCSOA). These devices are similar in structure to, and share many features with, vertical-cavity surface-emitting lasers (VCSELs). The major difference when comparing VCSOAs and VCSELs is the reduced mirror reflectivities used in the amplifier cavity. With VCSOAs, reduced feedback is necessary to prevent the device from reaching lasing threshold. Due to the extremely short cavity length, and correspondingly thin gain medium, these devices exhibit very low single-pass gain (typically on the order of a few per cent) and also a very large free spectral range (FSR). The small single-pass gain requires relatively high mirror reflectivities to boost the total signal gain. In addition to boosting the total signal gain, the use of the resonant cavity structure results in a very narrow gain bandwidth; coupled with the large FSR of the optical cavity, this effectively limits operation of the VCSOA to single-channel amplification. Thus, VCSOAs can be seen as amplifying filters.

Given their vertical-cavity geometry, VCSOAs are resonant cavity optical amplifiers that operate with the input/output signal entering/exiting normal to the wafer surface. In addition to their small size, the surface normal operation of VCSOAs leads to a number of advantages, including low power consumption, low noise figure, polarization insensitive gain, and the ability to fabricate high fill factor two-dimensional arrays on a single semiconductor chip. These devices are still in the early stages of research, though promising preamplifier results have been demonstrated. Further extensions to VCSOA technology are the demonstration of wavelength tunable devices. These MEMS-tunable vertical-cavity SOAs utilize a microelectromechanical systems (MEMS) based tuning mechanism for wide and continuous tuning of the peak gain wavelength of the amplifier.

Raman Amplifier

In a Raman amplifier, the signal is intensified by Raman amplification. Unlike the EDFA and SOA the amplification effect is achieved by a non-linear interaction between the signal and a pump laser within an optical fibre. There are two types of Raman amplifier: distributed and lumped.

A distributed Raman amplifier is one in which the transmission fibre is utilised as the gain medium by multiplexing a pump wavelength with signal wavelength, while a lumped Raman amplifier utilises a dedicated, shorter length of fibre to provide amplification. In the case of a lumped Raman amplifier highly non-linear fibre with a small core is utilised to increase the interaction between signal and pump wavelengths and thereby reduce the length of fibre required.

The pump light may be coupled into the transmission fibre in the same direction as the signal (co-directional pumping), in the opposite direction (contra-directional pumping) or both.

Contra-directional pumping is more common as the transfer of noise from the pump to the signal is reduced.

The pump power required for Raman amplification is higher than that required by the EDFA, with in excess of 500 mW being required to achieve useful levels of gain in a distributed amplifier. Lumped amplifiers, where the pump light can be safely contained to avoid safety implications of high optical powers, may use over 1W of optical power.

The principal advantage of Raman amplification is its ability to provide distributed amplification within the transmission fibre, thereby increasing the length of spans between amplifier and regeneration sites. The amplification bandwidth of Raman amplifiers is defined by the pump wavelengths utilised and so amplification can be provided over wider, and different, regions than may be possible with other amplifier types which rely on dopants and device design to define the amplification 'window'.

COMMON EMITTER AMPLIFIER

Then to summarize. The Common Emitter Amplifier circuit has a resistor in its Collector circuit. The current flowing through this resistor produces the voltage output of the amplifier. The value of this resistor is chosen so that at the amplifiers quiescent operating point, Q–point this output voltage lies half way along the transistors load line.

The Base of the transistor used in a common emitter amplifier is biased using two resistors as a potential divider network. This type of biasing arrangement is commonly used in the design of bipolar transistor amplifier circuits and greatly reduces the effects of varying Beta, (β) by holding the Base bias at a constant steady voltage. This type of biasing produces the greatest stability.

A resistor can be included in the emitter leg in which case the voltage gain becomes–R_L/R_E. If there is no external Emitter resistance, the voltage gain of the amplifier is not infinite as there is a very small internal resistance, R_e in the Emitter leg. The value of this internal resistance is equal to 25mV/ I_E

In the next tutorial about Amplifiers we will look at the Junction Field Effect Amplifier commonly called theJFET Amplifier. Like the transistor, the JFET is used in a single stage amplifier circuit making it easier to understand. There are several different kinds of field effect transistor that we could use but the easiest to understand is the junction field effect transistor, or JFET which has a very high input impedance making it ideal for amplifier circuits.

CIRCUIT

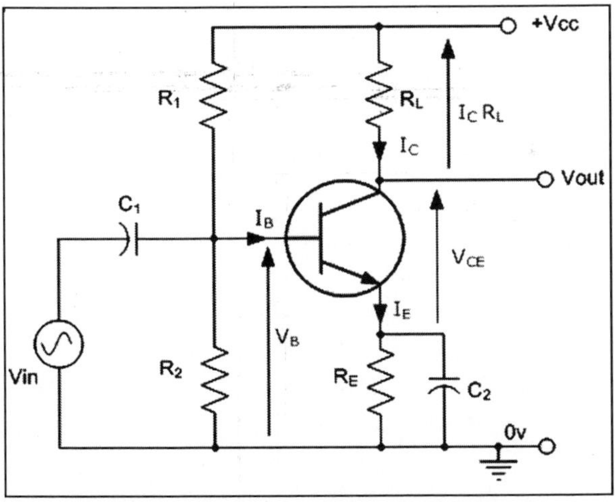

The single stage common emitter amplifier circuit shown above uses what is commonly called "Voltage Divider Biasing". This type of biasing arrangement uses two resistors as a potential divider network and is commonly used in the design of bipolar transistor amplifier circuits. This method of biasing the transistor greatly reduces the effects of varying Beta, (b) by holding the Base bias at a constant steady voltage level allowing for best stability.

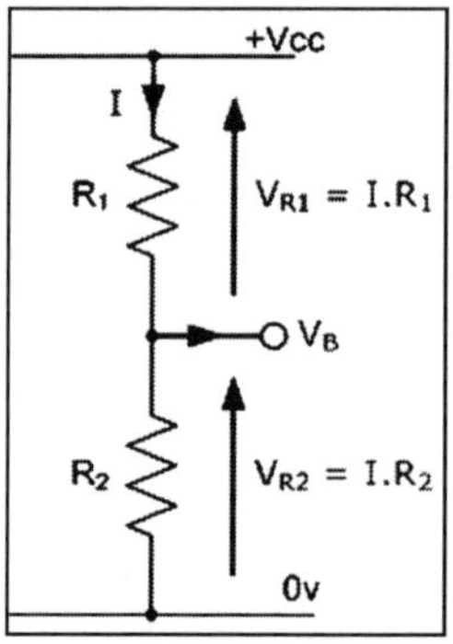

The quiescent Base voltage (Vb) is determined by the potential divider network formed by the two resistors, R_1, R_2 and the power supply voltage Vcc as shown with the current flowing through both resistors. Then the total resistance RT will be equal to $R_1 + R_2$ giving the current as $i = V_{CC}/RT$. The voltage level generated at the junction of resistors R_1 and R_2 holds the Base voltage (V_b) constant at a value below the supply voltage. Then the potential divider network used in the common emitter amplifier circuit divides the input signal in proportion to the resistance. This bias reference voltage can be easily calculated using the simple voltage divider formula below:

$$V_B = \frac{V_{CC} R_2}{R_1 + R_2}$$

The same supply voltage, (V_{CC}) also determines the maximum Collector current, I_c when the transistor is switched fully "ON" (saturation), $V_{ce} = 0$. The Base current Ib for the transistor is found from the Collector current, Ic and the DC current gain Beta, β of the transistor.

$$\beta = \frac{\Delta I_C}{\Delta I_B}.$$

Beta is sometimes referred to as hFE which is the transistors forward current gain in the common emitter configuration. Beta has no units as it is a fixed ratio of the two currents, I_c and I_b so a small change in the Base current will cause a large change in the Collector current. One final point about Beta. Transistors of the same type and part number will have large variations in their Beta value for example, the BC107 NPN Bipolar transistor has a DC current gain Beta value of between 110 and 450 (data sheet value) this is because Beta is a characteristic of their construction and not their operation.

As the Base/Emitter junction is forward-biased, the Emitter voltage, Ve will be one junction voltage drop different to the Base voltage. If the voltage across the Emitter resistor is known then the Emitter current,Ie can be easily calculated using Ohm's Law. The Collector current, I_c can be approximated, since it is almost the same value as the Emitter current.

COUPLING CAPACITORS

In Common Emitter Amplifier circuits, capacitors C_1 and C_2 are used as Coupling Capacitors to separate the AC signals from the DC biasing voltage. This ensures that the bias condition set up for the circuit to operate correctly is not effected by any additional amplifier stages, as the capacitors will only pass AC signals and block any DC component. The output AC signal is then superimposed on the biasing of the following stages. Also a bypass capacitor, CE is included in the Emitter leg circuit.

This capacitor is an open circuit component for DC bias meaning that the biasing currents and voltages are not affected by the addition of the capacitor maintaining a good Q-point stability. However, this bypass capacitor short circuits the Emitter resistor at high frequency signals and only RL plus a very small internal resistance acts as the transistors load increasing the voltage gain to its maximum. Generally, the value of the bypass capacitor, CE is chosen to provide a reactance of at most, 1/10th the value of REat the lowest operating signal frequency.

OUTPUT CHARACTERISTICS CURVES

Ok, so far so good. We can now construct a series of curves that show the Collector current, I_c against the Collector/Emitter voltage, Vce with different values of Base current, Ib for our simple common emitter amplifier circuit. These curves are known as the "Output Characteristic Curves" and are used to show how the transistor will operate over its dynamic range.

A static or DC load line is drawn onto the curves for the load resistor RL of 1.2kW to show all the transistors possible operating points. When the transistor is switched "OFF", Vce equals the supply voltage Vcc and this is point B on the line. Likewise when the transistor is fully "ON" and saturated the Collector current is determined by the load resistor, RL and this is point A on the line.

We calculated before from the DC gain of the transistor that the Base current required for the mean position of the transistor was 45.8mA and this is marked as point Q on the load line which represents the Quiescent point or Q-point of the amplifier. We could quite easily make life easy for ourselves and round off this value to 50μA exactly, without any effect to the operating point. Point Q on the load line gives us the Base current Q-point of Ib = 45.8μA or 46μA. We need to find the maximum and minimum peak swings of Base current that will result in a proportional change to the Collector current, Ic without any distortion to the output signal.

As the load line cuts through the different Base current values on the DC characteristics curves we can find the peak swings of Base current that are equally spaced along the load line.

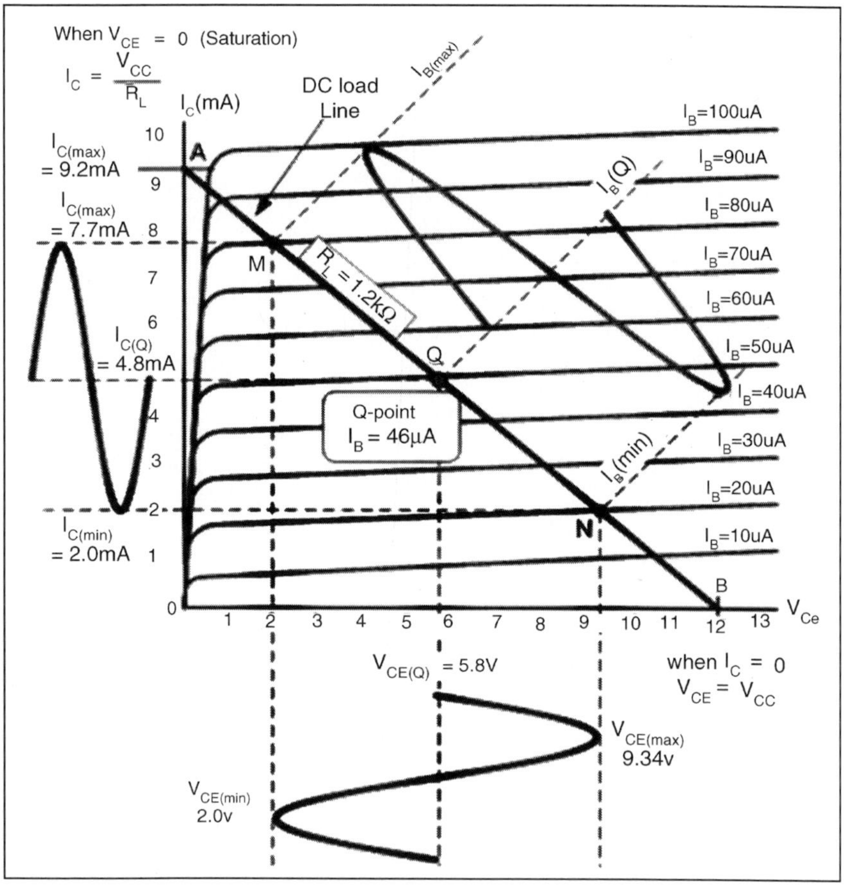

Fig. Output Characteristics Curves

These values are marked as points N and M on the line, giving a minimum and a maximum Base current of 20μA and 80μA respectively. These points, N and M can be anywhere along the load line that we choose as long as they are equally spaced from Q. This then gives us a theoretical maximum input signal to the Base terminal of 60μA peak-to-peak, (30ìμA peak) without producing any distortion to the output signal. Any input signal giving a Base current greater than this value will drive the transistor to go beyond point N and into its Cut-off region or beyond point M and into its Saturation region thereby resulting in distortion to the output signal in the form of "clipping".

Using points N and M as an example, the instantaneous values of Collector current and corresponding values of Collector-emitter voltage can be projected from the load line. It can be seen that the Collector-emitter voltage is in anti-phase (–180°) with the collector current.

As the Base current Ib changes in a positive direction from 50μA to 80μA, the Collector-emitter voltage, which is also the output voltage decreases from its steady state value of 5.8v to 2.0v. Then a single stage Common Emitter Amplifier is also an "Inverting Amplifier" as an increase in Base voltage causes a decrease in Vout and a decrease in Base voltage produces an increase in Vout.

VOLTAGE GAIN

The Voltage Gain of the common emitter amplifier is equal to the ratio of the change in the input voltage to the change in the amplifiers output voltage. Then ΔVL is Vout and ΔVB is Vin. But voltage gain is also equal to the ratio of the signal resistance in the Collector to the signal resistance in the Emitter and is given as:

$$Voltage\,Gain = \frac{V_{out}}{V_{in}} = \frac{\Delta V_L}{\Delta V_B}$$

$$= -\frac{R_L}{R_E}$$

We mentioned earlier that as the signal frequency increases the bypass capacitor, CE starts to short out the Emitter resistor. Then at high frequencies RE = 0, making the gain infinite.

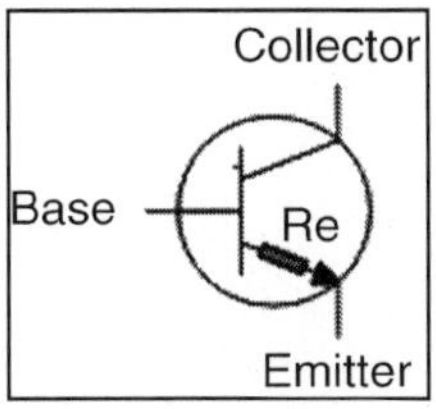

However, bipolar transistors have a small internal resistance built into their Emitter region called Re. The transistors semiconductor material offers an internal resistance to the flow of current through it and is generally represented by a small resistor symbol shown inside the main transistor symbol.

Transistor data sheets tell us that for a small signal bipolar transistors this internal resistance is the product of 25mV ÷ Ie (25mV being the internal volt drop across the Base/Emitter junction depletion layer), then for our common Emitter amplifier circuit above this resistance value will be equal to:

$$R_e = \frac{25mV}{I_E}$$
$$= \frac{25mV}{4.58mA}$$
$$= 5.5\Omega.$$

This internal Emitter leg resistance will be in series with the external Emitter resistor, RE, then the equation for the transistors actual gain will be modified to include this internal resistance and is given as:

$$Voltage\,Gain = -\frac{R_L}{(R_E + R_e)}$$

At low frequency signals the total resistance in the Emitter leg is equal to $R_E + R_e$. At high frequency, the bypass capacitor shorts out the Emitter resistor leaving only the internal resistance R_e in the Emitter leg resulting in a high gain. Then for our common emitter amplifier circuit above, the gain of the circuit at both low and high signal frequencies is given as:

At Low Frequencies,

$$Gain = -\frac{R_L}{(R_E + R_e)}$$
$$= -\frac{1200}{220 + 5.5}$$
$$= -5.32$$

At High Frequencies

$$Gain = -\frac{R_L}{R_e}$$
$$= -\frac{1200}{5.5}$$
$$= -2.18$$

One final point, the voltage gain is dependent only on the values of the Collector resistor, R_L and the Emitter resistance, ($R_E + R_e$) it is not affected by the current gain Beta, b (hFE) of the transistor.

So, for our simple example above we can now summarise all the values we have calculated for our common emitter amplifier circuit and these are:

	Minimum	**Mean**	**Maximum**
Base Current	20μA	50μA	80μA
Collector Current	2.0mA	4.8mA	7.7mA
Output Voltage Swing	2.0V	5.8V	9.3V
Amplifier Gain	-5.32		-218

ALTERNATIVE AC EQUIVALENT CIRCUITS FOR THE AMPLIFIER

An *operational amplifier* ("op-amp") is a DC-coupled high-gain electronic voltage amplifier with a differential input and, usually, a single-ended output. An op-amp produces an output voltage that is typically hundreds of thousands times larger than the voltage *difference* between its input terminals.

Operational amplifiers are important building blocks for a wide range of electronic circuits. They had their origins in analog computers where they were used in many linear, non-linear and frequency-dependent circuits. Their popularity in circuit design largely stems from the fact that characteristics of the final op-amp circuits with negative feedback (such as their gain) are set by external components with little dependence on temperature changes and manufacturing variations in the op-amp itself.

Op-amps are among the most widely used electronic devices today, being used in a vast array of consumer, industrial, and scientific devices. Many standard IC op-amps cost only a few cents in moderate production volume; however some integrated or hybrid operational amplifiers with special performance specifications may cost over $100 US in small quantities.Op-amps may be packaged as components, or used as elements of more complex integrated circuits.

The op-amp is one type of differential amplifier. Other types of differential amplifier include the fully differential amplifier (similar to the op-amp, but with two outputs), the instrumentation amplifier (usually built from three op-amps), the isolation amplifier (similar to the instrumentation amplifier, but with tolerance to common-mode voltages that would destroy an ordinary op-amp), and negative feedback amplifier (usually built from one or more op-amps and a resistive feedback network).

EQUIVALENT CIRCUITS OF TRANSISTOR AMPLIFIER

A *port* is a pair of terminals of a network (circuit). Across the port is a voltage, v, and through it flows a current, i, as shown below.

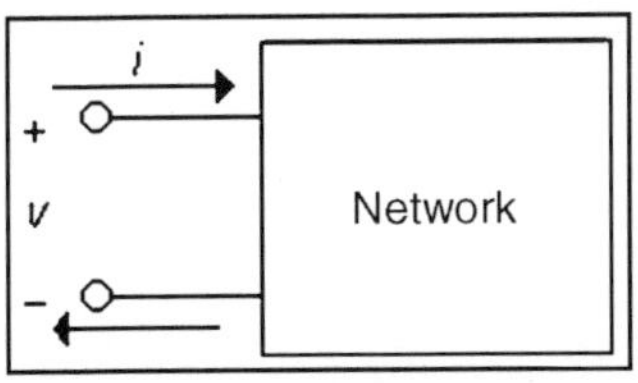

Amplifiers have two ports, input and output. An electrical *waveform* is a voltage or current as a function of time. A waveform to be amplified is applied to the input port and another waveform appears at the output port that is larger than the input waveform. Input and output quantities can be either voltages or currents, resulting in four basic kinds of amplifiers:

Amplifier Type	**Input Quantity**	**Output Quantity**
Voltage, A_v	voltage, v_i	voltage, v_o
Current, A_i	current, i_i	current, i_o
Transresistance, R_m	current, i_i	voltage, v_o
Transconductance, G_m	voltage, v_i	current, i_o

In the table under amplifier type is the expression for amplification or *gain* (or *transfer function*), which is the output quantity divided by the input quantity. In general, $A = x_o/x_i$, where x is either a voltage or current. An ideal input port is not affected by input source resistance nor is an ideal output port affected by output load resistance. The general amplifier is shown below:

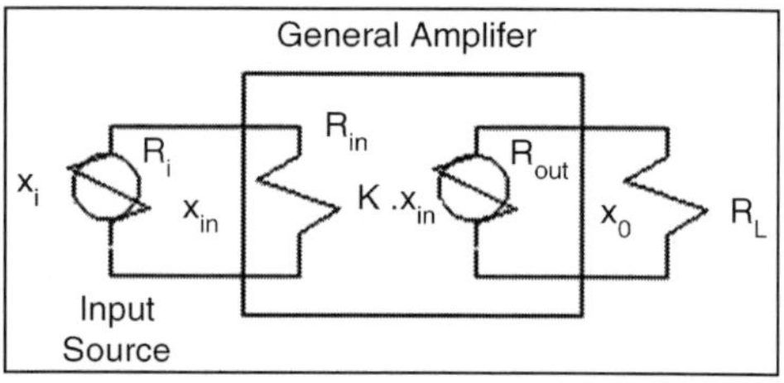

The combination source/resistance symbol is a generalized source: either a Thevenin or Norton equivalent circuit. The amplifier has an input resistance R_{in} and output resistance, R_{out}. The input source, x_i (where x_i is v_i or i_i), has resistance R_i. It forms a divider (voltage or current) with R_{in} so that $x_i \neq x_{in}$. Similarly, output resistance R_{out} forms a divider with output port load resistance R_L so that the output $x_{out} = K \times x_{in} \neq x_0$. The amplification of x_{in} by K results in x_{out} that is K times larger. K is the gain, and it scales x_{in}. (Gain less than 1 is called *attenuation*.) If source or load resistance is unknown or varies with K, then error in the overall amount of gain results. An accurate (or at least unchanging) gain is required for calibrated sensor circuits, so that the transducer output is multiplied by a known (and constant) amount.

An example of a voltage amplifier is shown below:

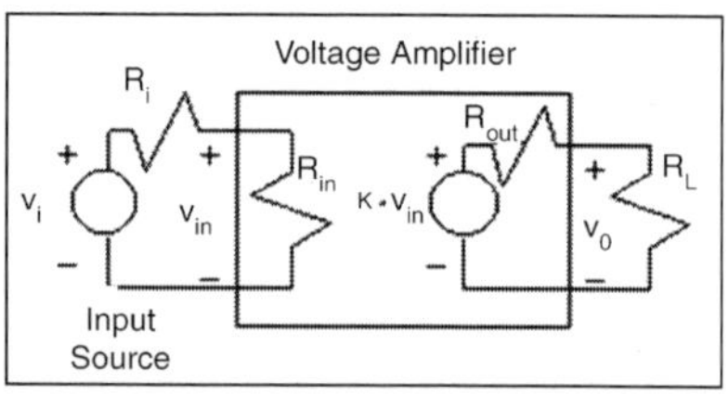

The overall voltage *gain* is:

$$A_v = \frac{v_0}{v_i} = \left(\frac{v_{in}}{v_i}\right)\left(\frac{v_{out}}{v_{in}}\right)\left(\frac{v_0}{v_{out}}\right)$$

$$= \left(\frac{R_{in}}{R_i + R_{in}}\right)(k)\left(\frac{R_L}{R_0 + R_L}\right)$$

The first factor is the input voltage divider attenuation, the second is the amplifier voltage gain and the third is the output voltage divider attenuation. For the ideal voltage amplifier, $A_v = K$. This is achieved when R_{in} approaches infinity (open-circuit input) and $R_0 = 0$.

The ideal port resistances are given in the following table:

Port Type	Ideal Resistance
Voltage input	infinite (open)
Current input	zero
Voltage output	zero
Current output	infinite (open)

In practice, good amplifier design approaches the ideal so that input and output loading does not affect the overall amplifier gain accuracy.

TRANSISTOR CONFIGURATIONS

Transistors have three terminals connected to input and output circuit loops. One-transistor amplifiers are two-port networks; one of the three transistor terminals must be shared by both input and output ports as the common terminal. This results in three possibilities. The first is the *common emitter* (CE) amplifier. The emitter is common to both input and output, as shown below.

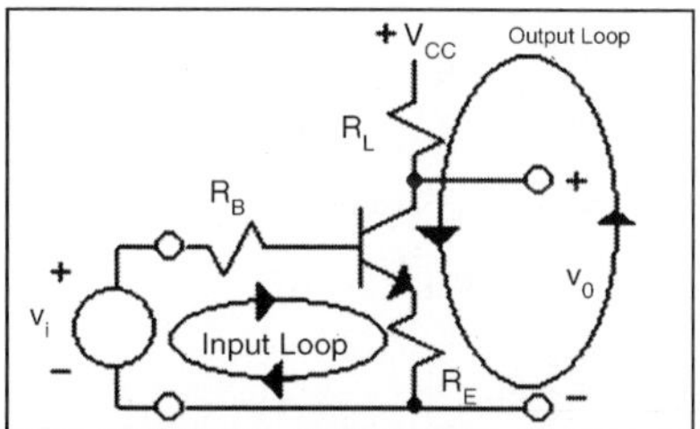

The emitter is part of both input and output loops. It is the common terminal of the transistor that is connected to both an input and output port terminal. With a series emitter resistor R_E the emitter terminal is still common to both loops. The output loop current is shown flowing from the power supply ($+V_{CC}$), through R_L and the BJT, through R_E to ground, which is connected to the negative terminal of the supply. The closure of the output loop from ground to $+V_{CC}$ implies flow through the voltage-source, $+V_{CC}$.

The common-base (CB) configuration is shown below:

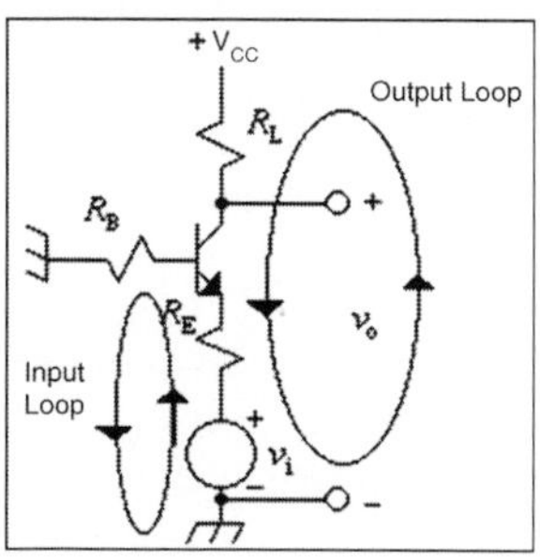

The common-collector (CE) configuration, also known as the *emitter-follower*, is shown below.

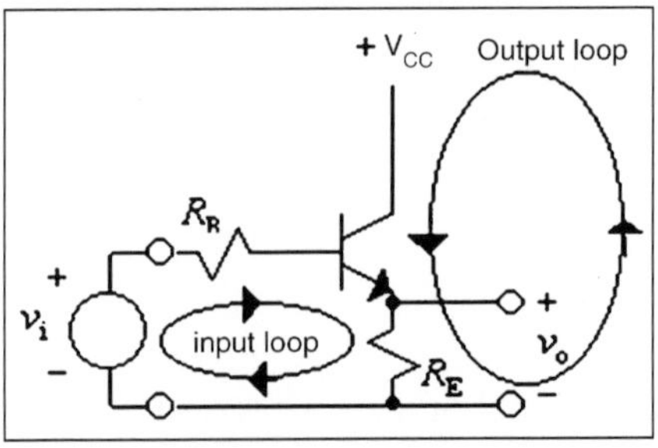

POWER GAINS AND VOLTAGE

Then the gain of an amplifier can be said to be the relationship that exists between the signal measured at the output with the signal measured at the input. There are three different kinds of Amplifier Gain, Voltage Gain, (Av), Current Gain (Ai) and Power Gain (Ap) and examples of these are given below.

AMPLIFIER GAIN OF THE INPUT SIGNAL

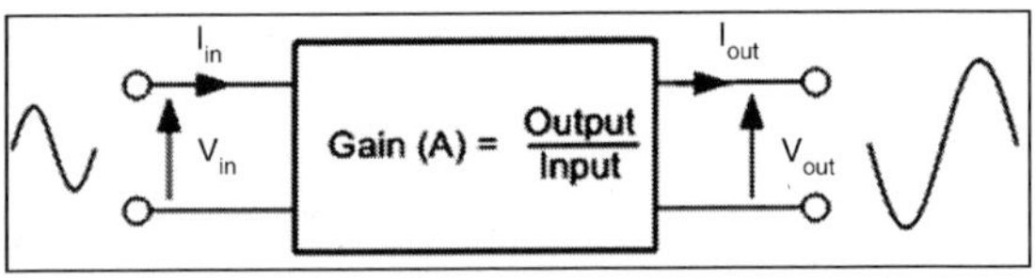

Fig. Amplification Stage

VOLTAGE AMPLIFIER GAIN

$$Voltage\,Gain(A_v) = \frac{Output\,Voltage}{Input\,Voltage} = \frac{V_{out}}{V_{in}}$$

CURRENT AMPLIFIER GAIN

$$Current\,Gain(A_i) = \frac{Output\,Current}{Input\,Current} = \frac{I_{out}}{I_{in}}.$$

POWER AMPLIFIER GAIN

$$Power\,Gain(A_p) = A_v \times A_i.$$

Note that for the Power Gain you can also divide the power obtained at the output with the power obtained at the input. Also when calculating the gain of an amplifier, the subscripts v, i and p are used to denote the type of signal gain being used.

The power Gain or power level of the amplifier can also be expressed in Decibels, (dB). The Bel is a logarithmic unit (base 10) of measurement that has no units. Since the Bel is too large a unit of measure, it is prefixed with *deci* making it Decibels instead with one decibel being one tenth (1/10th) of a Bel. To calculate the gain of the amplifier in Decibels or dB, we can use the following expressions.

- Voltage Gain in dB: $a_v = 20 \log Av$
- Current Gain in dB: $a_i = 20 \log Ai$
- Power Gain in dB: $a_p = 10 \log Ap$

Note that the DC power gain of an amplifier is equal to ten times the common log of the output to input ratio, where as voltage and current gains are 20 times the common log of the ratio. Note however, that 20dB is not twice as much power as 10dB because of the log scale. Also, a positive value of dB represents a Gain and a negative value of dB represents a Loss within the amplifier. For example, an amplifier gain of +3dB indicates that the amplifiers output signal has "doubled", (x2) while an amplifier gain of -3dB indicates that the signal has "halved", (x0.5) or in other words a loss.

The -3dB point of an amplifier is called the half-power point which is -3dB down from maximum, taking 0dB as the maximum output value.

POWER AMPLIFIERS

Small signal amplifiers are generally referred to as "Voltage" amplifiers as they convert a small input voltage into a much larger output voltage. Sometimes an amplifier is required to drive a motor or feed a loudspeaker and for these types of applications where high switching currents are needed Power Amplifiers are required.

The main function of Power amplifiers (also known as large signal amplifiers) is to deliver power, which as we know from above, is the product of the voltage and current applied to the load with the output signal power being greater than the input signal power. In other words, a power amplifier amplifies the power of the input signal which is why they are used in amplifier output stages.

The power amplifier works on the basic principle of converting the DC power drawn from the power supply into an AC voltage signal delivered to the load. Although the amplification is high the efficiency of the conversion from the DC power supply input to the AC voltage signal output is usually poor.

The perfect or ideal amplifier would give us an efficiency rating of 100 per cent or at least the power IN would be equal to the power OUT. However, this can never happen as some of its power is lost in the form of heat and also, the amplifier itself consumes power during the amplification process. Then the efficiency of an amplifier is given as:

Amplifier Efficiency:

$$\text{Efficiency}(\eta) = \frac{\text{Power delivered to the load}}{\text{D.C power taken from the supply}} = \frac{\text{point}}{\text{pin}}$$

BJT AMPLIFIER CONFIGURATIONS

There are plenty of texts around on basic electronics, so this is a very brief look at the three basic ways in which a bipolar junction transistor (BJT) can be used. In each case, one terminal is common to both the input and output signal. All the circuits shown here are without bias circuits and power supplies for clarity.

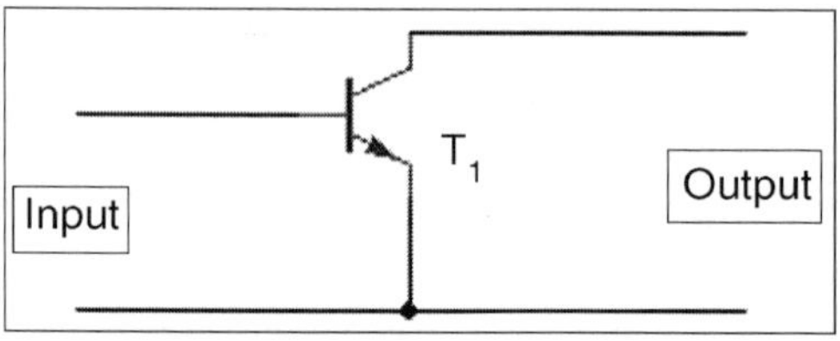

Fig. Common Emitter Configuration

Here the emitter terminal is common to both the input and output signal. The arrangement is the same for a PNP transistor. Used in this way the transistor has the advantages of a medium input impedance, medium output impedance, high voltage gain and high current gain.

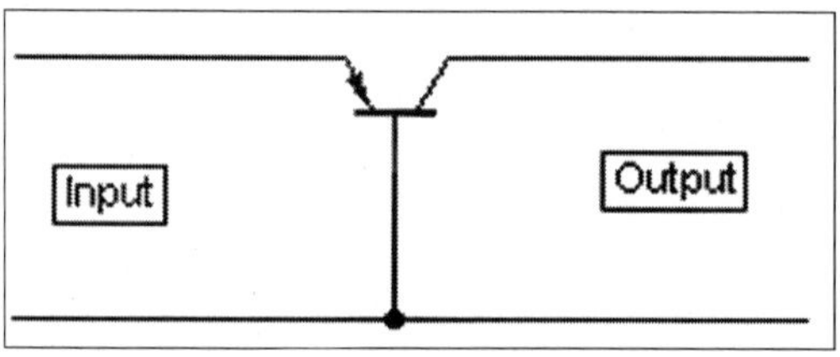

Fig. Common Base Configuration

Here the base is the common terminal. Used frequently for RF applications, this stage has the following properties. Low input impedance, high output impedance, unity (or less) current gain and high voltage gain.

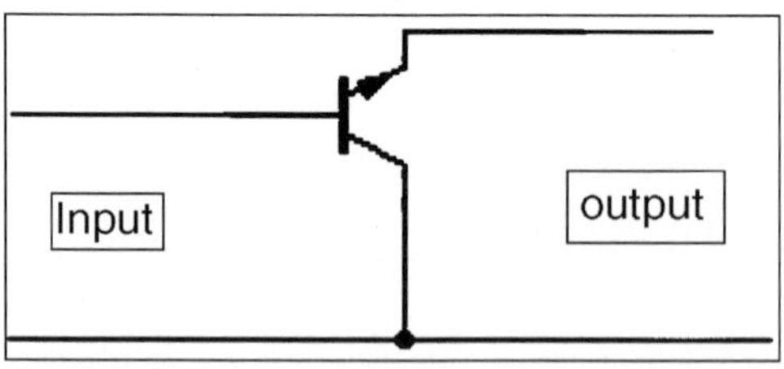

Fig. Common Collector Configuration

This last configuration is also more commonly known as the emitter follower. This is because the input signal applied at the base is "followed" quite closely at the emitter with a voltage gain close to unity.

The properties are a high input impedance, a very low output impedance, a unity (or less) voltage gain and a high current gain. This circuit is also used extensively as a "buffer" converting impedances or for feeding or driving long cables or low impedance loads.

Amplifier Type	**Common Base**	**Common Emitter**	**Common Emitter (Emitter Resistor)**	**Common Collector (Emitter Follower)**
Input/ Output Phase Relationship	0°	180°	180°	0°
Voltage Gain	High $\frac{\alpha R_C}{R_s + r_e}$	Medium $\frac{\beta (R_C \| r_o)}{R_s + r_\pi}$	Medium $\frac{\beta R_C}{R_s + (\beta+1)(r_e + R_E)}$	Low $\frac{(\beta+1)(R_L \| r_o)}{R_s + (\beta+1)[r_e + (R_L \| r}$
Current Gain	Low α	Medium $\beta \frac{r_o}{R_C + r_o}$	Medium β	High $(\beta+1)\frac{r_o}{r_o + R_L}$
Power Gain	Low	High	High	Medium
Input Resistance	Low r_e	Medium $r_\pi = (\beta+1) r_e$	Medium $(\beta+1)(r_e + R_E)$	High $(\beta+1)[r_e + (r_o \| R_L)]$
Output Resistance	High R_C	Medium $R_C \| r_o$	Medium R_C	Low $r_o \| \left[r_e + \frac{R_s}{(\beta+1)}\right]$

6

Circuits Transformers System

USING OF CIRCUITS TRANSFORMERS

The circuit will light three automobile headlight bulbs in parallel. If these bulbs draw 5 amperes each, the current labeled I2 will, of course, be 15 amperes a-c. But the value of I1 is not so obvious. For any transformer the ratio of the input to output current is the inverse of the ratio of the input to output voltage. If the input voltage of the transformer is twice the output voltage, the input current will always be half the output current. This is expressed as

$$\frac{V_{\text{input}}}{V_{\text{output}}} = \frac{I_{\text{output}}}{I_{\text{input}}}$$

Note that this equation can also be expressed as

Vinput × Iinput = Voutput × Ioutput

(*i.e.*)

Input watts = output watts

In other words, if a certain amount of power is coming out of a transformer, the same amount will be going in. This makes sense as long as the transformer itself does not use up much power. Most transformers operate so efficiently that errors introduced by using these equations are negligible. In the example, we are given the output current (15 amperes), the output voltage (6 volts), and the input voltage (115 volts). Therefore $I_{\text{input}} = \frac{15 \times 6}{115} = 0.78$ ampere

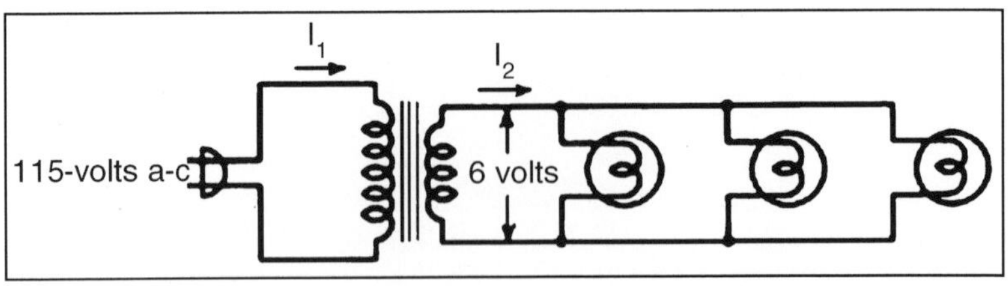

Fig. Circuit to Light Three Automobile Headlight Bulbs in Parallel. I_1, is

the Input Current and I_2 t he Output Current of the Transformer.

To extend this treatment, consider what happens when there is no connection (an infinite resistance) between the two output terminals of the transformer. The output voltage will still be 6 volts, but now the output current is zero. The output power is zero watts, and therefore the input power must also be zero watts. Since the input voltage is not zero, the input current must be zero. In other words, according to this equation, a transformer plugged into the wall will not draw any current in its primary as long as nothing is connected across its secondary. This conclusion looks peculiar but is true for all practical purposes.

TRANSFORMER RATINGS

When buying a transformer you should specify:

- Input (primary) voltage and frequency (usually 115 volts, 60 cycles).
- Output (secondary) voltage. Output (secondary) current.
- Other special requirements such as size, etc.

INPUT (PRIMARY) VOLTAGE

The transformer is most commonly used to provide a convenient source of a-c power at a voltage other than the one provided at the wall socket. There is only one type of transformer used at all frequently in the laboratory that has a rated input voltage or frequency other than 115 volts, 60 cycles. It is called an output transformer and is used in certain electronic circuits which will not be discussed here.

Many transformers found in surplus stores are rated for 400-cycle voltages. They are common because the standard frequency for certain military devices, such as airplanes, is 400 cycles rather than 60 cycles. A 400-cycle transformer will not operate properly on a 60cycle voltage.

OUTPUT (SECONDARY) VOLTAGE

The output, or secondary voltage, is simply the voltage that the transformer will deliver across its secondary coil terminals when the rated input voltage is applied to its primary coil. If you wish to light a 6-volt bulb from a wall socket, order a transformer with an output voltage of 6 volts. Many transformers are built to give a variety of output voltages all at the same time.

This is sometimes accomplished by winding two or more electrically separate output coils on the same core. That arrangement is very common in transformers designed to supply power for electronic devices. Most radios, for instance, require a low voltage, *e.g.*, 6 volts, to light the tubes, and a high voltage, *e.g.*, 300 volts, to drive electrons through the vacuum in the tubes.

Multiple-output voltages are also commonly obtained by the method. Connections are made at different places along a single output coil. It has already been pointed out that the more turns there are on the secondary winding, the higher the voltage that is generated.

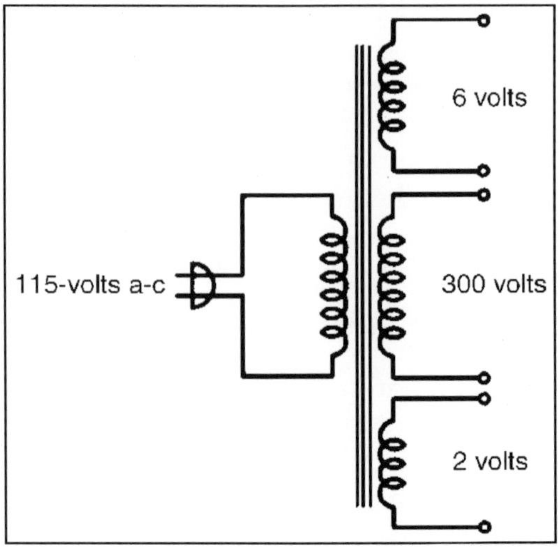

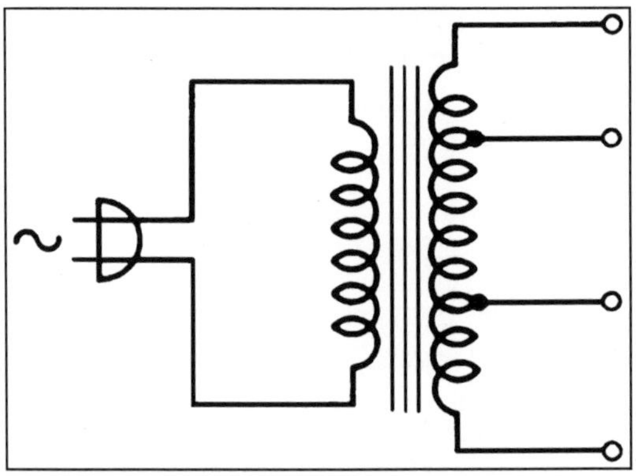

Fig. (*a*) Diagram of a Transformer with Three Different Output Windings.
(*b*) Diagram of a Transformer with a Multiple-tapped Output Winding.

It should be evident that different pairs of terminals have different numbers of turns between them and will, therefore, deliver different voltages. The wires connected along the coil are called taps, and this kind of transformer is said to have a multiple-tapped output winding. The most common tap is one halfway between the two ends of the output coil and such a transformer is said to be centre-tapped. Thus, if you want a transformer that will deliver either 6 or 12 volts, ask for a 12-volt-output, centre-tapped transformer.

OUTPUT CURRENT

Suppose that you wish to light three 6-volt, 5-ampere bulbs in parallel. You must order a transformer that gives a 6-volt output with an output current rating of at least 15 amperes; that is, the output coil must be able to carry 15 amperes without getting too hot. Note that this is a maximum current rating. If a transformer has a 6-volt, 15ampere output rating, it will always put out about 6 volts, but it will not necessarily deliver 15 amperes at all times. If there is nothing connected between the two terminals of the output coil, it will not deliver any current at all (6 volts/infinite ohms = 0 ampere). But if there is less than 0.4 ohm between the output terminals, more than the rated current will flow, and the transformer will get too hot.

ESTIMATING THE RATINGS

Transformers which have no ratings engraved on them can usually be found in the dark corners of the laboratory. This section will describe several ways in which you may start to classify them. First of all, if what you think might be a transformer has only two wires coming out of it, it is a choke rather than a transformer. A choke looks like a transformer but it consists of only a single coil of wire on an iron core. If there are either three of four terminals on the unknown transformer and they are labeled with letters such as *b*, *p*, or *g*, it is an output transformer and should not be used except in certain electronic applications which will not be discussed here. But if the transformer has at least four terminals or wires, and if they are not labeled with the letters, *b*, *p*, or *g*, it may be usable.

The easiest way to determine its output voltage is to apply 115 volts, 60 cycles per second, to the primary coil and measure the voltages resulting in the secondary or secondaries. On some transformers the primary is labeled "primary," or "input," or "115 volts a-c," and then this procedure is straightforward, but usually there are no such markings and you have to guess which wires connect to the input coil. When there are only four wires or terminals and none is labeled, the four must be divided into two pairs, one pair from the input and the other from the output coil. If the pairing is not obvious from the geography of the transformer, it can be determined simply by using an ohmmeter to find out which pairs of wires are connected together. Now examine the leads themselves.

If one pair consists of fairly heavy wires, it is very likely that the unit is a step-down transformer and that the heavy pair are the secondary leads. If one pair of terminals is mounted on a large ceramic holder, that pair is probably the output of a very-high-voltage transformer (*e.g.*, 5000 volts or more). If neither is true, either coil can be used as the primary and the other as the secondary. In one case you will have a step-up transformer, and in the other a step-down transformer. To find out the ratio of input to output voltage, one pair of terminals may be connected across the 115-volt wall voltage, and the voltage between the others measured with an a-c voltmeter. However, since it is possible that the application of 115 volts to one of the coils will damage it, a safer procedure is to increase the input voltage gradually from zero to 115 volts, stopping if the transformer begins to heat up.

If there are five wires on an unknown transformer, it is probable that the fifth wire is a centre tap on the output coil. If so, an ohmmeter will indicate that there is one pair and one trio of connected wires. The output voltages (two of them) can then be measured by applying a known a-c voltage across the pair and measuring the voltages between pairs of the three remaining wires. When there are more than five wires or terminals on a transformer, the job becomes a lot more difficult. In general, the best procedure is to find out with an ohmmeter which wires are connected to each other, and then try to reconstruct the circuit diagram of the transformer from this information.

Suppose you find a transformer with eight wires coming out of its casing, and an ohmmeter indicates that there are one pair and two groups of three. Because primaries rarely have centre taps, the pair is very likely the primary. Then examine the wires themselves. If in one of the groups of three, two of the wires are thick, the chances are very good that the group of three is a low-voltage output coil with a centre tap (probably 6 volts), and that the other group of three is a high-voltage output coil with a centre tap (*e.g.*, 300 volts).

It is very difficult for an unpracticed worker to judge the output current rating of an unknown transformer. The only real way to determine this is to let the transformer draw some current and see if it gets hot. This is done by putting the rated voltage across the input terminals and connecting a variable resistor across the output. Then gradually increase the current by decreasing the resistance until the transformer either gets hot or is delivering the required current. A good criterion for "hot" is this: If you can hold your hand against the transformer, it is not overheating. If you have to pull your hand away, the transformer is overloaded. Any transformer running near its rated load will be hot, but it should not be so hot that you cannot hold your hand against it.

TRANSFORMERS ELECTRICAL ENERGY

A transformer is a device that transfers electrical energy from one circuit to another through inductively coupled conductors — the transformer's coils or "windings". Except for air-core transformers, the conductors are commonly wound around a single iron-rich core, or around separate but magnetically-coupled cores. A varying current in the first or "primary" winding creates a varying magnetic field in the core (or cores) of the transformer. This varying magnetic field induces a varying electromotive force (EMF) or "voltage" in the "secondary" winding. This effect is called mutual induction.

If a load is connected to the secondary, an electric current will flow in the secondary winding and electrical energy will flow from the primary circuit through the transformer to the load. In an ideal transformer, the induced voltage in the secondary winding (V_S) is in proportion to the primary voltage (V_P), and is given by the ratio of the number of turns in the secondary to the number of turns in the prima

$$\frac{V_S}{V_P} = \frac{N_S}{N_P}$$

By appropriate selection of the ratio of turns, a transformer thus allows an alternating current (AC) voltage to be "stepped up" by making N_S greater than N_P, or "stepped down" by making N_S less than N_P.

Transformers come in a range of sizes from a thumbnail-sized coupling transformer hidden inside a stage microphone to huge units weighing hundreds of tons used to interconnect portions of national power grids.

All operate with the same basic principles, although the range of designs is wide. While new technologies have eliminated the need for transformers in some electronic circuits, transformers are still found in nearly all electronic devices designed for household ("mains") voltage. Transformers are essential for high voltage power transmission, which makes long distance transmission economically practical.

First Steps: Experiments with Induction Coils

What would become the "transformer principle" was revealed in 1831 by Michael Faraday in his demonstration of electromagnetic induction, but without recognition of its future role in manipulating EMF. The first "induction coils" to see wide use were invented by Rev. Nicholas Callan of Maynooth College, Ireland in 1836, one of the first researchers to realize that the more turns the secondary winding has in relation to the primary winding, the larger the increase in EMF. Induction coils evolved from scientists' and inventors' efforts to get higher voltages from batteries.

Rather than alternating current (AC), their action relied upon a vibrating "make-and-break" mechanism that regularly interrupted the flow of direct current (DC) from the batteries. Between the 1830s and the 1870s, efforts to build better induction coils, mostly by trial and error, slowly revealed the basic principles of transformers. Efficient, practical designs did not appear until the 1880s, but within a decade the "transformer" would be instrumental in the "War of Currents", and in seeing AC distribution systems triumph over their DC counterparts, a position in which they have remained dominant ever since.

In 1876, Russian engineer Pavel Yablochkov invented a lighting system based on a set of induction coils where the primary windings were connected to a source of alternating current and the secondary windings could be connected to several "electric candles" (arc lamps) of his own design. The coils used in the system behaved as primitive transformers. The patent claimed the system could "provide separate supply to several lighting fixtures with different luminous intensities from a single source of electric power".

In 1878, the engineers of the Ganz Company in Hungary assigned part of its extensive engineering works to the manufacture of electric lighting apparatus for Austria-Hungary, and by 1883 made over fifty installations. It offered an entire system consisting of both arc and incandescent lamps, generators, and other accessories. Lucien Gaulard and John Dixon Gibbs first exhibited a device with an open iron core called a "secondary generator" in London in 1882, then sold the idea to the Westinghouse company in the United States. They also exhibited the invention in Turin, Italy in 1884, where it was adopted for an electric lighting system.

Induction coils with open magnetic circuits are inefficient for transfer of power to loads. Various methods of adjusting the cores or bypassing magnetic flux around part of a coil were developed, since until about 1880 the paradigm for AC power transmission from a high voltage supply to a low voltage load was a series circuit.

In practice, several coils with a ratio near 1:1 were connected with their primaries in series to allow use of a high voltage for transmission while presenting a low voltage to the lamps. The inherent flaw in this method was that turning off a single lamp affected all the others on the circuit, and many adjustable coil designs were introduced in an effort to accommodate this problematic characteristic of the series circuit.

Between 1884 and 1885, Hungarian engineers Zipernowsky, Bláthy and Déri from the Ganz company in Budapest created the efficient "ZBD" closed-core model, which were based on the design by Gaulard and Gibbs. (Gaulard and Gibbs designed just an open core model). They discovered that all former (coreless or open-core) devices were incapable of regulating voltage, and were therefore impracticable.

Their joint patent described a transformer with no poles and comprised two versions of it, the "closed-core transformer" and the "shell-core transformer. In the closed-core transformer the iron core is a closed ring around which the two coils are arranged uniformly. In the shell type transformer, the copper induction cables are passed through the core. In both designs, the magnetic flux linking the primary and secondary coils travels (almost entirely) in the iron core, with no intentional path through air.

The core consists of iron cables or plates. Based on this invention, it became possible to provide economical and cheap lighting for industry and households." Zipernowsky, Bláthy and Déri discovered the mathematical formula of transformers: $V_s/V_p = N_s/N_p$. With this formula, transformers became calculable and proportionable.

Their patent application made the first use of the word "transformer", a word that had been coined by Ottó Bláthy. George Westinghouse had bought both Gaulard and Gibbs' and the "ZBD" patents in 1885. He entrusted William Stanley with the building of a ZBD-type transformer for commercial use. Stanley built the core from interlocking E-shaped iron plates. This design was first used commercially in 1886.

Early Developments and Applications

Russian engineer Mikhail Dolivo-Dobrovolsky developed the first three-phase transformer in 1889. In 1891 Nikola Tesla invented the Tesla coil, an air-cored, dual-tuned resonant transformer for generating very high voltages at high frequency. Audio frequency transformers (at the time called repeating coils) were used by the earliest experimenters in the development of the telephone.

Basic Principles

The transformer is based on two principles: firstly, that an electric current can produce a magnetic field (electromagnetism) and secondly that a changing magnetic field within a coil of wire induces a voltage across the ends of the coil (electromagnetic induction). Changing the current in the primary coil changes the magnitude of the applied magnetic field. The changing magnetic flux extends to the secondary coil where a voltage is induced across its ends.

A simplified transformer design is shown to the left. A current passing through the primary coil creates a magnetic field. The primary and secondary coils are wrapped around a core of very high magnetic permeability, such as iron; this ensures that most of the magnetic field lines produced by the primary current are within the iron and pass through the secondary coil as well as the primary coil.

Induction law

The voltage induced across the secondary coil may be calculated from Faraday's law of induction, which states that:

$$V_S = N_S \frac{d\Phi}{dt}$$

where V_S is the instantaneous voltage, N_S is the number of turns in the secondary coil and Ö equals the magnetic flux through one turn of the coil. If the turns of the coil are oriented perpendicular to the magnetic field lines, the flux is the product of the magnetic field strength B and the area A through which it cuts.

The area is constant, being equal to the cross-sectional area of the transformer core, whereas the magnetic field varies with time according to the excitation of the primary. Since the same magnetic flux passes through both the primary and secondary coils in an ideal transformer, the instantaneous voltage across the primary winding equals

$$V_P = N_P \frac{d\Phi}{dt}$$

Taking the ratio of the two equations for V_S and V_P gives the basic equation for stepping up or stepping down the voltage.

Ideal Power Equation

If the secondary coil is attached to a load that allows current to flow, electrical power is transmitted from the primary circuit to the secondary circuit. Ideally, the transformer is perfectly efficient; all the incoming energy is transformed from the primary circuit to the magnetic field and into the secondary circuit. If this condition is met, the incoming electric power must equal the outgoing power.

$$P_{\text{incoming}} = I_P V_P = P_{\text{outgoing}} = I_S V_S$$

giving the ideal transformer equation

If the voltage is increased (stepped up) ($V_S > V_P$), then the current is decreased (stepped down) ($I_S < I_P$) by the same factor. Transformers are efficient so this formula is a reasonable approximation.

The impedance in one circuit is transformed by the *square* of the turns ratio. For example, if an impedance Z_S is attached across the terminals of the secondary coil, it appears to the primary circuit to have an impedance of. This relationship is reciprocal, so that the impedance Z_P of the primary circuit appears to the secondary to be.

Detailed Operation

The simplified description above neglects several practical factors, in particular the primary current required to establish a magnetic field in the core, and the contribution to the field due to current in the secondary circuit.

Models of an ideal transformer typically assume a core of negligible reluctance with two windings of zero resistance. When a voltage is applied to the primary winding, a small current flows, driving flux around the magnetic circuit of the core. The current required to create the flux is termed the *magnetizing current*; since the ideal core has been assumed to have near-zero reluctance, the magnetizing current is negligible, although still required to create the magnetic field.

The changing magnetic field induces an electromotive force (EMF) across each winding. Since the ideal windings have no impedance, they have no associated voltage drop, and so the voltages V_P and V_S measured at the terminals of the transformer, are equal to the corresponding EMFs. The primary EMF, acting as it does in opposition to the primary voltage, is sometimes termed the "back EMF". This is due to Lenz's law which states that the induction of EMF would always be such that it will oppose development of any such change in magnetic field.

Practical Considerations

Leakage Flux

The ideal transformer model assumes that all flux generated by the primary winding links all the turns of every winding, including itself. In practice, some flux traverses paths that take it outside the windings. Such flux is termed *leakage flux*, and results in leakage inductance in series with the mutually coupled transformer windings.

Leakage results in energy being alternately stored in and discharged from the magnetic fields with each cycle of the power supply. It is not directly a power loss, but results in inferior voltage regulation, causing the secondary voltage to fail to be directly proportional to the primary, particularly under heavy load. Transformers are therefore normally designed to have very low leakage inductance. However, in some applications, leakage can be a desirable property, and long magnetic paths, air gaps, or magnetic bypass shunts may be deliberately introduced to a transformer's design to limit the short-circuit current it will supply.

Leaky transformers may be used to supply loads that exhibit negative resistance, such as electric arcs, mercury vapour lamps, and neon signs; or for safely handling loads that become periodically short-circuited such as electric arc welders. Air gaps are also used to keep a transformer from saturating, especially audio-frequency transformers in circuits that have a direct current flowing through the windings.

Effect of Frequency

The time-derivative term in Faraday's Law shows that the flux in the core is the integral of the applied voltage. Hypothetically an ideal transformer would work with direct-current excitation, with the core flux increasing linearly with time. In practice, the flux would rise to the point where magnetic saturation of the core occurs, causing a huge increase in the magnetizing current and overheating the transformer. All practical transformers must therefore operate with alternating (or pulsed) current.

TRANSFORMER UNIVERSAL EMF EQUATION

The EMF of a transformer at a given flux density increases with frequency. By operating at higher frequencies, transformers can be physically more compact because a given core is able to transfer more power without reaching saturation, and fewer turns are needed to achieve the same impedance. However properties such as core loss and conductor skin effect also increase with frequency. Aircraft and military equipment employ 400 Hz power supplies which reduce core and winding weight. Operation of a transformer at its designed voltage but at a higher frequency than intended will lead to reduced magnetizing current; at lower frequency, the magnetizing current will increase. Operation of a transformer at other than its design frequency may require assessment of voltages, losses, and cooling to establish if safe operation is practical.

For example, transformers may need to be equipped with "volts per hertz" over-excitation relays to protect the transformer from overvoltage at higher than rated frequency. Knowledge of natural frequencies of transformer windings is of importance for the determination of the transient response of the windings to impulse and switching surge voltages.

ENERGY LOSSES

An ideal transformer would have no energy losses, and would be 100 per cent efficient. In practical transformers energy is dissipated in the windings, core, and surrounding structures. Larger transformers are generally more efficient, and those rated for electricity distribution usually perform better than 98 per cent.

Experimental transformers using superconducting windings achieve efficiencies of 99.85 per cent, While the increase in efficiency is small, when applied to large heavily-loaded transformers the annual savings in energy losses are significant. A small transformer, such as a plug-in "wall-wart" or power adapter type used for low-power consumer electronics, may be no more than 85 per cent efficient, with considerable loss even when not supplying any load. Though individual power loss is small, the aggregate losses from the very large number of such devices are coming under increased scrutiny.

The losses vary with load current, and may be expressed as "no-load" or "full-load" loss. Winding resistance dominates load losses, whereas hysteresis and eddy currents losses contribute to over 99 per cent of the no-load loss. The no-load loss can be significant, meaning that even an idle transformer constitutes a drain on an electrical supply, which encourages development of low-loss transformers.

Transformer losses are divided into losses in the windings, termed copper loss, and those in the magnetic circuit, termed iron loss. Losses in the transformer arise from:

Winding Resistance

Current flowing through the windings causes resistive heating of the conductors. At higher frequencies, skin effect and proximity effect create additional winding resistance and losses.

Hysteresis Losses

Each time the magnetic field is reversed, a small amount of energy is lost due to hysteresis within the core. For a given core material, the loss is proportional to the frequency, and is a function of the peak flux density to which it is subjected.

Eddy Currents

Ferromagnetic materials are also good conductors, and a solid core made from such a material also constitutes a single short-circuited turn throughout its entire length. Eddy currents therefore circulate within the core in a plane normal to the flux, and are responsible for resistive heating of the core material. The eddy current loss is a complex function of the square of supply frequency and inverse square of the material thickness.

Magnetostriction

Magnetic flux in a ferromagnetic material, such as the core, causes it to physically expand and contract slightly with each cycle of the magnetic field, an effect known as magnetostriction. This produces the buzzing sound commonly associated with transformers, and in turn causes losses due to frictional heating in susceptible cores.

Mechanical Losses

In addition to magnetostriction, the alternating magnetic field causes fluctuating electromagnetic forces between the primary and secondary windings. These incite vibrations within nearby metalwork, adding to the buzzing noise, and consuming a small amount of power.

Stray Losses

Leakage inductance is by itself lossless, since energy supplied to its magnetic fields is returned to the supply with the next half-cycle. However, any leakage flux that intercepts nearby conductive materials such as the transformer's support structure will give rise to eddy currents and be converted to heat.

Equivalent Circuit

The physical limitations of the practical transformer may be brought together as an equivalent circuit model built around an ideal lossless transformer. Power loss in the windings is current-dependent and is represented as in-series resistances R_P and R_S. Flux leakage results in a fraction of the applied voltage dropped without contributing to the mutual coupling, and thus can be modeled as reactances of each leakage inductance X_P and X_S in series with the perfectly-coupled region.

Iron losses are caused mostly by hysteresis and eddy current effects in the core, and are proportional to the square of the core flux for operation at a given frequency. Since the core flux is proportional to the applied voltage, the iron loss can be represented by a resistance R_C in parallel with the ideal transformer.

A core with finite permeability requires a magnetizing current I_M to maintain the mutual flux in the core. The magnetizing current is in phase with the flux; saturation effects cause the relationship between the two to be non-linear, but for simplicity this effect tends to be ignored in most circuit equivalents.

With a sinusoidal supply, the core flux lags the induced EMF by 90° and this effect can be modeled as a magnetizing reactance (reactance of an effective inductance) X_M in parallel with the core loss component. R_C and X_M are sometimes together termed the *magnetizing branch* of the model. If the secondary winding is made open-circuit, the current I_0 taken by the magnetizing branch represents the transformer's no-load current.

The secondary impedance R_S and X_S is frequently moved (or "referred") to the primary side after multiplying the components by the impedance scaling factor. The resulting model is sometimes termed the "exact equivalent circuit", though it retains a number of approximations, such as an assumption of linearity.

Analysis may be simplified by moving the magnetizing branch to the left of the primary impedance, an implicit assumption that the magnetizing current is low, and then summing primary and referred secondary impedances, resulting in so-called equivalent impedance. The parameters of equivalent circuit of a transformer can be calculated from the results of two transformer tests: open-circuit test and short-circuit test.

Types

A wide variety of transformer designs are used for different applications, though they share several common features. Important common transformer types include:

Autotransformer

An autotransformer has only a single winding with two end terminals, plus a third at an intermediate tap point. The primary voltage is applied across two of the terminals, and the secondary voltage taken from one of these and the third terminal. The primary and secondary circuits therefore have a number of windings turns in common. Since the volts-per-turn is the same in both windings, each develops a voltage in proportion to its number of turns. An adjustable autotransformer is made by exposing part of the winding coils and making the secondary connection through a sliding brush, giving a variable turns ratio.

Polyphase Transformers

For three-phase supplies, a bank of three individual single-phase transformers can be used, or all three phases can be incorporated as a single three-phase transformer. In this case, the magnetic circuits are connected together, the core thus containing a three-phase flow of flux.

A number of winding configurations are possible, giving rise to different attributes and phase shifts. One particular polyphase configuration is the zigzag transformer, used for grounding and in the suppression of harmonic currents.

Leakage Transformers

A leakage transformer, also called a stray-field transformer, has a significantly higher leakage inductance than other transformers, sometimes increased by a magnetic bypass or shunt in its core between primary and secondary, which is sometimes adjustable with a set screw.

This provides a transformer with an inherent current limitation due to the loose coupling between its primary and the secondary windings. The output and input currents are low enough to prevent thermal overload under all load conditions – even if the secondary is shorted.

Leakage transformers are used for arc welding and high voltage discharge lamps (neon lamps and cold cathode fluorescent lamps, which are series-connected up to 7.5 kV AC). It acts then both as a voltage transformer and as a magnetic ballast. Other applications are short-circuit-proof extra-low voltage transformers for toys or doorbell installations.

Resonant Transformers

A resonant transformer is a kind of the leakage transformer. It uses the leakage inductance of its secondary windings in combination with external capacitors, to create one or more resonant circuits. Resonant transformers such as the Tesla coil can generate very high voltages, and are able to provide much higher current than electrostatic high-voltage generation machines such as the Van de Graaff generator.

One of the applications of the resonant transformer is for the CCFL inverter. Another application of the resonant transformer is to couple between stages of a superheterodyne receiver, where the selectivity of the receiver is provided by tuned transformers in the intermediate-frequency amplifiers.

Audio Transformers

Audio transformers are those specifically designed for use in audio circuits. They can be used to block radio frequency interference or the DC component of an audio signal, to split or combine audio signals, or to provide impedance matching between high and low impedance circuits, such as between a high impedance tube (valve) amplifier output and a low impedance loudspeaker, or between a high impedance instrument output and the low impedance input of a mixing console.

Such transformers were originally designed to connect different telephone systems to one another while keeping their respective power supplies isolated, and are still commonly used to interconnect professional audio systems or system components.

Being magnetic devices, audio transformers are susceptible to external magnetic fields such as those generated by AC current-carrying conductors. "Hum" is a term commonly used to describe unwanted signals originating from the "mains" power supply (typically 50 or 60 Hz). Audio transformers used for low-level signals, such as those from microphones, often included shielding to protect against extraneous magnetically-coupled signals.

Instrument Transformers

Instrument transformers are used for measuring voltge, current, power and energy in electrical systems, and for protection and control. Where a voltage or current is too large to be conveniently measured by an instrument, it can be scaled down to a standardized low value. Instrument transformers isolate measurement and control circuitry from the high currents or voltages present on the circuits being measured or controlled.

A current transformer is a transformer designed to provide a current in its secondary coil proportional to the current flowing in its primary coil. Voltage transformers (VTs), also referred to as "potential transformers" (PTs), are used in high-voltage circuits. They are designed to present a negligible load to the supply being measured, to allow protective relay equipment to be operated at lower voltages, and to have a precise winding ratio for accurate metering.

Classification

Transformers can be classified in different ways:

- *By power capacity*: from a fraction of a volt-ampere (VA) to over a thousand MVA;
- *By frequency range*: power-, audio-, or radio frequency;
- *By voltage class*: from a few volts to hundreds of kilovolts;
- *By cooling type*: air cooled, oil filled, fan cooled, or water cooled;
- *By application*: such as power supply, impedance matching, output voltage and current stabilizer, or circuit isolation;
- *By end purpose*: distribution, rectifier, arc furnace, amplifier output;
- *By winding turns ratio*: step-up, step-down, isolating (equal or near-equal ratio), variable.

Construction

Cores

Laminated Steel Cores

Transformers for use at power or audio frequencies typically have cores made of high permeability silicon steel. The steel has a permeability many times that of free space, and the core thus serves to greatly reduce the magnetizing current, and confine the flux to a path which closely couples the windings.

Early transformer developers soon realized that cores constructed from solid iron resulted in prohibitive eddy-current losses, and their designs mitigated this effect with cores consisting of bundles of insulated iron wires. Later designs constructed the core by stacking layers of thin steel laminations, a principle that has remained in use. Each lamination is insulated from its neighbours by a thin non-conducting layer of insulation. The universal transformer equation indicates a minimum cross-sectional area for the core to avoid saturation.

The effect of laminations is to confine eddy currents to highly elliptical paths that enclose little flux, and so reduce their magnitude. Thinner laminations reduce losses, but are more laborious and expensive to construct. Thin laminations are generally used on high frequency transformers, with some types of very thin steel laminations able to operate up to 10 kHz.

One common design of laminated core is made from interleaved stacks of E-shaped steel sheets capped with I-shaped pieces, leading to its name of "E-I transformer". Such a design tends to exhibit more losses, but is very economical to manufacture. The cut-core or C-core type is made by winding a steel strip around a rectangular form and then bonding the layers together. It is then cut in two, forming two C shapes, and the core assembled by binding the two C halves together with a steel strap. They have the advantage that the flux is always oriented parallel to the metal grains, reducing reluctance.

A steel core's remanence means that it retains a static magnetic field when power is removed. When power is then reapplied, the residual field will cause a high inrush current until the effect of the remaining magnetism is reduced, usually after a few cycles of the applied alternating current. Overcurrent protection devices such as fuses must be selected to allow this harmless inrush to pass. On transformers connected to long, overhead power transmission lines, induced currents due to geomagnetic disturbances during solar storms can cause saturation of the core and operation of transformer protection devices.

Distribution transformers can achieve low no-load losses by using cores made with low-loss high-permeability silicon steel or amorphous (non-crystalline) metal alloy. The higher initial cost of the core material is offset over the life of the transformer by its lower losses at light load.

Solid Cores

Powdered iron cores are used in circuits (such as switch-mode power supplies) that operate above main frequencies and up to a few tens of kilohertz. These materials combine high magnetic permeability with high bulk electrical resistivity. For frequencies extending beyond the VHF band, cores made from non-conductive magnetic ceramic materials called ferrites are common. Some radio-frequency transformers also have movable cores (sometimes called 'slugs') which allow adjustment of the coupling coefficient (and bandwidth) of tuned radio-frequency circuits.

Toroidal Cores

Toroidal transformers are built around a ring-shaped core, which, depending on operating frequency, is made from a long strip of silicon steel or permalloy wound into a coil, powdered iron, or ferrite. A strip construction ensures that the grain boundaries are optimally aligned, improving the transformer's efficiency by reducing the core's reluctance.

The closed ring shape eliminates air gaps inherent in the construction of an E-I core. The cross-section of the ring is usually square or rectangular, but more expensive cores with circular cross-sections are also available. The primary and secondary coils are often wound concentrically to cover the entire surface of the core. This minimizes the length of wire needed, and also provides screening to minimize the core's magnetic field from generating electromagnetic interference.

Toroidal transformers are more efficient than the cheaper laminated E-I types for a similar power level. Other advantages compared to E-I types, include smaller size (about half), lower weight (about half), less mechanical hum (making them superior in audio amplifiers), lower exterior magnetic field (about one tenth), low off-load losses (making them more efficient in standby circuits), single-bolt mounting, and greater choice of shapes.

The main disadvantages are higher cost and limited power capacity. Ferrite toroidal cores are used at higher frequencies, typically between a few tens of kilohertz to a megahertz, to reduce losses, physical size, and weight of switch-mode power supplies.

A drawback of toroidal transformer construction is the higher cost of windings. As a consequence, toroidal transformers are uncommon above ratings of a few kVA. Small distribution transformers may achieve some of the benefits of a toroidal core by splitting it and forcing it open, then inserting a bobbin containing primary and secondary windings.

Air Cores

A physical core is not an absolute requisite and a functioning transformer can be produced simply by placing the windings in close proximity to each other, an arrangement termed an "air-core" transformer. The air which comprises the magnetic circuit is essentially lossless, and so an air-core transformer eliminates loss due to hysteresis in the core material.

The leakage inductance is inevitably high, resulting in very poor regulation, and so such designs are unsuitable for use in power distribution. They have however very high bandwidth, and are frequently employed in radio-frequency applications, for which a satisfactory coupling coefficient is maintained by carefully overlapping the primary and secondary windings.

Windings

The conducting material used for the windings depends upon the application, but in all cases the individual turns must be electrically insulated from each other to ensure that the current travels throughout every turn. For small power and signal transformers, in which currents are low and the potential difference between adjacent turns is small, the coils are often wound from enameled magnet wire, such as Formvar wire. Larger power transformers operating at high voltages may be wound with copper rectangular strip conductors insulated by oil-impregnated paper and blocks of pressboard.

High-frequency transformers operating in the tens to hundreds of kilohertz often have windings made of braided Litz wire to minimize the skin-effect and proximity effect losses. Large power transformers use multiple-stranded conductors as well, since even at low power frequencies non-uniform distribution of current would otherwise exist in high-current windings.

Each strand is individually insulated, and the strands are arranged so that at certain points in the winding, or throughout the whole winding, each portion occupies different relative positions in the complete conductor. The transposition equalizes the current flowing in each strand of the conductor, and reduces eddy current losses in the winding itself. The stranded conductor is also more flexible than a solid conductor of similar size, aiding manufacture.

For signal transformers, the windings may be arranged in a way to minimize leakage inductance and stray capacitance to improve high-frequency response. This can be done by splitting up each coil into sections, and those sections placed in layers between the sections of the other winding. This is known as a stacked type or interleaved winding.

Both the primary and secondary windings on power transformers may have external connections, called taps, to intermediate points on the winding to allow selection of the voltage ratio. The taps may be connected to an automatic on-load tap changer for voltage regulation of distribution circuits. Audio-frequency transformers, used for the distribution of audio to public address loudspeakers, have taps to allow adjustment of impedance to each speaker.

A centre-tapped transformer is often used in the output stage of an audio power amplifier in a push-pull circuit. Modulation transformers in AM transmitters are very similar.

Certain transformers have the windings protected by epoxy resin. By impregnating the transformer with epoxy under a vacuum, one can replace air spaces within the windings with epoxy, thus sealing the windings and helping to prevent the possible formation of corona and absorption of dirt or water. This produces transformers more suited to damp or dirty environments, but at increased manufacturing cost.

Coolant

High temperatures will damage the winding insulation. Small transformers do not generate significant heat and are cooled by air circulation and radiation of heat. Power transformers rated up to several hundred kVA can be adequately cooled by natural convective air-cooling, sometimes assisted by fans.

In larger transformers, part of the design problem is removal of heat. Some power transformers are immersed in transformer oil that both cools and insulates the windings. The oil is a highly refined mineral oil that remains stable at transformer operating temperature. Indoor liquid-filled transformers must use a non-flammable liquid, or must be located in fire resistant rooms. Air-cooled dry transformers are preferred for indoor applications even at capacity ratings where oil-cooled construction would be more economical, because their cost is offset by the reduced building construction cost.

The oil-filled tank often has radiators through which the oil circulates by natural convection; some large transformers employ forced circulation of the oil by electric pumps, aided by external fans or water-cooled heat exchangers. Oil-filled transformers undergo prolonged drying processes to ensure that the transformer is completely free of water vapour before the cooling oil is introduced. This helps prevent electrical breakdown under load. Oil-filled transformers may be equipped with Buchholz relays, which detect gas evolved during internal arcing and rapidly de-energize the transformer to avert catastrophic failure.

Polychlorinated biphenyls have properties that once favoured their use as a coolant, though concerns over their environmental persistence led to a widespread ban on their use. Today, non-toxic, stable silicone-based oils, or fluorinated hydrocarbons may be used where the expense of a fire-resistant liquid offsets additional building cost for a transformer vault. Before 1977, even transformers that were nominally filled only with mineral oils may also have been contaminated with polychlorinated biphenyls at 10-20 ppm. Since mineral oil and PCB fluid mix, maintenance equipment used for both PCB and oil-filled transformers could carry over small amounts of PCB, contaminating oil-filled transformers.

Some "dry" transformers (containing no liquid) are enclosed in sealed, pressurized tanks and cooled by nitrogen or sulfur hexafluoride gas. Experimental power transformers in the 2 MVA range have been built with superconducting windings which eliminates the copper losses, but not the core steel loss. These are cooled by liquid nitrogen or helium.

Terminals

Very small transformers will have wire leads connected directly to the ends of the coils, and brought out to the base of the unit for circuit connections. Larger transformers may have heavy bolted terminals, bus bars or high-voltage insulated bushings made of polymers or porcelain. A large bushing can be a complex structure since it must provide careful control of the electric field gradient without letting the transformer leak oil.

Applications

A major application of transformers is to increase voltage before transmitting electrical energy over long distances through wires. Wires have resistance and so dissipate electrical energy at a rate proportional to the square of the current through the wire. By transforming electrical power to a high-voltage (and therefore low-current) form for transmission and back again afterwards, transformers enable economic transmission of power over long distances. Consequently, transformers have shaped the electricity supply industry, permitting generation to be located remotely from points of demand. All but a tiny fraction of the world's electrical power has passed through a series of transformers by the time it reaches the consumer.

Transformers are also used extensively in electronic products to step down the supply voltage to a level suitable for the low voltage circuits they contain. The transformer also electrically isolates the end user from contact with the supply voltage.

Signal and audio transformers are used to couple stages of amplifiers and to match devices such as microphones and record player s to the input of amplifiers. Audio transformers allowed telephone circuits to carry on a two-way conversation over a single pair of wires. Transformers are also used when it is necessary to couple a differential-mode signal to a ground-referenced signal, and for between external cables and internal circuits.

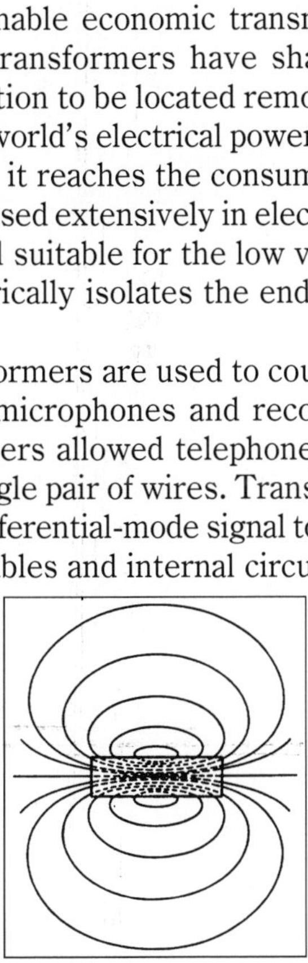

Fig. Schematic Representation of a Bar Magnet and its Magnetic Field. All the Lines are Actually Loops, but some of the Loops are too Large to be Represented on the Diagram.

For instance, if the current should suddenly be doubled, the field would double in strength, as represented in Fig. After doubling the current, there will be twice as many lines of force spread out over a larger volume than before. If one of these hypothetical lines of force were visible while the current were in the process of doubling, it would be seen to move outward.

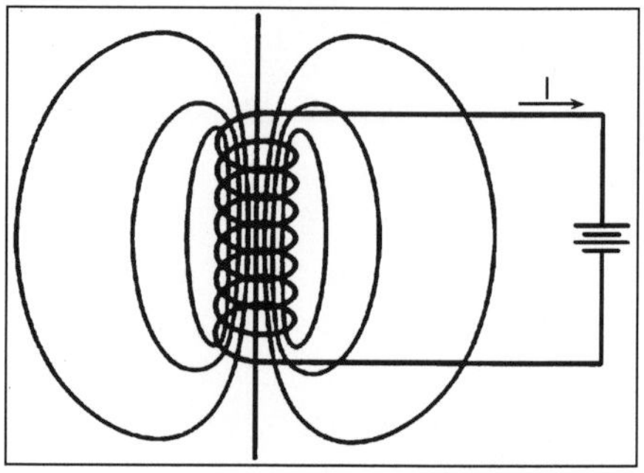

Fig. Schematic Representation of the Magnetic Field Generated by the Passage of Current I Through a Coil of Wire.

In other words, increasing the current through the coil not only adds more lines to the field but also causes the field to expand. Conversely, if the current were reduced, the field would collapse. Now store that concept for a moment and consider the opposite sort of phenomenon, namely, the effect that a magnetic field has on a coil of wire. If you have a permanent magnet at hand, you can demonstrate certain phenomena very readily. Connect your meter directly across the two ends of the coil on your sensitive relay, and set the meter to its most sensitive scale (*e.g.*, 100 microamperes, 1 volt, etc.).

Then put the magnet down next to the relay coils. Some of the lines of force from the magnet will now be passing through the relay coil but you will notice that nothing happens to the meter. In general, a stationary magnetic field has no effect on a coil of wire. But a moving field does affect the coil. Any time that lines of force move with respect to a coil of wire, they induce a voltage into it. (There is one particular direction of motion of a field with respect to a coil which will not induce a voltage into the coil, but that special case will be ignored for the present discussion.) It is convenient to think of this process as one in which each line of magnetic force tends to push some of the free electrons in the wire along in front of it as it moves.

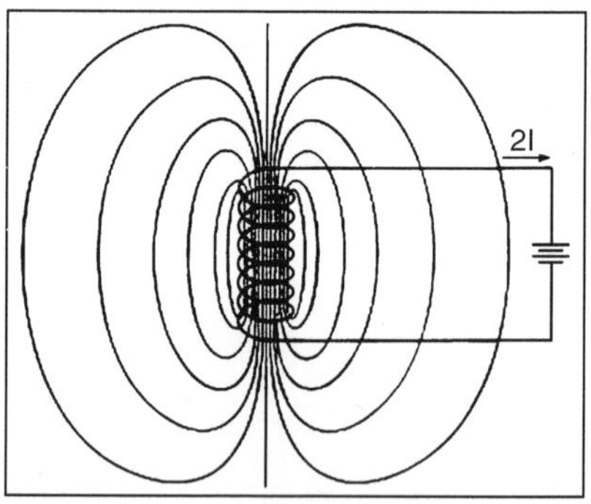

Fig. The Magnetic Field of the Coil when the Current Through it is Doubled. The Field is Represented as having Twice as many Lines, Distributed Over a Larger Area (Volume).

You can observe this by moving the magnet rapidly past the relay coil so that its lines of force move through the coil. If you have a reasonably strong magnet and if the meter is set at a sensitive scale, the pointer will jump to the right when the magnet passes in one direction, and to the left when it moves in the opposite direction. The lines of force are pushing electrons one way or the other. Note, though, that the needle reads non-zero only when the field is actually moving.

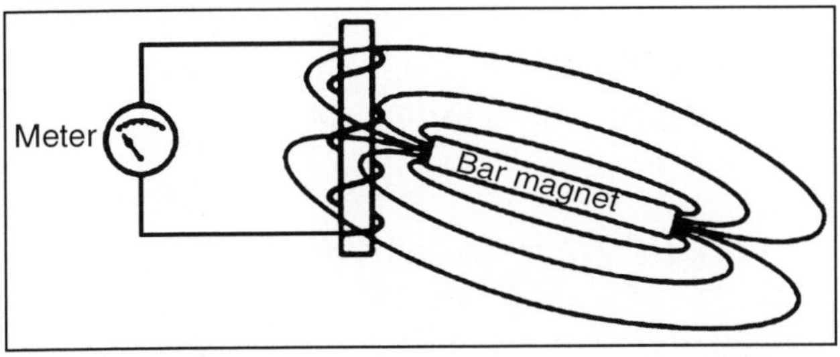

Fig. Representation of a Bar Magnet Close to a Relay Coil. Some of the Lines of Force Intersect the Coil. So Long as the Magnet is Stationary with Respect to the Coil, no Current will Flow Around the Loop Containing the Meter.

These two phenomena, the generation of a magnetic field by current flowing in a coil and the induction of current in a coil by a moving magnetic field, are combined in an extensively used device called the transformer. Its principle of operation is fairly simple. The switch is open and no current flows in either circuit.

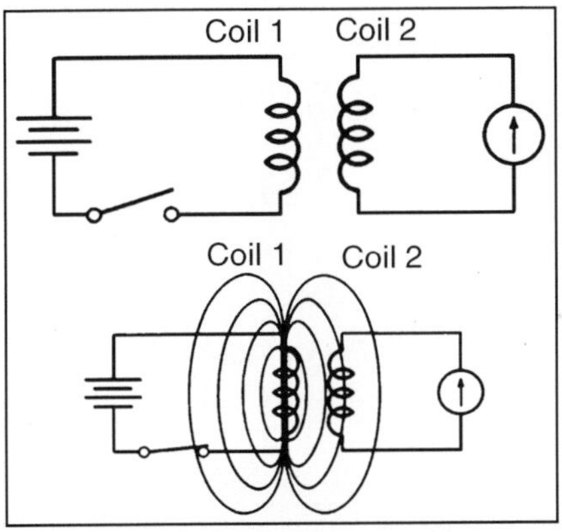

Fig. (*a*) A Circuit Containing two Coils of Wire, Near Each Other but not Connected Together Electrically. (*b*) Representation of the Circuit after the Switch has been Closed. Some of the Lines of Force from Coil I Intersect Coil 2, but so Long as the Current Through Coil I is Constant, no Current will Flow Through Coil 2.

Figure *b* shows the state of affairs when the switch has been closed and current is flowing steadily through coil 1. Although the lines of force from coil 1 are passing through coil 2, no voltage is generated in coil 2 because the lines are not moving. However, between the time when no current flows through coil 1 and when the full current is flowing, that is the lines of force must have multiplied and expanded through coil 2 from zero lines and volume to their positions. During the time when the field was being built up, a voltage must have been induced into coil 2. Similarly, if the switch were opened again, the field would collapse and, during that short time, a voltage would be induced into coil 2 as well. The voltage in coil 2 would have the opposite polarity in the two cases because the lines of force would be traveling in opposite directions during the buildup and collapse of the field. The buildup and the collapse do not occur instantaneously but take finite times which depend upon the characteristics of the circuit. Now suppose the battery and switch were replaced by an a-c source, for instance, the 115-volt a-c wall socket supply. In this case, the current through coil 1 will build up from zero to a maximum, then back through zero again, etc., repeating this complete cycle 60 times a second.

In turn, the magnetic field will also build up and collapse over and over again, continuously generating a changing and reversing voltage in coil 2. This kind of combination of coils is called a transformer. For more efficient transfer of power from coil I to coil 2, both coils are typically wound on a single iron core, because such a core tends to channel the magnetic fields, keeping the lines of force in the close vicinity of the coils. The schematic diagram for a transformer, together with photographs of actual transformers.

If coil 1 consists of the same number of turns of wire as coil 2, the voltage generated in coil 2 will be the same as that put into coil 1. However, if there are, say, ten times as many turns on coil 2 as on coil 1, the lines of force from coil 1 will cut across coil 2 at ten times as many places.

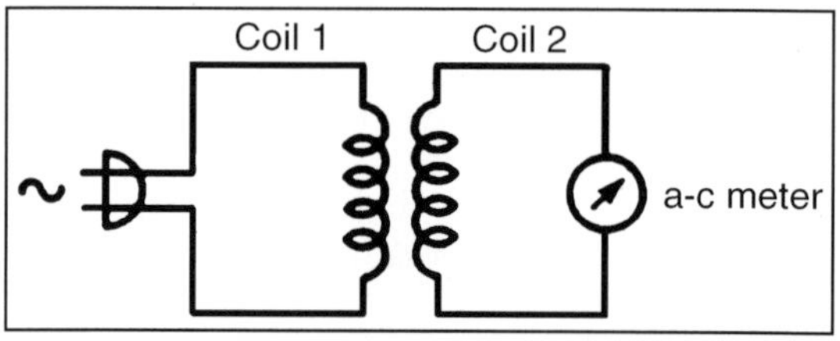

Fig. Diagram of a Simple Transformer Circuit.
The Meter will Indicate a Continuous Voltage.

Therefore the voltage generated in coil 2 will be ten times as great as the input voltage to coil 1. In general:

$$\frac{V_i}{V_o} = \frac{N_i}{N_o}$$

where Vi = the input voltage Vo = the output voltage Ni = the number of turns on the input coil No = the number of turns on the output coil

It is this phenomenon that makes the transformer so useful. For example, suppose that you were designing a circuit to give a rat a strong constant-current shock. For the shock to have a relatively constant current, a large resistance must be in series with the rat. In fact, the resistance must be large enough that almost all of the voltage from the source is lost across the resistor, and only about one-tenth of it actually affects the rat.

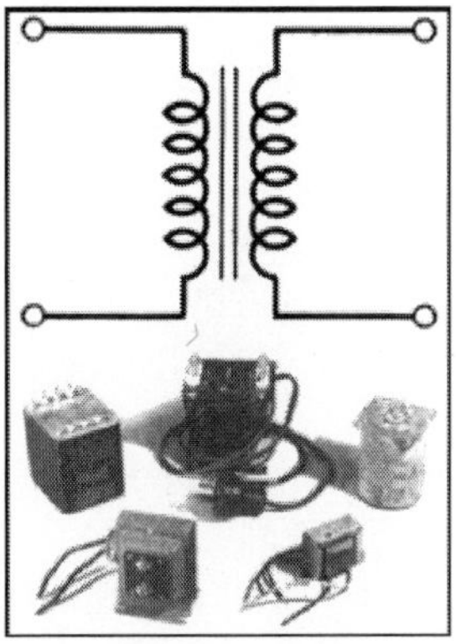

Fig. (*a*) Circuit Symbol for a Transformer, and (*b*) a Photograph of Various Transformers.

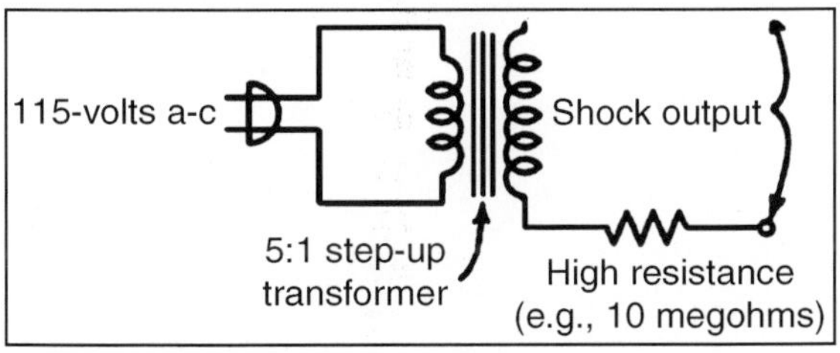

Fig. Constant-current a-c Shock-Delivery Circuit.

To give the rat a good strong shock, the source voltage must be higher than the 115 volts from the wall. A 50-volt shock, for example, requires a 500-volt source. The solution is to put between the wall socket and the shock box a transformer which has about five times as many turns on the output (or secondary) coil as it has on the input (or primary) coil. The transformer is called a step-up transformer.

Electric-eye systems which detect the presence of a rat in an alley require a small source of light such as a flashlight bulb or an automobile taillight bulb. These bulbs are typically rated at 6 volts.

Flashlight batteries could be used to light the bulbs, but they burn out quickly, and storage batteries are inconvenient. However, a small *step-down* transformer which takes the 115 volts from a wall socket and transforms it to 6 volts is cheap, available, and never wears out. Such a transformer has approximately 20 times as many turns on its input coil as on its output coil.

Note that a transformer is so named because it transforms one alternating voltage into another alternating voltage. It does not transform direct current into alternating current (that is accomplished by a device called a converter) nor alternating current into direct current (that device is called a rectifier). A transformer is strictly an a-c device, and as such can be quickly and irreversibly damaged if it is connected across a d-c source.

VOLTAGE TRANSFORMERS SYSTEM

The voltage at the wall socket is usually said to be 115 volts. Actually, that voltage varies widely from time to time. It is a common experience to see the room lights dim when the refrigerator goes on. The refrigerator motor, when it first turns on, draws a large current which flows through the wires leading into the house and to the wall socket. The increased current running through the resistance of those wires causes an increased voltage drop in the wires, and less Voltage is left to hear the lights.

In and around the laboratory, and all along the circuit that includes the power station and the laboratory, devices turn on and off repeatedly, so that it is not unusual for the voltage at the wall to vary from 90 to 130 volts during the day. (The voltage can be higher than 115 because, to overcome the average voltage loss in the wires, the power company delivers more than 115 volts to the system. If everyone but you should suddenly turn off everything electrical, all of your light bulbs would burn out.)

For many devices, such changes in voltage do not matter very much (*e.g.*, fan motors, relays), but several types of apparatus are very sensitive to voltage changes and must be protected from them. All laboratory apparatus in which the intensity of a light bulb is important fall into this category because relatively small changes in line voltage will produce large changes in any lamp's brightness. For example, variations in the line voltage can be a source of great variance in tachistoscopic studies.

Devices which take variable a-c inputs and deliver constant-voltage a-c outputs are called voltage regulators or voltage stabilizers and may be ordered at any electronics supply store. There are a number of different types of voltage regulators, some considerably more costly than others. The cheapest, and one which meets most requirements, is called a constant-voltage transformer. It has no moving parts and is essentially a standard transformer built around a special core. The molecules of metal in any transformer core are involved in the transfer of energy from the input to the output coil. In a constant-voltage transformer, all of the core molecules are brought into action when the instantaneous input voltage is, say, 90 volts. The coils are wound so that the output voltage is 115 volts under this condition.

Any further increase in the input voltage cannot affect the output voltage appreciably because all of the core molecules are already involved (the core is said to be saturated). This transformer will, therefore, give almost constant output voltage for input voltages ranging from 90 to some higher voltage where the system begins to break down. Typically, the upper limit is approximately 130 volts. Transformers of this type are built with various output-current capacities from about 1/10 ampere up to 10 or 20 amperes. The rating is usually stated in units, called volt-amperes, which are very closely related to watts. If you want 115 volts at about 4 amperes, buy a 500-volt-ampere unit. (For a-c circuits, power—in watts—is sometimes different from the simple product of volts times amperes.)

The constant-voltage transformers described above have several defects which may be of consequence in certain applications. First, they only hold the voltage stable to about 11 per cent of its nominal level. For the rare situation in which something better is needed, a more expensive unit must be bought. Second, these transformers usually hum and, because of this, should not be used when absolute quiet is needed. A third defect is that, since the core saturates in the normal range of input voltages, the output wave form is a flat-topped wave that approaches the shape of a square wave. For the rare occasions when a very low harmonic content is required of the supply voltage, regulators which give pure sine-wave outputs can be purchased.

Voltage regulators are not to be confused with the very common piece of electronic equipment called a regulated power supply. The term "power supply," when it is applied to a single unit of equipment, usually refers to a unit that supplies d-c power. A regulated, or stabilized, power supply unit is one that delivers a constant d-c output voltage independent of changes in the line voltage.

TRANSFORMER TYPES

A variety of types of electrical transformer are made for different purposes. Despite their design differences, the various types employ the same basic principle as discovered in 1831 by Michael Faraday, and share several key functional parts.

Power transformers

Laminated core

This is the most common type of transformer, widely used in appliances to convert mains voltage to low voltage to power electronics

- Widely available in power ratings ranging from mW to MW
- Insulated laminations minimize eddy current losses
- Small appliance and electronic transformers may use a split bobbin, giving a high level of insulation between the windings
- Rectangular core
- Core laminate stampings are usually in EI shape pairs. Other shape pairs are sometimes used.
- Mumetal shields can be fitted to reduce EMI (electromagnetic interference)

- A screen winding is occasionally used between the 2 power windings
- Small appliance and electronics transformers may have a thermal cut out built in
- Occasionally seen in low profile format for use in restricted spaces
- Laminated core made with silicon steel with high permeability

Toroidal

Doughnut shaped toroidal transformers are used to save space compared to EI cores, and sometimes to reduce external magnetic field. These use a ring shaped core, copper windings wrapped round this ring (and thus threaded through the ring during winding), and tape for insulation.

Toroidals compared to EI core transformers:

- Lower external magnetic field
- Smaller for a given power rating
- Higher cost in most cases, as winding requires more complex and slower equipment
- Less robust
- Central fixing is either
  - Bolt, large metal washers and rubber pads
  - Bolt and potting resin
- Overtightening the central fixing bolt may short the windings
- Greater inrush current at switch-on.

Autotransformer

An autotransformer has only a single winding, which is tapped at some point along the winding. AC or pulsed voltage is applied across a portion of the winding, and a higher (or lower) voltage is produced across another portion of the same winding. The higher voltage will be connected to the ends of the winding, and the lower voltage from one end to a tap. For example, a transformer with a tap at the centre of the winding can be used with 230 volts across the entire winding, and 115 volts between one end and the tap.

It can be connected to a 230-volt supply to drive 115-volt equipment, or reversed to drive 230-volt equipment from 115 volts. Since the current in the windings is lower, the transformer is smaller, lighter cheaper and more efficient. For voltage ratios not exceeding about 3:1, an autotransformer is cheaper, lighter, smaller and more efficient than an isolating (two-winding) transformer of the same rating. Large three-phase autotransformers are used in electric power distribution systems, for example, to interconnect 33 kV and 66 kV sub-transmission networks.

In practice, transformer losses mean that autotransformers are not perfectly reversible; one designed for stepping down a voltage will deliver slightly less voltage than required if used to step up. The difference is usually slight enough to allow reversal where the actual voltage level is not critical. This is true of isolated winding transformers too.

Variac

By exposing part of the winding coils of an autotransformer, and making the secondary connection through a sliding carbon brush, an autotransformer with a near-continuously variable turns ratio can be obtained, allowing for wide voltage adjustment in very small increments.

Stray Field Transformer

A Stray field transformer has a significant stray field or a (sometimes adjustable) magnetic bypass in its core. It can act as a transformer with inherent current limitation due to its lower tight coupling between the primary and the secondary winding, which is unwanted in much other cases. The output and input currents are low enough to prevent thermal overload under each load condition - even if the secondary is shortened.

Stray field transformers are used for arc welding and high voltage discharge lamps (cold cathode fluorescent lamps, series connected up to 7,5 kV AC working voltage). It acts both as voltage transformer and magnetic ballast.

Polyphase Transformers

For three-phase power, three separate single-phase transformers can be used, or all three phases can be connected to a single polyphase transformer. The three primary windings are connected together and the three secondary windings are connected together.

The most common connections are Y-Delta, Delta-Y, Delta-Delta and Y-Y. A vector group indicates the configuration of the windings and the phase angle difference between them.

If a winding is connected to earth (grounded), the earth connection point is usually the centre point of a Y winding. If the secondary is a Delta winding, the ground may be connected to a centre tap on one winding (high leg delta) or one phase may be grounded (corner grounded delta). A special purpose polyphase transformer is the zigzag transformer. There are many possible configurations that may involve more or fewer than six windings and various tap connections.

Resonant Transformers

A resonant transformer operates at the resonant frequency of one or more of its coils and (usually) an external capacitor. The resonant coil, usually the secondary, acts as an inductor, and is connected in series with a capacitor. When the primary coil is driven by a periodic source of alternating current, such as a square or sawtooth wave at the resonant frequency, each pulse of current helps to build up an oscillation in the secondary coil.

Due to resonance, a very high voltage can develop across the secondary, until it is limited by some process such as electrical breakdown. These devices are used to generate high alternating voltages, and the current available can be much larger than that from electrostatic machines such as the Van de Graaff generator or Wimshurst machine.

Examples:

- Tesla coil
- Oudin coil (or Oudin resonator; named after its inventor Paul Oudin)
- D'Arsonval apparatus
- Ignition coil or induction coil used in the ignition system of a petrol engine
- Flyback transformer of a CRT television set or video monitor.
- Electrical breakdown and insulation testing of high voltage equipment and cables. In the latter case, the transformer's secondary is resonated with the cable's capacitance.

Other applications of resonant transformers are as coupling between stages of a superheterodyne receiver, where the selectivity of the receiver is provided by the tuned transformers of the intermediate-frequency amplifiers.

Constant Voltage Transformer

By arranging particular magnetic properties of a transformer core, and installing a ferro-resonant tank circuit (a capacitor and an additional winding), a transformer can be arranged to automatically keep the secondary winding voltage relatively constant for varying primary supply without additional circuitry or manual adjustment.

CVA transformers run hotter than standard power transformers, because regulating action depends on core saturation, which reduces efficiency somewhat. The output waveform is heavily distorted unless careful measures are taken to prevent this. Saturating transformers provide a simple rugged method to stabilize an AC power supply.

Ferrite Core

Ferrite core power transformers are widely used in switched mode power supplies (SMPSUs). The powder core enables high frequency operation, and hence much smaller size to power ratio than laminated iron transformers. Ferrite transformers are not usable as power transformers at mains frequency.

Planar Transformer

Manufacturers etch spiral patterns on a printed circuit board to form the "windings" of a planar transformer. (Manufacturers literally wind pieces of wire on some core or bobbin to form the windings of other kinds of transformers).

Some planar transformers are commercially sold as discrete components

— the transformer is the only thing on that printed circuit board. Other planar transformers are one of many components on one large printed circuit board.

- Much thinner than other transformers, for low-profile applications (even when several PCBs are stacked)
- Almost all use a ferrite planar core

Oil Cooled Transformer

For large transformers used in power distribution or electrical substations, the core and coils of the transformer are immersed in oil which cools and insulates. Oil circulates through ducts in the coil and around the coil and core assembly, moved by convection. The oil is cooled by the outside of the tank in small ratings, and in larger ratings an air-cooled radiator is used.

Where a higher rating is required, or where the transformer is used in a building or underground, oil pumps are used to circulate the oil and an oil-to-water heat exchanger may also be used. Formerly, indoor transformers required to be fire-resistant used PCB liquids; since these are now banned, substitute fire-resistant liquids such as silicone oils are instead used.

Isolating Transformer

Most transformers isolate, meaning the secondary winding is not connected to the primary. But this isn't true of all transformers. However the term 'isolating transformer' is normally applied to mains transformers providing isolation rather than voltage transformation. They are simply 1:1 laminated core transformers. Extra voltage tappings are sometimes included, but to earn the name 'isolating transformer' it is expected that they will usually be used at 1:1 ratio.

Instrument Transformers

Current Transformers

A current transformer (CT) is a measurement device designed to provide a current in its secondary coil proportional to the current flowing in its primary. Current transformers are commonly used in metering and protective relaying in the electrical power industry where they facilitate the safe measurement of large currents, often in the presence of high voltages. The current transformer safely isolates measurement and control circuitry from the high voltages typically present on the circuit being measured.

Current transformers are often constructed by passing a single primary turn (either an insulated cable or an uninsulated bus bar) through a well-insulated toroidal core wrapped with many turns of wire. The CT is typically described by its current ratio from primary to secondary. For example, a 4000:5 CT would provide an output current of 5 amperes when the primary was passing 4000 amperes.

The secondary winding can be single ratio or have several tap points to provide a range of ratios. Care must be taken that the secondary winding is not disconnected from its load while current flows in the primary, as this will produce a dangerously high voltage across the open secondary and may permanently affect the accuracy of the transformer.

Specially constructed wideband CTs are also used, usually with an oscilloscope, to measure high frequency waveforms or pulsed currents within pulsed power systems. One type provides a voltage output that is proportional to the measured current; another, called a Rogowski coil, requires an external integrator in order to provide a proportional output.

Voltage Transformers

Voltage transformers (VTs) or potential transformers (PTs) are another type of instrument transformer, used for metering and protection in high-voltage circuits. They are designed to present negligible load to the supply being measured and to have a precise voltage ratio to accurately step down high voltages so that metering and protective relay equipment can be operated at a lower potential. Typically the secondary of a voltage transformer is rated for 69 or 120 Volts at rated primary voltage, to match the input ratings of protection relays. The transformer winding high-voltage connection points are typically labelled as H1, H2 (sometimes H0 if it is internally grounded) and X1, X2, and sometimes an X3 tap may be present. Sometimes a second isolated winding (Y1, Y2, Y3) may also be available on the same voltage transformer. The high side (primary) may be connected phase to ground or phase to phase. The low side (secondary) is usually phase to ground. The terminal identifications (H1, X1, Y1, etc.) are often referred to as polarity. This applies to current transformers as well. At any instant terminals with the same suffix numeral have the same polarity and phase. Correct identification of terminals and wiring is essential for proper operation of metering and protection relays.

While VTs were formerly used for all voltages greater than 240V primary, modern meters eliminate the need VTs for most secondary service voltages. VTs are typically used in circuits where the system voltage level is above 600 V. Modern meters eliminate the need of VT's since the voltage remains constant and it is measured in the incoming supply.

Pulse Transformers

A pulse transformer is a transformer that is optimised for transmitting rectangular electrical pulses (that is, pulses with fast rise and fall times and a relatively constant amplitude). Small versions called *signal* types are used in digital logic and telecommunications circuits, often for matching logic drivers to transmission lines.

Medium-sized *power* versions are used in power-control circuits such as camera flash controllers. Larger *power* versions are used in the electrical power distribution industry to interface low-voltage control circuitry to the high-voltage gates of power semiconductors. Special high voltage pulse transformers are also used to generate high power pulses for radar, particle accelerators, or other high energy pulsed power applications.

To minimise distortion of the pulse shape, a pulse transformer needs to have low values of leakage inductance and distributed capacitance, and a high open-circuit inductance. In power-type pulse transformers, a low coupling capacitance (between the primary and secondary) is important to protect the circuitry on the primary side from high-powered transients created by the load.

For the same reason, high insulation resistance and high breakdown voltage are required. A good transient response is necessary to maintain the rectangular pulse shape at the secondary, because a pulse with slow edges would create switching losses in the power semiconductors.

The product of the peak pulse voltage and the duration of the pulse (or more accurately, the voltage-time integral) is often used to characterise pulse transformers. Generally speaking, the larger this product, the larger and more expensive the transformer. Pulse transformers by definition have a duty cycle of less than 1, whatever energy stored in the coil during the pulse must be "dumped" out before the pulse is fired again.

RF Transformers

There are several types of transformer used in radio frequency (RF) work. Steel laminations are not suitable for RF.

Air-Core Transformers

These are used for high frequency work. The lack of a core means very low inductance. Such transformers may be nothing more than a few turns of wire soldered onto a printed circuit board.

Ferrite-Core Transformers

Widely used in intermediate frequency (IF) stages in superheterodyne radio receivers. These are mostly tuned transformers, containing a threaded ferrite slug that is screwed in or out to adjust IF tuning. The transformers are usually canned for stability and to reduce interference.

Transmission-Line Transformers

For radio frequency use, transformers are sometimes made from configurations of transmission line, sometimes bifilar or coaxial cable, wound around ferrite or other types of core. This style of transformer gives an extremely wide bandwidth but only a limited number of ratios (such as 1:9, 1:4 or 1:2) can be achieved with this technique.

The core material increases the inductance dramatically, thereby raising its Q factor. The cores of such transformers help improve performance at the lower frequency end of the band. RF transformers sometimes used a third coil (called a tickler winding) to inject feedback into an earlier (detector) stage in antique regenerative radio receivers.

Baluns

Baluns are transformers designed specifically to connect between balanced and unbalanced circuits. These are sometimes made from configurations of transmission line and sometimes bifilar or coaxial cable and are similar to transmission line transformers in construction and operation.

Audio Transformers

Audio transformers are usually the factor which limit sound quality when used; electronic circuits with wide frequency response and low distortion are relatively simple to design.

Transformers are also used in DI boxes to convert high-impedance instrument signals (*e.g.* bass guitar) to low impedance signals to enable them to be connected to a microphone input on the mixing console.

A particularly critical component is the output transformer of an audio power amplifier. Valve circuits for quality reproduction have long been produced with no other (inter-stage) audio transformers, but an output transformer is needed to couple the relatively high impedance (up to a few hundred ohms depending upon configuration) of the output valve(s) to the low impedance of a loudspeaker. (The valves can deliver a low current at a high voltage; the speakers require high current at low voltage.) Most solid-state power amplifiers need no output transformer at all.

For good low-frequency response a relatively large iron core is required; high power handling increases the required core size. Good high-frequency response requires carefully designed and implemented windings without excessive leakage inductance or stray capacitance. All this makes for an expensive component. Early transistor audio power amplifiers often had output transformers, but they were eliminated as designers discovered how to design amplifiers without them.

Loudspeaker Transformers

In the same way that transformers are used to create high voltage power transmission circuits that minimize transmission losses, loudspeaker transformers can be used allow many individual loudspeakers to be powered from a single audio circuit operated at higher-than normal loudspeaker voltages. This application is common in industrial public address applications. Such circuits are commonly referred to as constant voltage speaker systems, although the audio waveform is a changing voltage. Such systems are also known by other terms such as 25-, 70- and 100-volt speaker systems, referring to the nominal voltage of the loudspeaker line.

At the audio amplifier, a large audio transformer may be used to step-up the low impedance, low-voltage output of the amplifier to the designed line voltage of the loudspeaker circuit. At the distant loudspeaker location, a smaller step-down transformer returns the voltage and impedance to ordinary loudspeaker levels. The loudspeaker transformers commonly have multiple primary taps, allowing the volume at each speaker to be adjusted in discrete steps.

Output Transformer (valve)

Valve (tube) amplifiers almost always use an output transformer to match the high load impedance requirement of the valves (several kilohms) to a low impedance speaker.

Small Signal Transformers

Moving coil phonograph cartridges produce a very small voltage. In order for this to be amplified with a reasonable signal-noise ratio, a transformer is usually used to convert the voltage to the range of the more common moving-magnet cartridges.

Microphones may also be matched to their load with a small transformer, which is mumetal shielded to minimise noise pickup. These transformers are less widely used today, as transistorized buffers are now cheaper.

'Interstage' and Coupling Transformers

A use for interstage transformers is in the case of push-pull amplifiers where an inverted signal is required. Here two secondary windings wired in opposite polarities may be used to drive the output devices. These phase splitting transformers are not much used today.

Cast Resin Transformers

Cast-resin power transformers have been widely used for a long time. These transformers have the advantage of easy installation and improved fire behaviour in case of class. This indoor type transformer is totally dry, without cooling oil.

Homemade and Obsolete Transformers

Transformer Kits

Transformers may be wound at home using commercial transformer kits, which contain laminations and bobbin. Or ready made transformers may be disassembled and rewound. These approaches are occasionally used by home constructors, but are usually avoided where possible due to the number of hours required to hand wind a transformer.

100 per cent Homemade

It is possible to make the transformer laminations by hand too. Such transformers are encountered at times in 3rd world countries, using laminations cut from scrap sheet steel, paper slips between the laminations, and string to tie the assembly together. The result works, but is usually noisy due to poor clamping of laminations.

- Picture
- Device in use

Hedgehog

Hedgehog transformers are occasionally encountered in homemade 1920s radios. They are homemade audio interstage coupling transformers. Enamelled copper wire is wound round the central half of the length of a bundle of insulated iron wire (eg florists' wire), to make the windings.

The ends of the iron wires are then bent around the electrical winding to complete the magnetic circuit, and the whole is wrapped with tape or string to hold it together. These were sometimes used when the cost of a ready made transformer could not be justified. Inductance tends to be on the low side, with consequent loss of bass. With the speakers of the day this was no bad thing.

Variocouplers

Variocouplers are rf transformers with 2 windings and variable coupling between the windings. They were standard equipment in 1920s radio sets. Pancake coil variocouplers were common in 1920s radios for variable rf coupling. The 2 planar coils were arranged to swing away from each other and for the angle between them to increase to 90 degrees, thus giving wide variation in coupling. No core was used.

These were mostly used to control reaction. The pancake structure was a means to minimize stray capacitance. In another design of variocoupler, 2 coils were wound on a 2 circular bands, and housed one inside the other, with provision for rotating the inner coil. Coupling varies as one coil is rotated between 0 and 90 degrees from the other. These had higher stray capacitance than the pancake type.

Not Transformers

Finally there are some items often mistaken for transformers, but which are not always transformers. Wall warts: small power supplies with integral mains plug. These can contain a transformer and other circuitry. Most use a laminated iron transformer, but an increasing number now contain a small SMPSU. These are smaller and much lighter. Halogen lighting transformers: Toroidal transformers are sometimes used for this task, but most halogen 'transformers' are SMPSUs.

7

Digital Circuits and Electronics System

CIRCUIT DIAGRAM

A circuit diagram (also known as an electrical diagram, wiring diagram, elementary diagram, or electronic schematic) is a simplified conventional pictorial representation of an electrical circuit. It shows the components of the

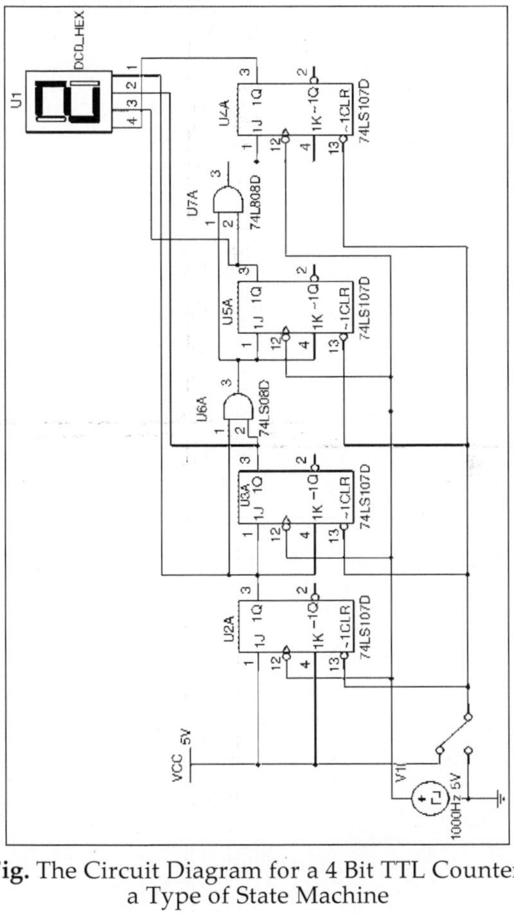

Fig. The Circuit Diagram for a 4 Bit TTL Counter, a Type of State Machine

circuit as simplified standard symbols, and the power and signal connections between the devices. Arrangement of the components interconnections on the diagram does not correspond to their physical locations in the finished device.

Unlike a block diagram or layout diagram, a circuit diagram shows the actual wire connections being used. The diagram does not show the physical arrangement of components. A drawing meant to depict what the physical arrangement of the wires and the components they connect is called "artwork" or "layout" or the "physical design."

Circuit diagrams are used for the design (circuit design), construction (such as PCB layout), and maintenance of electrical and electronic equipment.

LEGENDS

On a circuit diagram, the symbols for components are labelled with a descriptor (or reference designator) matching that on the list of parts. For example, C1 is the first capacitor, L1 is the first inductor, Q1 is the first transistor, and R1 is the first resistor (note that this is not written as a subscript, as in R_1, L_1,...). The letters that precede the numbers were chosen in the early days of the electrical industry, even before the vacuum tube (thermionic valve), so "Q" was the only one available for semiconductor devices in the mid-twentieth century. Often the value or type designation of the component is given on the diagram beside the part, but detailed specifications would go on the parts list.

SYMBOLS

Circuit diagram symbols have differed from country to country and have changed over time, but are now to a large extent internationally standardized. Simple components often had symbols intended to represent some feature of the physical construction of the device. For example, the symbol for a resistor shown here dates back to the days when that component was made from a long piece of wire wrapped in such a manner as to not produce inductance, which would have made it a coil.

These wirewound resistors are now used only in high-power applications, smaller resistors being cast from *carbon composition* (a mixture of carbon and filler) or fabricated as an insulating tube or chip coated with a metal film. The internationally standardized symbol for a resistor is therefore now simplified to an oblong, sometimes with the value in ohms written inside, instead of the zig-zag symbol. A less common symbol is simply a series of peaks on one side of the line representing the conductor, rather than back-and-forth as shown here.

STANDARDS

There are several national and international standards for graphical symbols in circuit diagrams, in particular:

- IEC 60617 (also known as British Standard BS 3939)
- ANSI standard Y32 (also known as IEEE Std 315)

IEC 60617 originally consisted of 13 parts, from resistors and capacitors to logic symbols and even a generalised drawing standard of connections and bus line widths. It is now published as a subscription online database IEC 60617-DB.

Different symbols may be used depending on the discipline using the drawing; for example, lighting and power symbols used as part of architectural drawings may be different from symbols for devices used in electronics.

LINKAGES

Schematic wire junctions:

- Old style: (a) connection, (b) no connection.
- One CAD style: (a) connection, (b) no connection.
- Alternative CAD Style: (a) connection, (b) no connection.

The linkages between leads were once simple crossings of lines; one wire insulated from and "jumping over" another was indicated by it making a little semicircle over the other line. With the arrival of computerized drafting, a connection of two intersecting wires was shown by a crossing with a dot or "blob", and a crossover of insulated wires by a simple crossing without a dot.

However, there was a danger of confusing these two representations if the dot was drawn too small or omitted. Modern practice is to avoid using the "crossover with dot" symbol, and to draw the wires meeting at two points instead of one. It is also common to use a hybrid style, showing connections as a cross with a dot while insulated crossings use the semicircle.

The following codes which vary slightly from the American codes are in common use in European and Australian standard electrical circuit diagrams. These codes are used for the "reference designators" printed on PCBs (which match the corresponding ones written on the corresponding schematic).

- A: Assemblies
- B: Transducers (photo cells, inductive proximity, thermocouple, flame detection)
- C: Capacitors
- D: Storage devices
- E: Miscellaneous
- F: Fuses
- G: Generator, battery pack
- H: Indicators, lamps (not for illumination), signalling devices
- K: Relays, contactors
- L: Inductors and filters
- M: Motors
- N: Analogue devices

- P: Measuring/test equipment
- Q: Circuit breakers, isolators, re-closers
- R: Resistors, brake resistors
- S: Switches, push buttons, emergency stops and limit switches
- T: Transformers
- U: Power converters, variable speed drives, soft starters, DC power supplies
- V: Semiconductors
- W: Wires, conductors, power, neutral and earthing busses
- X: Terminal strips, terminations, joins
- Y: Solenoids, electrical actuators
- Z: Filters

Detailed rules for reference designations are provided in the International standard IEC 61346.

ORGANIZATION OF DRAWINGS

It is a usual although not universal convention that schematic drawings are organized on the page from left to right and top to bottom in the same sequence as the flow of the main signal or power path. For example, a schematic for a radio receiver might start with the antenna input at the left of the page and end with the loudspeaker at the right.

Positive power supply connections for each stage would be shown towards the top of the page, with grounds, negative supplies, or other return paths towards the bottom. Schematic drawings intended for maintenance may have the principle signal paths highlighted to assist in understanding the signal flow through the circuit. More complex devices have multi-page schematics and must rely on cross-reference symbols to show the flow of signals between the different sheets of the drawing.

Detailed rules for the preparation of circuit diagrams (and other document kinds used in electrotechnology) are provided in the International standard IEC 61082-1. Relay logic line diagrams (also called ladder logic diagrams) use another common standardized convention for organizing schematic drawings, with a vertical power supply "rail" on the left and another on the right, and components strung between them like the rungs of a ladder.

ART WORK

Once the schematic has been made, it is converted into a layout that can be fabricated onto a Printed Circuit Board (PCB). The layout is usually prepared by the process of schematic capture. The result is what is known as a Rat's Nest. The Rat's Nest is a jumble of wires (lines) criss crossing each other to their destination nodes. These wires are routed either manually or by the use of Electronics Design Automation (EDA) tools. The EDA tools arrange and rearrange the placement of components and finds paths for tracks to connect various nodes. This results into an Art Work.

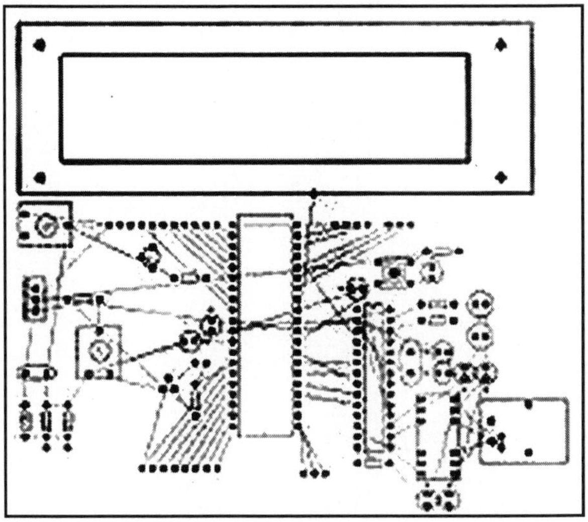

Fig. A Rat's Nest

A generalized design flow would be as:

Schematic → Schematic Capture → Rat's Nest → Routing → Art Work → PCB Development and etching → Component Mounting → Testing

Diode Bridge

A diode bridge or bridge rectifier is an arrangement of four diodes in a bridge configuration that provides the same polarity of output voltage for either polarity of input voltage. When used in its most common application, for conversion of alternating current (AC) input into direct current (DC) output, it is known as a bridge rectifier.

Fig. Three Bridge Rectifiers. The Size is Generally Related to the Current Handling Capability.

A bridge rectifier provides full-wave rectification from a two-wire AC input, resulting in lower cost and weight as compared to a centre-tapped transformer design, but has two diode drops rather than one, thus exhibiting reduced efficiency over a centre-tapped design for the same output voltage.

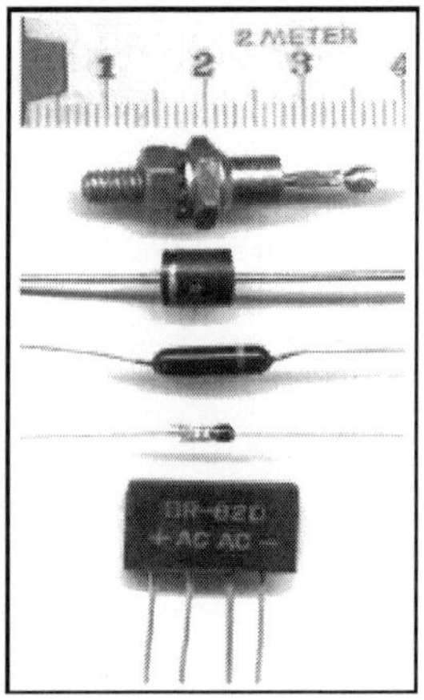

Fig. Diodes; the One on the Bottom is a Diode Bridge

The essential feature of a diode bridge is that the polarity of the output is the same regardless of the polarity at the input. The diode bridge circuit is also known as the *Graetz circuit* after its inventor, physicist Leo Graetz.

BASIC OPERATION

When the input connected at the left corner of the diamond is positive with respect to the one connected at the right hand corner, current flows to the right along the upper coloured path to the output, and returns to the input supply via the lower one. When the right hand corner is positive relative to the left hand corner, current flows along the upper coloured path and returns to the supply via the lower coloured path.

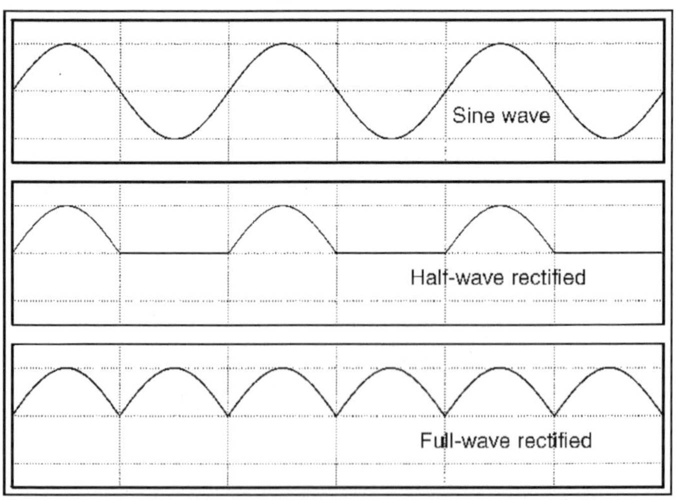

Fig. AC, Half-wave and Full Wave Rectified Signals

In each case, the upper right output remains positive with respect to the lower right one. Since this is true whether the input is AC or DC, this circuit not only produces DC power when supplied with AC power: it also can provide what is sometimes called "reverse polarity protection". That is, it permits normal functioning when batteries are installed backwards or DC input-power supply wiring "has its wires crossed" (and protects the circuitry it powers against damage that might occur without this circuit in place).

Prior to availability of integrated electronics, such a bridge rectifier was always constructed from discrete components. Since about 1950, a single four-terminal component containing the four diodes connected in the bridge configuration became a standard commercial component and is now available with various voltage and current ratings.

OUTPUT SMOOTHING

For many applications, especially with single phase AC where the full-wave bridge serves to convert an AC input into a DC output, the addition of a capacitor may be important because the bridge alone supplies an output voltage of fixed polarity but pulsating magnitude.

The function of this capacitor, known as a reservoir capacitor (aka smoothing capacitor) is to lessen the variation in (or 'smooth') the rectified AC output voltage waveform from the bridge. One explanation of 'smoothing' is that the capacitor provides a low impedance path to the AC component of the output, reducing the AC voltage across, and AC current through, the resistive load. In less technical terms, any drop in the output voltage and current of the bridge tends to be cancelled by loss of charge in the capacitor. This charge flows out as additional current through the load. Thus the change of load current and voltage is reduced relative to what would occur without the capacitor.

Increases of voltage correspondingly store excess charge in the capacitor, thus moderating the change in output voltage/current. Also see rectifier output smoothing. The simplified circuit shown has a well deserved reputation for being dangerous, because, in some applications, the capacitor can retain a *lethal* charge after the AC power source is removed. If supplying a dangerous voltage, a practical circuit should include a reliable way to safely discharge the capacitor.

If the normal load can not be guaranteed to perform this function, perhaps because it can be disconnected, the circuit should include a bleeder resistor connected as close as practical across the capacitor. This resistor should consume a current large enough to discharge the capacitor in a reasonable time, but small enough to minimise unnecessary power waste. Because a bleeder sets a minimum current drain, the regulation of the circuit, defined as percentage voltage change from minimum to maximum load, is improved. However in many cases the improvement is of insignificant magnitude.

The capacitor and the load resistance have a typical time constant ô = *RC* where *C* and *R* are the capacitance and load resistance respectively. As long as the load resistor is large enough so that this time constant is much longer than the time of one ripple cycle, the above configuration will produce a smoothed DC voltage across the load. In some designs, a series resistor at the load side of the capacitor is added.

The smoothing can then be improved by adding additional stages of capacitor–resistor pairs, often done only for sub-supplies to critical high-gain circuits that tend to be sensitive to supply voltage noise.

The idealized waveforms shown above are seen for both voltage and current when the load on the bridge is resistive. When the load includes a smoothing capacitor, both the voltage and the current waveforms will be greatly changed. While the voltage is smoothed, as described above, current will flow through the bridge only during the time when the input voltage is greater than the capacitor voltage.

For example, if the load draws an average current of n Amps, and the diodes conduct for 10 per cent of the time, the average diode current during conduction must be 10n Amps.

This non-sinusoidal current leads to harmonic distortion and a poor power factor in the AC supply.

In a practical circuit, when a capacitor is directly connected to the output of a bridge, the bridge diodes must be sized to withstand the current surge that occurs when the power is turned on at the peak of the AC voltage and the capacitor is fully discharged.

Sometimes a small series resistor is included before the capacitor to limit this current, though in most applications the power supply transformer's resistance is already sufficient. Output can also be smoothed using a choke and second capacitor. The choke tends to keep the current (rather than the voltage) more constant. Due to the relatively high cost of an effective choke compared to a resistor and capacitor this is not employed in modern equipment.

Some early console radios created the speaker's constant field with the current from the high voltage ("B +") power supply, which was then routed to the consuming circuits, (permanent magnets were then too weak for good performance) to create the speaker's constant magnetic field. The speaker field coil thus performed 2 jobs in one: it acted as a choke, filtering the power supply, and it produced the magnetic field to operate the speaker.

POLYPHASE DIODE BRIDGES

This construction can be generalized to rectify Polyphase system AC inputs. For instance, for *three*-phase AC, a full wave bridge rectifier consists of six diodes.

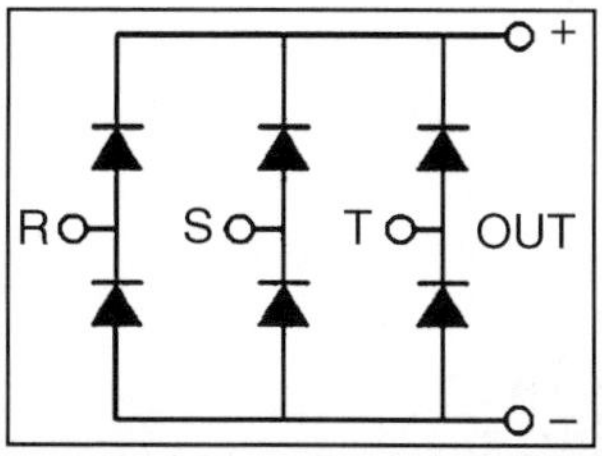

Fig. Three Phase Bridge Rectifier for a Wind Turbine.

Fig. Three Phase Bridge Rectifier for a Wind Turbine.

AUTOMATED DESIGN TOOLS

To save costly engineering effort, much of the effort of designing large logic machines has been automated. The computer programmes are called "electronic design automation tools" or just "EDA." Simple truth table-style descriptions of logic are often optimized with EDA that automatically produces reduced systems of logic gates or smaller lookup tables that still produce the desired outputs. The most common example of this kind of software is the Espresso heuristic logic minimizer.

Most practical algorithms for optimizing large logic systems use algebraic manipulations or binary decision diagrams, and there are promising experiments with genetic algorithms and annealing optimizations. To automate costly engineering processes, some EDA can take state tables that describe state machines and automatically produce a truth table or a function table for the combinatorial part of a state machine. The state table is a piece of text that lists each state, together with the conditions controlling the transitions between them and the belonging output signals.

It is common for the function tables of such computer-generated state-machines to be optimized with logic-minimization software such as Minilog. Often, real logic systems are designed as a series of sub-projects, which are combined using a "tool flow." The tool flow is usually a "script," a simplified computer language that can invoke the software design tools in the right order.

Tool flows for large logic systems such as microprocessors can be thousands of commands long, and combine the work of hundreds of engineers. Writing and debugging tool flows is an established engineering specialty in companies that produce digital designs. The tool flow usually terminates in a detailed computer file or set of files that describe how to physically construct the logic. Often it consists of instructions to draw the transistors and wires on an integrated circuit or a printed circuit board.

Parts of tool flows are "debugged" by verifying the outputs of simulated logic against expected inputs. The test tools take computer files with sets of inputs and outputs, and highlight discrepancies between the simulated behaviour and the expected behaviour. Once the input data is believed correct, the design itself must still be verified for correctness. Some tool flows verify designs by first producing a design, and then scanning the design to produce compatible input data for the tool flow. If the scanned data matches the input data, then the tool flow has probably not introduced errors.

The functional verification data are usually called "test vectors." The functional test vectors may be preserved and used in the factory to test that newly constructed logic works correctly. However, functional test patterns don't discover common fabrication faults. Production tests are often designed by software tools called "test pattern generators." These generate test vectors by examining the structure of the logic and systematically generating tests for particular faults. This way the fault coverage can closely approach 100 per cent, provided the design is properly made testable.

Once a design exists, and is verified and testable, it often needs to be processed to be manufacturable as well. Modern integrated circuits have features smaller than the wavelength of the light used to expose the photoresist. Manufacturability software adds interference patterns to the exposure masks to eliminate open-circuits, and enhance the masks' resolution and contrast.

Design for Testability

A large logic machine (say, with more than a hundred logical variables) can have an astronomical number of possible states. Obviously, in the factory, testing every state is impractical if testing each state takes a microsecond, and there are more states than the number of microseconds since the universe began. Unfortunately, this ridiculous-sounding case is typical. Fortunately, large logic machines are almost always designed as assemblies of smaller logic machines. To save time, the smaller sub-machines are isolated by permanently-installed "design for test" circuitry, and are tested independently.

One common test scheme known as "scan design" moves test bits serially (one after another) from external test equipment through one or more serial shift registers known as "scan chains". Serial scans have only one or two wires to carry the data, and minimize the physical size and expense of the infrequently-used test logic. After all the test data bits are in place, the design is reconfigured to be in "normal mode" and one or more clock pulses are applied, to test for faults (*e.g.* stuck-at low or stuck-at high) and capture the test result into flip-flops and/or latches in the scan shift register(s). Finally, the result of the test is shifted out to the block boundary and compared against the predicted "good machine" result. In a board-test environment, serial to parallel testing has been formalized with a standard called "JTAG" (named after the "Joint Test Action Group" that proposed it). Another common testing scheme provides a test mode that forces some part of the logic machine to enter a "test cycle." The test cycle usually exercises large independent parts of the machine.

Trade-offs

Several numbers determine the practicality of a system of digital logic. Engineers explored numerous electronic devices to get an ideal combination of fanout, speed, low cost and reliability. The cost of a logic gate is crucial. In the 1930s, the earliest digital logic systems were constructed from telephone relays because these were inexpensive and relatively reliable. After that, engineers always used the cheapest available electronic switches that could still fulfill the requirements.

The earliest integrated circuits were a happy accident. They were constructed not to save money, but to save weight, and permit the Apollo Guidance Computer to control an inertial guidance system for a spacecraft. The first integrated circuit logic gates cost nearly $50 (in 1960 dollars, when an engineer earned $10,000/year). To everyone's surprise, by the time the circuits were mass-produced, they had become the least-expensive method of constructing digital logic. Improvements in this technology have driven all subsequent improvements in cost.

With the rise of integrated circuits, reducing the absolute number of chips used represented another way to save costs. The goal of a designer is not just to make the simplest circuit, but to keep the component count down. Sometimes this results in slightly more complicated designs with respect to the underlying digital logic but nevertheless reduces the number of components, board size, and even power consumption. For example, in some logic families, NAND gates are the simplest digital gate to build. All other logical operations can be implemented by NAND gates.

If a circuit already required a single NAND gate, and a single chip normally carried four NAND gates, then the remaining gates could be used to implement other logical operations like logical and. This could eliminate the need for a separate chip containing those different types of gates. The "reliability" of a logic gate describes its mean time between failure (MTBF). Digital machines often have millions of logic gates. Also, most digital machines are "optimized" to reduce their cost. The result is that often, the failure of a single logic gate will cause a digital machine to stop working.

Digital machines first became useful when the MTBF for a switch got above a few hundred hours. Even so, many of these machines had complex, well-rehearsed repair procedures, and would be non-functional for hours because a tube burned-out, or a moth got stuck in a relay. Modern transistorized integrated circuit logic gates have MTBFs of nearly a trillion (1 × 10) hours, and need them because they have so many logic gates. Fanout describes how many logic inputs can be controlled by a single logic output.

The minimum practical fanout is about five. Modern electronic logic using CMOS transistors for switches have fanouts near fifty, and can sometimes go much higher. The "switching speed" describes how many times per second an inverter (an electronic representation of a "logical not" function) can change from true to false and back. Faster logic can accomplish more operations in less time.

Digital logic first became useful when switching speeds got above fifty hertz, because that was faster than a team of humans operating mechanical calculators. Modern electronic digital logic routinely switches at five gigahertz (5 × 10 hertz), and some laboratory systems switch at more than a terahertz (1 × 10 hertz).

LOGIC FAMILIES

Design started with relays. Relay logic was relatively inexpensive and reliable, but slow. Occasionally a mechanical failure would occur. Fanouts were typically about ten, limited by the resistance of the coils and arcing on the contacts from high voltages. Later, vacuum tubes were used. These were very fast, but generated heat, and were unreliable because the filaments would burn out. Fanouts were typically five to seven, limited by the heating from the tubes' current. In the 1950s, special "computer tubes" were developed with filaments that omitted volatile elements like silicon.

These ran for hundreds of thousands of hours. The first semiconductor logic family was Resistor-transistor logic. This was a thousand times more reliable than tubes, ran cooler, and used less power, but had a very low fan-in of three. Diode-transistor logic improved the fanout up to about seven, and reduced the power. Some DTL designs used two power-supplies with alternating layers of NPN and PNP transistors to increase the fanout.

Transistor transistor logic (TTL) was a great improvement over these. In early devices, fanout improved to ten, and later variations reliably achieved twenty. TTL was also fast, with some variations achieving switching times as low as twenty nanoseconds. TTL is still used in some designs. Another contender was emitter coupled logic. This is very fast but uses a lot of power. It's now used mostly in radio-frequency circuits. Modern integrated circuits mostly use variations of CMOS, which is acceptably fast, very small and uses very little power. Fanouts of forty or more are possible, with some speed penalty.

NON-ELECTRONIC LOGIC

It is possible to construct non-electronic digital mechanisms. In principle, any technology capable of representing discrete states and representing logic operations could be used to build mechanical logic. Danny Hillis, co-author of The Connection Machine, once built a working computer from Tinker toys, string, a brick, and a sharpened pencil, which is supposed to be in the Houston Museum of Natural Science. Hydraulic, pneumatic and mechanical versions of logic gates exist and are used in situations where electricity cannot be used.

The first two types are considered under the heading of fluidics. One application of fluidic logic is in military hardware that is likely to be exposed to a nuclear electromagnetic pulse (nuclear EMP, or NEMP) that would destroy electrical circuits. Mechanical logic is frequently used in inexpensive controllers, such as those in washing machines. Famously, the first computer design, by Charles Babbage, was designed to use mechanical logic. Mechanical logic might also be used in very small computers that could be built by nanotechnology. Another example is that if two particular enzymes are required to prevent the construction of a particular protein, this is the equivalent of a biological "NAND" gate.

Recent Developments

The discovery of superconductivity has enabled the development of Rapid Single Flux Quantum (RSFQ) circuit technology, which uses Josephson junctions instead of transistors. Most recently, attempts are being made to construct purely optical computing systems capable of processing digital information using non-linear optical elements.

DIGITAL ELECTRONICS

Digital electronics are electronics systems that use digital signals. Digital electronics are representations of Boolean algebra and are used in computers, mobile phones, and other consumer products. In a digital circuit, a signal is represented in one of two states or logic levels. The advantages of digital techniques stem from the fact it is easier to get an electronic device to switch into one of two states, than to accurately reproduce a continuous range of values.

Digital electronics or any digital circuit are usually made from large assemblies of logic gates, simple electronic representations of Boolean logic functions. To most electronic engineers, the terms "digital circuit", "digital system" and "logic" are interchangeable in the context of digital circuits.

ADVANTAGES

One advantage of digital circuits when compared to analog circuits is that signals represented digitally can be transmitted without degrading because of noise. For example, a continuous audio signal, transmitted as a sequence of 1s and 0s, can be reconstructed without error provided the noise picked up in transmission is not enough to prevent identification of the 1s and 0s. An hour of music can be stored on a compact disc as about 6 billion binary digits.

In a digital system, a more precise representation of a signal can be obtained by using more binary digits to represent it. While this requires more digital circuits to process the signals, each digit is handled by the same kind of hardware. In an analog system, additional resolution requires fundamental improvements in the linearity and noise charactersitics of each step of the signal chain. Computer-controlled digital systems can be controlled by software, allowing new functions to be added without changing hardware.

Often this can be done outside of the factory by updating the product's software. So, the product's design errors can be corrected after the product is in a customer's hands. Information storage can be easier in digital systems than in analog ones. The noise-immunity of digital systems permits data to be stored and retrieved without degradation. In an analog system, noise from aging and wear degrade the information stored. In a digital system, as long as the total noise is below a certain level, the information can be recovered perfectly.

DISADVANTAGES

In some cases, digital circuits use more energy than analog circuits to accomplish the same tasks, thus producing more heat as well. In portable or battery-powered systems this can limit use of digital systems. For example, battery-powered cellular telephones often use a low-power analog front-end to amplify and tune in the radio signals from the base station.

However, a base station has grid power and can use power-hungry, but very flexible software radios. Such base stations can be easily reprogrammed to process the signals used in new cellular standards. Digital circuits are sometimes more expensive, especially in small quantities. The sensed world is analog, and signals from this world are analog quantities.

For example, light, temperature, sound, electrical conductivity, electric and magnetic fields are analog. Most useful digital systems must translate from continuous analog signals to discrete digital signals. This causes quantization errors. Quantization error can be reduced if the system stores enough digital data to represent the signal to the desired degree of fidelity. The Nyquist-Shannon sampling theorem provides an important guideline as to how much digital data is needed to accurately portray a given analog signal.

In some systems, if a single piece of digital data is lost or misinterpreted, the meaning of large blocks of related data can completely change. Because of the cliff effect, it can be difficult for users to tell if a particular system is right on the edge of failure, or if it can tolerate much more noise before failing. Digital fragility can be reduced by designing a digital system for robustness. For example, a parity bit or other error management method can be inserted into the signal path.

These schemes help the system detect errors, and then either correct the errors, or at least ask for a new copy of the data. In a state-machine, the state transition logic can be designed to catch unused states and trigger a reset sequence or other error recovery routine. Embedded software designs that employ Immunity Aware Programming, such as the practice of filling unused programme memory with interrupt instructions that point to an error recovery routine. This helps guard against failures that corrupt the microcontroller's instruction pointer which could otherwise cause random code to be executed. Digital memory and transmission systems can use techniques such as error detection and correction to use additional data to correct any errors in transmission and storage. On the other hand, some techniques used in digital systems make those systems more vulnerable to single-bit errors. These techniques are acceptable when the underlying bits are reliable enough that such errors are highly unlikely.

- A single-bit error in audio data stored directly as linear pulse code modulation (such as on a CD-ROM causes, at worst, a single click). Instead, many people use audio compression to save storage space and download time, even though a single-bit error may corrupt the entire song.

ELECTRICAL ENGINEERING IN SIGNAL PROCESSING

In the fields of communications, signal processing, and in electrical engineering more generally, a signal is any time-varying or spatial-varying quantity. In the physical world, any quantity measurable through time or over space can be taken as a signal. Within a complex society, any set of human information or machine data can also be taken as a signal.

Such information or machine data (for example, the dots on a screen, the ink making up text on a paper page, or the words now flowing into the reader's mind) must all be part of systems existing in the physical world – either living or non-living. Despite the complexity and even mystery – in the case of the reader's mind – of such systems, their outputs and inputs can often be represented with great fidelity as simple quantities measurable through time or across space.

In the latter half of the 20th century, electrical engineering itself separated into several disciplines, specializing in the design and analysis of physical signals and systems, on the one hand, and in the functional behaviour and conceptual structure of the complex human and machine systems, on the other. These engineering disciplines have led the way in the design, study, and implementation of systems that take advantage of signals as simple measurable quantities in order to facilitate the transmission, storage, and manipulation of information.

SOME DEFINITIONS

Definitions specific to subfields are common. For example, in information theory, a *signal* is a codified message, that is, the sequence of states in a communication channel that encodes a message. In a *communication system*, a *transmitter* encodes a *message* into a signal, which is carried to a *receiver* by the communications *channel*. For example, the words "Mary had a little lamb" might be the message spoken into a telephone. The telephone transmitter converts the sounds into an electrical voltage signal.

The signal is transmitted to the receiving telephone by wires; and at the receiver it is reconverted into sounds. Signals can be categorized in various ways. The most common distinction is between discrete and continuous spaces that the functions are defined over, for example discrete and continuous time domains.

Discrete-time signals are often referred to as *time series* in other fields. Continuous-time signals are often referred to as *continuous signals* even when the signal functions are not continuous; an example is a square-wave signal. A second important distinction is between discrete-valued and continuous-valued. Digital signals are discrete-valued, but are often derived from an underlying continuous-valued physical process.

Discrete-time and Continuous-time Signals

If for a signal, the quantities are defined only on a discrete set of times, we call it a discrete-time signal. In other words, a discrete-time real (or complex) signal can be seen as a function from the set of integers to the set of real (or complex) numbers. A continuous-time real (or complex) signal is any real-valued (or complex-valued) function which is defined for all time t in an interval, most commonly an infinite interval.

Analog and Digital Signals

Less formally than the theoretical distinctions mentioned above, two main types of signals encountered in practice are *analog* and *digital*. In short, the difference between them is that digital signals are *discrete* and *quantized*, as defined below, while analog signals possess neither property.

DISCRETIZATION

One of the fundamental distinctions between different types of signals is between continuous and discrete time. In the mathematical abstraction, the domain of a continuous-time (CT) signal is the set of real numbers (or some interval thereof), whereas the domain of a discrete-time (DT) signal is the set of integers (or some interval). What these integers represent depends on the nature of the signal.

DT signals often arise via sampling of CT signals. An audio signal, for example consists of a continually fluxuating voltage on a line that can be digitized by an ADC circuit, wherein the circuit will read the voltage level on the line, say, every 50 μs. The resulting stream of numbers is stored as digital data on a discrete-time signal. Computers and other digital devices are restricted to discrete time.

A discrete signal or discrete-time signal is a time series, perhaps a signal that has been sampled from a continuous-time signal. Unlike a continuous-time signal, a discrete-time signal is not a function of a continuous-time argument, but is a sequence of quantities; that is, a function over a domain of discrete integers. Each value in the sequence is called a sample.

When a discrete-time signal is a sequence corresponding to uniformly spaced times, it has an associated sampling rate; the sampling rate is not apparent in the data sequence, so may be associated as a separate data item.

DIGITAL SIGNALS

A digital signal is a discrete-time signal that takes on only a discrete set of values. It typically derives from a discrete signal that has been quantized.

Common practical digital signals are represented as 8-bit (256 levels), 16-bit (65,536 levels), 32-bit (4.3 billion levels), and so on, though any number of quantization levels is possible, not just powers of two.

QUANTIZATION

If a signal is to be represented as a sequence of numbers, it is impossible to maintain arbitrarily high precision - each number in the sequence must have a finite number of digits.

As a result, the values of such a signal are restricted to belong to a finite set; in other words, it is quantized. In digital signal processing, quantization is the process of approximating a continuous range of values (or a very large set of possible discrete values) by a relatively small set of discrete symbols or integer values.

More specifically, a signal can be multi-dimensional and quantization need not be applied to all dimensions. Discrete signals (a common mathematical model) need not be quantized, which can be a point of confusion. A common use of quantization is in the conversion of a discrete signal (a sampled continuous signal) into a digital signal by quantizing. Both of these steps (sampling and quantizing) are performed in analog-to-digital converters with the quantization level specified in bits.

A specific example would be compact disc (CD) audio which is sampled at 44,100 Hz and quantized with 16 bits (2 bytes) which can be one of 65,536 (*i.e.* 2) possible values per sample. In electronics, adaptive quantization is a quantization process that varies the step size based on the changes of the input signal, as a means of efficient compression.Two approaches commonly used are forward adaptive quantization and backward adaptive quantization.

Mathematical Description

The simplest and best-known form of quantization is referred to as scalar quantization, since it operates on scalar (as opposed to multi-dimensional vector) input data. The integer-valued quantization index i is the representation that is typically stored or transmitted, and then the final interpretation is constructed using $g(i)$ when the data is later interpreted.

In computer audio and most other applications, a method known as *uniform quantization* is the most common. There are two common variations of uniform quantization, called *mid-rise* and *mid-tread* uniform quantizers.

In this case the $f(x)$ and $g(i)$ operators are just multiplying scale factors (one multiplier being the inverse of the other) along with an offset in $g(i)$ function to place the representation value in the middle of the input region for each quantization index.

The value $2^{-(M-)}$ is often referred to as the *quantization step size*. Using this quantization law and assuming that quantization noise is approximately uniformly distributed over the quantization step size (an assumption typically accurate for rapidly varying x or high M) and further assuming that the input signal x to be quantized is approximately uniformly distributed over the entire interval from -1 to 1, the signal to noise ratio (SNR) of the quantization.

From this equation, it is often said that the SNR is approximately 6 dB per bit. For mid-tread uniform quantization, the offset of 0.5 would be added within the floor function instead of outside of it. Sometimes, mid-rise quantization is used without adding the offset of 0.5. This reduces the signal to noise ratio by approximately 6.02 dB, but may be acceptable for the sake of simplicity when the step size is small.

In digital telephony, two popular quantization schemes are the 'A-law' (dominant in Europe) and 'ì-law' (dominant in North America and Japan). These schemes map discrete analog values to an 8-bit scale that is nearly linear for small values and then increases logarithmically as amplitude grows. Because the human ear's perception of loudness is roughly logarithmic, this provides a higher signal to noise ratio over the range of audible sound intensities for a given number of bits.

QUANTIZATION AND DATA COMPRESSION

Quantization plays a major part in lossy data compression. In many cases, quantization can be viewed as the fundamental element that distinguishes lossy data compression from lossless data compression, and the use of quantization is nearly always motivated by the need to reduce the amount of data needed to represent a signal. In some compression schemes, like MP3 or Vorbis, compression is also achieved by selectively discarding some data, an action that can be analyzed as a quantization process (*e.g.*, a vector quantization process) or can be considered a different kind of lossy process.

·One example of a lossy compression scheme that uses quantization is JPEG image compression. During JPEG encoding, the data representing an image (typically 8-bits for each of three colour components per pixel) is processed using a discrete cosine transform and is then quantized and entropy coded. By reducing the precision of the transformed values using quantization, the number of bits needed to represent the image can be reduced substantially. For example, images can often be represented with acceptable quality using JPEG at less than 3 bits per pixel (as opposed to the typical 24 bits per pixel needed prior to JPEG compression). Even the original representation using 24 bits per pixel requires quantization for its PCM sampling structure.

In modern compression technology, the entropy of the output of a quantizer matters more than the number of possible values of its output (the number of values being 2^M in the above example). In order to determine how many bits are necessary to effect a given precision, algorithms are used. Suppose, for example, that it is necessary to record six significant digits, that is to say, millionths. The number of values that can be expressed by N bits is equal to two to the Nth power.

To express six decimal digits, the required number of bits is determined by rounding (6/log 2)—where log refers to the base ten, or common, logarithm—up to the nearest integer. Since the logarithm of 2, base ten, is approximately 0.30102, the required number of bits is then given by (6/0.30102), or 19.932, rounded up to the nearest integer, *viz.*, 20 bits. This type of quantization—where a set of binary digits, *e.g.*, an arithmetic register in a CPU, are used to represent a quantity—is called Vernier quantization. It is also possible, although rather less efficient, to rely upon equally spaced quantization levels.

This is only practical when a small range of values is expected to be captured: for example, a set of eight possible values requires eight equally spaced quantization levels—which is not unreasonable, although obviously less efficient than a mere trio of binary digits (bits)—but a set of, say, sixty-four possible values, requiring sixty-four equally spaced quantization levels, can be expressed using only six bits, which is obviously far more efficient.

RELATION TO QUANTIZATION IN NATURE

At the most fundamental level, some physical quantities are quantized. This is a result of quantum mechanics). Signals may be treated as continuous for mathematical simplicity by considering the small quantizations as negligible.

In any practical application, this inherent quantization is irrelevant for two reasons. First, it is overshadowed by signal noise, the intrusion of extraneous phenomena present in the system upon the signal of interest.

The second, which appears only in measurement applications, is the inaccuracy of instruments. Thus, although all physical signals are intrinsically quantized, the error introduced by modeling them as continuous is vanishingly small.

Examples of Signals

- *Motion*. The motion of a particle through some space can be considered to be a signal, or can be represented by a signal. The domain of a motion signal is one-dimensional (time), and the range is generally three-dimensional. Position is thus a 3-vector signal; position and orientation is a 6-vector signal.
- *Sound*. Since a sound is a vibration of a medium (such as air), a sound signal associates a pressure value to every value of time and three space coordinates. A microphone converts sound pressure at some place to just a function of time, using a voltage signal as an analog of the sound signal.
- *Compact discs* (CDs). CDs contain discrete signals representing sound, recorded at 44,100 samples per second. Each sample contains data for a left and right channel, which may be considered to be a 2-vector (since CDs are recorded in stereo).
- *Pictures*. A picture assigns a colour value to each of a set of points. Since the points lie on a plane, the domain is two-dimensional. If the picture is a physical object, such as a painting, it's a continuous signal. If the picture is a digital image, it's a discrete signal. It's often convenient to represent colour as the sum of the intensities of three primary colours, so that the signal is vector-valued with dimension three.
- *Videos*. A video signal is a sequence of images. A point in a video is identified by its position (two-dimensional) and by the time at which it occurs, so a video signal has a three-dimensional domain. Analog video has one continuous domain dimension (across a scan line) and two discrete dimensions (frame and line).
- *Biological membrane potentials*. The value of the signal is a straightforward electric potential ("voltage"). The domain is more difficult to establish. Some cells or organelles have the same membrane potential throughout; neurons generally have different potentials at different points. These signals have very low energies, but are enough to make nervous systems work; they can be measured in aggregate by the techniques of electrophysiology.

FREQUENCY ANALYSIS

Signals are often analyzed or modeled in terms of their frequency spectrum. Frequency domain techniques are applicable to all signals, both continuous-time and discrete-time.

If a signal is passed through an LTI system, the frequency spectrum of the resulting output signal is the product of the frequency spectrum of the original input signal and the frequency response of the system.

Frequency domain is a term used to describe the analysis of mathematical functions or signals with respect to frequency. Speaking non-technically, a time-domain graph shows how a signal changes over time, whereas a frequency-domain graph shows how much of the signal lies within each given frequency band over a range of frequencies.

A frequency-domain representation can also include information on the phase shift that must be applied to each sinusoid in order to be able to recombine the frequency components to recover the original time signal. The frequency domain relates to the Fourier transform or Fourier series by decomposing a function into an infinite or finite number of frequencies.

This is based on the concept of Fourier series that any waveform can be expressed as a sum of sinusoids (sometimes infinitely many.) A spectrum analyzer is the tool commonly used to visualize real-world signals in the frequency domain.

MAGNITUDE AND PHASE

In using the Laplace, Z-, or Fourier transforms, the frequency spectrum is complex, describing the magnitude and phase of a signal, or of the response of a system, as a function of frequency. In many applications, phase information is not important. By discarding the phase information it is possible to simplify the information in a frequency domain representation to generate a frequency spectrum or spectral density.

A spectrum analyzer is a device that displays the spectrum. The power spectral density is a frequency-domain description that can be applied to a large class of signals that are neither periodic nor square-integrable; to have a power spectral density a signal needs only to be the output of a wide-sense stationary random process.

PARTIAL FREQUENCY-DOMAIN EXAMPLE

Due to popular simplifications of the hearing process and titles such as Plomp's "The Ear as a Frequency Analyzer," the inner ear is often thought of as converting time-domain sound waveforms to frequency-domain spectra. The frequency domain is not actually a very accurate or useful model for hearing, but a time/frequency space or time/place space can be a useful description.

BRIDGE CIRCUIT

A bridge circuit is a type of electrical circuit in which the current in a conductor splits into two parallel paths and then recombines into a single conductor, thereby enclosing a loop. It was originally used for measurement purposes, but can also be used in power supplies.

The best-known bridge circuit, the Wheatstone bridge, was invented by Samuel Hunter Christie and popularized by Charles Wheatstone, and is used for measuring resistance. It is constructed from four resistors, one of which has an unknown value (R_x), one of which is variable (R_2), and two of which are fixed and equal (R_1 and R_3), connected as the sides of a square. Two opposite corners of the square are connected to a source of electrical current, such as a battery. A galvanometer is connected across the other two opposite corners. The variable resistor is adjusted until the galvanometer reads zero.

It is then known that the ratio between the variable resistor and its neighbour is equal to the ratio between the unknown resistor and its neighbour, and this enables the value of the unknown resistor to be calculated. The Wheatstone bridge has also been developed to measure impedance in AC circuits, resulting in designs such as the Wien bridge, the Maxwell bridge and the Heaviside bridge. All are based on the same principle, which is to compare the output of two potentiometers sharing a common source.

In power supply design, a bridge circuit or bridge rectifier is an arrangement of diodes or similar devices used to rectify an electric current, *i.e.* to convert it from an unknown or alternating polarity to a direct current of known polarity. In some motor controllers, a H-bridge is used to control the direction the motor turns.

8

Alternating Currents and Direct Current

ALTERNATING CURRENTS

An alternating current (AC) is an electric current whose direction reverses cyclically, as opposed to direct current, whose direction remains constant. The usual waveform of an AC power circuit is a sine wave, as this results in the most efficient transmission of energy. However in certain applications different waveforms are used, such as triangular or square waves.

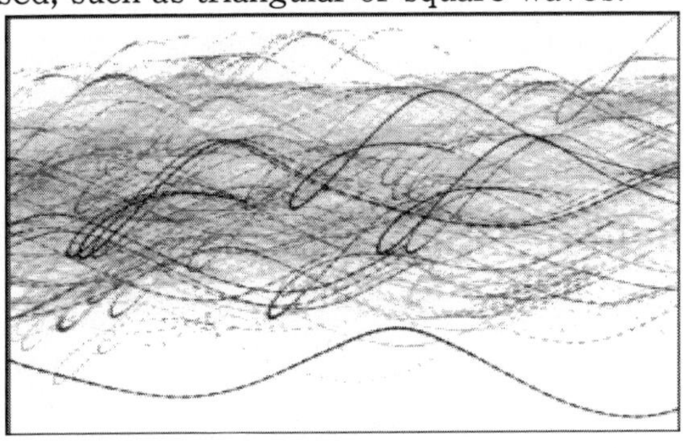

Fig. City lights viewed in a motion blurred exposure. The AC blinking causes the lines to be dotted rather than continuous.

Used generically, *AC* refers to the form in which electricity is delivered to businesses and residences. However, audio and radio signals carried on electrical wires are also examples of alternating current. In these applications, an important goal is often the recovery of information encoded (or modulated) onto the AC signal.

William Stanley, Jr. designed one of the first practical devices to transfer AC power efficiently between isolated circuits. Using pairs of coils wound on a common iron core, his design, called an induction coil, was an early transformer. The AC power system used today developed rapidly after 1886, and includes key concepts by Nikola Tesla, who subsequently sold his patent to George Westinghouse. Lucien Gaulard, John Dixon Gibbs, Carl Wilhelm Siemens and others contributed subsequently to this field.

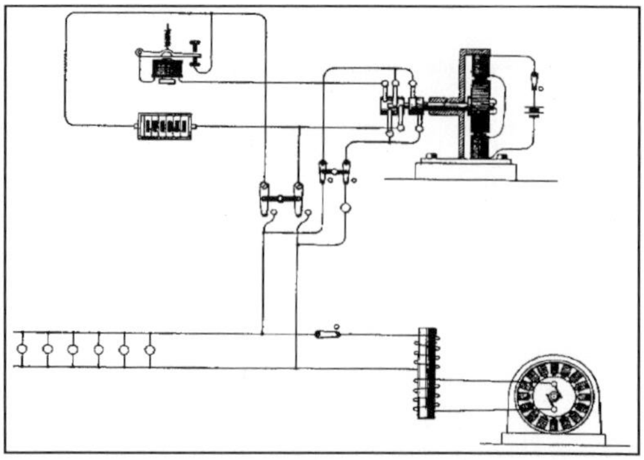

Fig. Westinghouse Early AC System 1887.

AC systems overcame the limitations of the direct current system used by Thomas Edison to distribute electricity efficiently over long distances even though Edison attempted to discredit alternating current as too dangerous during the War of Currents. The first commercial power plant using three-phase alternating current was at the Mill Creek hydroelectric plant near Redlands, California in 1893 designed by Almirian Decker. Decker's design incorporated 10,000 volt three-phase transmission and established the standards for the complete system of generation, transmission and motors used today.

Alternating current circuit theory evolved rapidly in the latter part of the 19th and early 20th century. Notable contributors to the theoretical basis of alternating current calculations include Charles Steinmetz, James Clerk Maxwell, Oliver Heaviside, and many others. Calculations in unbalanced three-phase systems were simplified by the symmetrical components methods discussed by Charles Legeyt Fortescue in 1918.

AC POWER SUPPLY FREQUENCIES

The frequency of the electrical system varies by country; most electric power is generated at either 50 or 60 Hz. Some countries have a mixture of 50 Hz and 60 Hz supplies, notably Japan.

A low frequency eases the design of low speed electric motors, particularly for hoisting, crushing and rolling applications, and commutator-type traction motors for applications such as railways, but also causes a noticeable flicker in incandescent lighting and objectionable flicker of fluorescent lamps. $16\frac{2}{3}$ Hz power is still used in some European rail systems, such as in Austria, Germany, Norway, Sweden and Switzerland.

The use of lower frequencies also provided the advantage of lower impedance losses, which are proportional to frequency. The original Niagara Falls generators were built to produce 25 Hz power, as a compromise between low frequency for traction and heavy induction motors, while still allowing incandescent lighting to operate (although with noticeable flicker); most of the

25 Hz residential and commercial customers for Niagara Falls power were converted to 60 Hz by the late 1950s, although some 25 Hz industrial customers still existed as of the start of the 21st century. Off-shore, military, textile industry, marine, computer mainframe, aircraft, and spacecraft applications sometimes use 400 Hz, for benefits of reduced weight of apparatus or higher motor speeds.s

EFFECTS AT HIGH FREQUENCIES

A direct, constant current flows uniformly throughout the cross-section of the (uniform) wire that carries it. With alternating current of any frequency, the current is forced towards the outer surface of the wire, and away from the centre. This is because an electric charge which accelerates (as is the case of an alternating current) radiates electromagnetic waves, and materials of high conductivity (the metal which makes up the wire) do not allow propagation of electromagnetic waves. This phenomenon is called skin effect.

At very high frequencies the current no longer flows *in* the wire, but effectively flows *on* the surface of the wire, within a thickness of a few skin depths. The skin depth is the thickness at which the current density is reduced by 63 per cent.

Even at relatively low frequencies used for high power transmission (50–60 Hz), non-uniform distribution of current still occurs in sufficiently thick conductors. For example, the skin depth of a copper conductor is approximately 8.57 mm at 60 Hz, so high current conductors are usually hollow to reduce their mass and cost.

Since the current tends to flow in the periphery of conductors, the effective cross-section of the conductor is reduced. This increases the effective *AC* resistance of the conductor, since resistance is inversely proportional to the cross-sectional area in which the current actually flows. The AC resistance often is many times higher than the *DC* resistance, causing a much higher energy loss due to ohmic heating (also called IR loss).

Techniques for Reducing AC Resistance

For low to medium frequencies, conductors can be divided into stranded wires, each insulated from one other, and the individual strands specially arranged to change their relative position within the conductor bundle. Wire constructed using this technique is called Litz wire.

This measure helps to partially mitigate skin effect by forcing more equal current flow throughout the total cross section of the stranded conductors. Litz wire is used for making high Q inductors, reducing losses in flexible conductors carrying very high currents at power frequencies, and in the windings of devices carrying higher radio frequency current (up to hundreds of kilohertz), such as switch-mode power supplies and radio frequency transformers.

Techniques for Reducing Radiation Loss

As written above, an alternating current is made of electric charge under periodic acceleration, which causes radiation of electromagnetic waves. Energy that is radiated represents a loss. Depending on the frequency, different techniques are used to minimize the loss due to radiation.

Twisted Pairs

At frequencies up to about 1 GHz, wires are paired together in cabling to form a twisted pair in order to reduce losses due to electromagnetic radiation and inductive coupling. A twisted pair must be used with a balanced signalling system, where the two wires carry equal but opposite currents. The result is that each wire in the twisted pair radiates a signa that is effectively cancelled by the other wire, resulting in almost no electromagnetic radiation.

Coaxial Cables

At frequencies above 1 GHz, unshielded wires of practical dimensions lose too much energy to radiation, so coaxial cables are used instead. A coaxial cable has a conductive wire inside a conductive tube. The current flowing on the inner conductor is equal and opposite to the current flowing on the inner surface of the outer tube. This causes the electromagnetic field to be completely contained within the tube, and (ideally) no energy is radiated or coupled outside the tube.

Coaxial cables have acceptably small losses for frequencies up to about 20 GHz. For microwave frequencies greater than 20 GHz, the dielectric losses (due mainly to the dissipation factor of the dielectric layer which separates the inner wire from the outer tube) become too large, making waveguides a more efficient medium for transmitting energy.

WAVEGUIDES

Waveguides are similar to coax cables, as both consist of tubes, with the biggest difference being that the waveguide has no inner conductor. Waveguides can have any arbitrary cross section, but rectangular cross sections are the most common. Because waveguides do not have an inner conductor to carry a return current, waveguides cannot deliver energy by means of an electric current, but rather by means of a *guided* electromagnetic field.

Although surface currents do flow on the inner walls of the waveguides, those surface currents do not carry power. Power is carried by the guided electromagnetic fields. The surface currents are set up by the guided electromagnetic fields and have the effect of keeping the fields inside the waveguide and preventing leakage of the fields to the space outside the waveguide.

Waveguides have dimensions comparable to the wavelength of the alternating current to be transmitted, so they are only feasible at microwave frequencies. In addition to this mechanical feasibility, electrical resistance of the non-ideal metals forming the walls of the waveguide cause dissipation of power (surface currents flowing on lossy conductors dissipate power). At higher frequencies, the power lost to this dissipation becomes unacceptably large.

Fibre Optics

At frequencies greater than 200 GHz, waveguide dimensions become impractically small, and the ohmic losses in the waveguide walls become large. Instead, fibre optics, which are a form of dielectric waveguides, can be used. For such frequencies, the concepts of voltages and currents are no longer used.

The existence of two types of electric current had been known since the dawn of the electrical age. Hippolyte Pixii's generator of 1832 had had a rotating magnetic field that cut the conductors first in one direction and then in the other. This produced a series of interrupted electrical pulses as the magnets swung back and forth; the flow of electricity reached a maximum in one direction, then decreased, moved towards a maximum in the other direction, and reversed again to begin a new cycle.

Such an "alternating" current did not seem useful to the early experimenters. Ampère had suggested a simple device called a commutator, which converted the oscillating or alternating current into a non-pulsating "direct" current, and commutators became standard features on the dynamos of the are light era. Edison's first generating machines were direct-current dynamos, but his early commitment to d.c. was accidental, to a large extent. For the purpose of providing incandescent lighting, it did not matter whether Edison used d.c. or a.c.

The dynamo technology of 1878 was almost exclusively a d.c. technology, and so he chose to make no departure from it. Before long, though, he was irrevocably wedded to the d.c. concept as an essential part of his system. What was important to Edison as he created his system was not the type of current he used, but the voltage. The Edison system was based on power that traveled at 110 volts all the way from the generating station to the consumer. The great advantage of this was its safety.

Current at 110 volts and at the low amperage sufficient for electric lighting could do no harm; a customer who came in contact with an unshielded line would receive a mild shock, no more. The danger of fire was low, too. Sensitive to the problems inherent in getting a new kind of illumination across to a suspicious public, Edison wished to avoid any possibility of spectacular electrical disasters, so he opted for the safest form of distribution.

However, low voltages had a built-in drawback: they limited the distance of transmission. Current moving with such a modest force could be carried no more than a mile or two from the generating station; then the drop in voltage became so great that the outlying lights dimmed. In addition, much of the power would be dissipated in the form of heat. Under Edison's low voltage transmission system, Edison's own great concept of the central station was seriously impaired, for 110-volt service required a chain of power stations at close intervals in order to supply any large city with electricity. At best an Edison station could serve some sixteen square miles.

Ideally, the solution was to have power leave the generating station at a high voltage, and then to step it down to a safe 110 volts before it reached the customer. But this was technically impossible on the Edison system. The young Edison would have looked for some way to *make* it possible; the middle-aged Edison simply accepted the situation as a regrettable innate flaw in his design, and left it at that.

But there was a way—with alternating current.

The germ of the long-distance a.c. transmission system lay in the original work of Faraday. His first successful experiment had made use of two coils wound on an iron core; by stopping or starting a current in one coil, he induced a current in the other. Although he did not realize it, the basis of the transformer, or voltage-changer, could be found in this experiment.

Seven years later, in 1838, the American Joseph Henry used Faraday's two-coil arrangement as a step-down transformer. One coil consisted of many turns of wire, the other of only a few; high-voltage power entered the first coil, and was stepped down to a low voltage in the second.

The transformer received little further attention until 1882, when Lucien Gaulard of France and John D. Gibbs of England produced a transformer specifically designed to reduce the voltage of a.c. power en route from the generator to the consumer. Gaulard and Gibbs obtained an English patent on their work in 1883, but in practice it had serious mechanical defects and could not be used without modification.

The modifications were quick in coming. In 1885 three Hungarian inventors found a way to employ the Gaulard-Gibbs transformer in long-distance power installations. Their device, called the "Z.B.D. transformer" after the initials of its inventors, went into immediate use in Europe. It permitted power first to be boosted to very high voltage as it left the generator, then to be stepped down again as it reached the consumer—with remarkably little energy loss over great distances. The higher the voltage used in transmission, the less the loss on the line.

Rights to the Z.B.D. system were offered to Edison in 1886. He sent Upton to Europe to inspect the equipment, and at Upton's suggestion Edison paid $5000 for a three-year option, against a total price of $20,000. But Edison could not bring himself to veer from his low-voltage d.c. transmission system. In November, 1886 he received a report from a Berlin engineering firm to the effect that the Z.B.D. system was troublesome, costly to install, and dangerous to work with. That was all Edison needed to hear. He allowed the option to lapse, and angrily rejected the suggestions of his associates that he give the Hun garian transformer a further trial.

Edison's adherence to his own original system had become something deeply personal, a matter for the emotions, not the reason. He clung to his 110-volt concept despite all advice to the contrary. Yet younger men all over Europe and America were excited by the new development. High-voltage transmission suddenly was the technological darling of the moment. Of course, it would require a changeover from d.c. to a.c. power. Since transformers functioned only with a changing current, the output of d.c. generators could be neither stepped up nor stepped down with transformers. High voltages required the use of a.c. equipment.

And that produced certain new problems. Alternating current could not be used for driving electric motors; it seemed to demand two or three times as much coal per kilowatt hour as direct current; and it rendered obsolete most of the existing capital plant of the power industry. Nevertheless, the forward looking men of that industry were confident that someone would invent a feasible a.c. electric motor and that the generating cost of a.c. could be made comparable to that of d.c. As for the obsolescence of existing equipment, that would cease to matter in time. Alternating current was sure to win universal acceptance.

Edison—nearly alone—fought the new trend.

His option on the Z.B.D. patents gave him the exclusive right to use the new transformer in the United States until 1889. Since that was the case, rival manufacturers had no choice but to go back to the defective Gaulard-Gibbs device and attempt to modify it themselves in a way that circumvented the Hungarians. The most eager of these rivals was George Westinghouse of Pittsburgh, who had gone into the electric business in 1885.

Westinghouse, one year Edison's senior, had founded his industrial fortunes on his invention of the air brake for railways. Bold, ambitious, even reckless, he dreamed of an electrical empire to rival Edison's, and by liberal use of cash had attracted some of the best engineers of the day away from the ever more conservative Edison organization.

The new Westinghouse Electrical and Manufacturing Company sought to steal a march on Edison by getting control of that segment of the electrical field that Edison scorned: high-voltage alternating current equipment of 1,000 volts or more. For that Westinghouse needed a practicable transformer.

Westinghouse bought the American rights to the Gaulard Gibbs transformer and assigned the job of making it work properly to a young engineering consultant, William A. Stanley. The flaw in the Gaulard-Gibbs system was that it employed a number of transformers to be connected in series, so that they could not work independently. Stanley found a patentable way of connecting transformers in parallel and making each transformer automatically regulate itself, thus permitting independent control.

On March 6, 1886, the first commercial alternating current lighting system went into operation in Great Barrington, Massachusetts, using the Stanley transformers. At Great Barrington Stanley generated power at 500 volts just outside town, stepped it up to 3000 volts for transmission, then lowered it to 500 volts again for use in lighting. The transmission distance was only 4000 feet, but the principle was established; Westinghouse saw no reason why the Stanley system could not be used to send power cheaply for distances of many miles.

Westinghouse, in 1886, also purchased the United States Electric Lighting Company, which had control of the contested Maxim incandescent lamp patents. Now he proposed to compete in both spheres of the electrical industry. Through Westinghouse Electric he would manufacture high-voltage equipment; through United States Electric he would set up power companies that, by taking advantage of the economies of long distance transmission, would outstrip the Edison companies. Where he could, Westinghouse would rely on the non-Edison technology of a.c. Where he could not, he intended to invade the Edison patents, whose validity was under attack in the courts. In an advertisement exceptional for its blandly self congratulatory tone Westinghouse declared:

We regard it as fortunate that we have deferred entering the electrical field until the present moment. Having thus profited by the public experience of others, we enter ourselves for competition, hampered by a minimum of expense for experimental outlay.... In short, our organization is free, in large measure, of the load with which [other] electrical enterprises seem to be encumbered. The fruit of this... we propose to share with the customer.

That is, having allowed Edison to bear the burden of developing the electrical industry, Westinghouse planned to undersell him by making use of his pioneering work. In March of 1887 Westinghouse began to move into New York City, Edison's home grounds, with an a.c. incandescent lighting system. He organized the Safety Electric Light and Power Company and licensed it to use the Westinghouse system of generation and distribution of electricity by alternating current.

Safety Electric began to buy into an existing are light company, United States Illuminating, which held several important street and park contracts. Soon Safety Electric controlled United States Illuminating; it changed its name in 1889 to United Electric Light and Power Company and began to spread the Westinghouse system into those parts of Manhattan that Edison Illuminating had not yet reached.

Naturally the older company fought back. In 1888 it denounced the Westinghouse system in a pamphlet that declared:

- They cannot make it safe.
- They cannot make it reliable.
- They cannot make it run twelve 16-candlepower lamps per horsepower.

- They cannot make its lamps even.
- They cannot make its lamps last a reasonable time.
- They cannot make it sell by meter.
- They cannot make it run motors.

Some of these charges were true at the time; others, particularly the accusation of unsafeness, were not. But the men who had made such a deep financial and emotional investment in d.c. could not abide the thought that their system was obsolescent.

Edison himself, though no longer directly concerned with the fortunes of the Edison Illuminating Company, was seriously perturbed by the spread of a.c. It was an attack on his technical judgement, he felt; he was worried, also, about Westinghouse's talk of using power at 5,000 or even 10,000 volts. Contemporary insulating techniques did not seem adequate for such voltages, Edison felt. To Edward Johnson he wrote:

Just as certain as death Westinghouse will kill a customer within 6 months after he puts in a system of any size. He has got a new thing and it will require a great deal of experimenting to get it working practically. It will never be free from danger....

None of his plans worry me in the least; only thing that disturbs me is that Westinghouse is a great man for flooding the country with agents and travelers. He is ubiquitous and will form numerous companies before we know anything about it.

Edison's earlier rival, Thomson-Houston Electric, was also toying with a.c. Elihu Thomson had constructed a rudimentary transformer in 1879, but he did nothing with it then because, like everyone else at the time, he was building d.c. dynamos. After the success of the three Hungarians provoked interest in a.c., Thomson hastily filed an application in November of 1885 for a patent on his transformer. But several other groups, among them the Gaulard-Gibbs team, contested the application and Thomson's patent was not granted for nearly twenty years, only to be thrown out anyway on its first court test.

Patent or not, Thomson saw the advantages of a long-distance a.c. transmission system, and continued his research. By the spring of 1887 Thomson-Houston was manufacturing a.c. dynamos—"alternators," as they were called—and transformers. However, Thomson was perturbed by the supposed danger of bringing high-voltage lines to the consumers, and so Thomson Houston held back from any really enthusiastic promotion of a.c. for several years, leaving a clear field to Westinghouse while it worked on protective devices.

The problem of the electric motor also hindered the spread of a.c. in 1887. All existing motors worked on d.c., and d.c. alone; no one yet knew a way of adapting them for a.c. use. Since the electrical industry was coming to depend for its revenues more and more on industrial power use, less and less on lighting, a practical a.c. motor was essential if the new system were to last. Elihu Thomson built one of the first a.c. motors in May of 1887. It had a laminated magnet consisting of many disks of metal pressed together, and a laminated armature with three coils. The Thomson motor attracted considerable attention when it was displayed before a scientific group, but it did not go into production at once, and the aggressive Westinghouse gained a leading position with the a.c. motor of Nikola Tesla.

Tesla, a towering, mystical genius out of Yugoslavia, had once been part of the Edison organization but had slipped away. As an electrical inventor he was extraordinary, very likely superior in vision even to Edison. But he was a romantic, brooding figure of little practical awareness who could never have become a permanent part of the Edison research team. There was room in Edison's entourage for only one genius at a time; besides, Tesla believed passionately in the alternating current.

In 1882, in Budapest, Tesla was seized with a vision of an alternating-current motor: a rotating magnetic field that carried the armature around with it. The Tesla motor employed two alternating currents at once; the currents were equal in the frequency of their cycle of minimum-to-maximum amplitude, but were out of phase relative to each other. With one current or the other always at a maximum, the motor worked continuously. It was a brilliant idea, a landmark in engineering. But in 1882 there was no a.c. power source to drive Tesla's motor.

He drifted on to Paris soon after and took a job with the new Société Continentale Edison, the French Edison company. There he dutifully worked on d.c. lighting systems while inwardly pursuing his own dream. Early in 1883, sent to Strasbourg as a troubleshooter, he collected some spare parts and built the first working model of his motor, and a small dynamo to drive it. Everything worked as he had envisioned it. Tesla was then twenty-seven.

Calling some wealthy industrialists of Strasbourg together, he tried to interest them in forming a company to back him, for he knew that the Edison company had no desire even to hear about an a.c. system. But the Strasbourgers could not understand what Tesla was telling them, and he returned to Paris without the capital he had hoped to raise.

There he came to the attention of Charles Batchelor in the spring of 1884. Batchelor, sensing Tesla's brilliance, invited him to go to the United States and work directly with Edison. Naively confident that he could convert Edison to his own system, Tesla sank his savings into a steamship passage to New York. He arrived with four cents in cash, the clothes on his back, and the plans for his a.c. motor; he also carried a letter of introduction from Batchelor to Edison, saying, "I know two great men and you are one of them; the other is this young man."

Tesla went to Edison's Fifth Avenue office, presented his letter, and was treated to a discourse on the problems of Pearl Street. Edison would hear no talk of alternating current; he put Tesla to work at once on routine electrical work. When Tesla solved a knotty problem with unexpected cleverness, Edison saw his merit and gave him harder assignments—without, however, raising his pay beyond that of a beginning technician. Feeling underpaid, and angered by Edison's refusal to consider his theories, Tesla came to the inevitable collision with his employer and resigned after a year. From the spring of 1886 to the spring of 1887 he supported himself as a day laborer, digging ditches and taking on occasional electrical jobs.

Then he found backers, and set up the Tesla Electric Company. Though their financial support was modest, Tesla was able to carry on his research and to apply for patents. Between November and December, 1887, Tesla filed for seven patents on his motor and on an electrical distribution system; so radically different were they from anything in existence that the applications were granted in only six months. Tesla was invited to explain his ideas at a meeting of the American Institute of Electrical Engineers on May 16, 1888. He had "arrived."

Now he held patents on a revolutionary power system. His high-voltage lines could carry power hundreds of miles with relatively little energy loss or voltage drop. His motors were far more efficient than the small d.c. motors then in use, and could handle much greater loads. But the Tesla Electric Company lacked the capital to make use of the patents. Inevitably, Tesla was summoned to meet George Westinghouse. Westinghouse offered the inventor a million dollars in cash for his patents, plus a generous royalty arrangement, and hired Tesla as a consultant.

Though the unpredictable Tesla soon quarreled with Westinghouse and returned to private research, his patents had passed into the manufacturer's control. Now it was Westinghouse, and not Edison, who held the technological advantage. Westinghouse could make use of the newest and most exciting ideas in the field; Edison, relegated to the role of grim defender of the past, was forced to rely on the great work of 1878-83 at a time when that work had passed into obsolescence. As often happens when a great man must occupy an untenable position, Edison grew more convinced that he was right each time Westinghouse added new proof that he was wrong. Thus began the famed "Battle of the Currents."

Edison set out to prove that a.c. was a menace to the public. He rigged an a.c. generator at West Orange that supplied current at 1,000 volts, and before invited audiences of reporters and other guests staged grisly demonstrations in which hapless dogs and cats were nudged out onto wired sheets of tin and electrocuted. The stray animals of upper New Jersey perished at a ghastly rate in this gruesome promotional campaign.

Then, in February of 1888, Edward Johnson issued a manifesto titled, "A Warning from the Edison Electric Light Company." It summarized the results of the animal executions and provided a long list of the fatalities caused by is lights, which used high voltages. The pamphlet sideswiped Westinghouse and Thomson-Houston by branding them as "patent pirates" who were out to gain quick wealth by introducing a newfangled and risky kind of electricity to the American home.

Westinghouse issued relatively restrained denials, pointing out that the mishaps of arc light companies in 1880 had nothing to do with his own incandescent system of 1888, and that in any event high voltages would be used only in transmission lines; current entering the home would be stepped down to a voltage no greater than that favoured by Edison.

Thomson-Houston stayed out of the battle. After its initial delay it had gone into full production of a.c. equipment, which Elihu Thomson at last regarded as perfectly safe. ThomsonHouston was selling hundreds of alternators and thousands of transformers now. As Elihu Thomson pointed out, a.c. was actually less deadly than d.c., given the same voltage and amperage.

The only real danger in using a.c. lay in the possibility that a short-circuit in the transformer might allow a high voltage to emerge from the secondary, or step-down coil; but Thomson found a means of protecting against this danger, and had no further fears of trouble. Nevertheless, Charles Coffin, the shrewd head of Thomson-Houston, preferred to let Westinghouse and Edison battle it out in the headlines while his company went ahead with steady output of the new equipment.

Edison's promotional campaign—sparked by the ingenuity of Samuel Insull—swung into a high level in the autumn of 1888. Edison lobbyists asked state legislatures for laws limiting electric circuits to 800 volts, and nearly succeeded in Ohio and Virginia. In New York, Edison took a different and much more bizarre approach: he persuaded the legislature to legalize the electric chair for the execution of condemned criminals—and saw to it that the fatal chair used Westinghouse equipment! What better way to dramatize the lethal nature of alternating current?

Harold P. Brown, a former laboratory assistant at West Orange, served as Edison's chief agent in this enterprise. Brown went before the legislators to preach the virtues of electrical execution—"instantaneous, painless, and humane"—and pushed through a bill, late in 1888, providing for the adoption of the electric chair in place of hanging. The New York State Legislature appointed Brown as a consultant to develop such a chair. With much fanfare he purchased three Westinghouse alternators and put on a display of their potency on March 2, 1889. The victims included several large dogs, four calves, and an elderly horse.

Ten seconds at 800 volts finished off the dogs; the calves were bigger, and had greater resistance, but they died in fifteen seconds. The horse required twenty-five seconds at 1000 volts. For human beings, Brown recommended

2000 volts for rapid and painless dispatch. A human being was available for this research: one William Kemmler, a condemned murderer. The state authorities were eager to try Mr. Edison's death chair, and so Kemmler was sentenced "to suffer death by electricity at Auburn Prison within the week beginning Monday, June 24, 1889."

Kemmler seemed pleased at the attention he would get and the place in history that would be his, but his lawyers objected. Hearings followed; objections were filed. Twice the prisoner was reprieved while more experiments were performed. At one hearing Edison was called to the witness stand to give his evidence. He was interrogated by Deputy Attorney General Poste, who asked, "What is your calling or profession?"

"Inventor," Edison replied.

"Have you devoted a great deal of attention to the subject of electricity?"

"Yes."

"How long have you been engaged in the work of an inventor or electrician?"

"Twenty-six years."

Edison was asked to explain the difference between continuous (direct) and alternating current. "A continuous current," he said, "is one that flows like water through a pipe. An alternating current is the same as if a body of water were allowed to flow through the pipe in one direction for a given time and then its direction reversed for a given time."

Had he measured the electrical resistance of human beings? Yes, he said, he had. Measurements of 250 people showed an average resistance of 1000 ohms, the highest being 1800, the lowest 600.

"In your judgment," Poste asked, "can an artificial electric current be generated and applied in such a manner as to produce death in human beings in every case?"

"Yes."

"Instantly?"

"Yes."

W. Bourke Cockran, one of Kemmler's lawyers, now crossexamined the inventor. Cockran was troubled by the point that different men had different resistances. What if Kemmler did not die at once, but met some hideous and lingering doom? "What would be the effect of the current on Kemmler in case the current was applied for five or six minutes?" Cockran asked. "Would he not be carbonized?"

"No," Edison replied. "He would be mummified. All the water in his body would evaporate."

Edison expressed the belief that 1000 volts of one-ampere current would be ten times as much as was needed to kill any man with the Westinghouse alternator.

"That is your belief, not from knowledge?" Cockran asked.

"From belief," said Edison quietly. "I never killed anybody."

After lengthy debate, Kemmler's execution date was set for August 6, 1890. The warden of Auburn Prison was empowered to invite twenty-one witnesses. Most of the guests selected were scientific men, among them Edison—who, however, did not care to be present at Kemmler's death.

An ordinary commercial Westinghouse machine capable of producing a current at 1500 volts was used. It was driven by a steam engine in the prison basement; the power lines were run out of a window of the dynamo room to the roof of the jail and along the roof to the death chamber. A current varying in force from 800 to 1300 volts gave Kemmler his place in the annals of penology; the witnesses unofficially reported that the death had been slower and less pleasant than Edison had predicted.

None the less, electrocution became the rule in New York State, and Edison, at some loss in dignity, managed to frighten many Americans into thinking that a.c. was a public menace. Some of his own men, particularly Frank Sprague, pleaded with him to halt his campaign, but Edison had gone too far now to back down. All his prestige was pledged to direct current. Out of bitterness and testiness he had elevated his original shortsightedness to the status of an inflexible policy; it was the worst mistake of his career.

Through 1888 and 1889 it became increasingly apparent to most men in the field that the future of electricity lay with a.c. By the time William Kemmler went to his death in the summer of 1890, that fact seemed certain to virtually everyone but Edison. Among the converts to the new creed were the executives of Edison General Electric, although they did not dare suggest to Edison that he had outlived his usefulness in electrical engineering. A series of complicated developments grew out of Edison's vested interest in the outmoded system of electrical generation and transmission, and when the dust settled, Edison had been quietly shown to the exit.

ELECTRICITY DISTRIBUTION

Electricity distribution is the penultimate stage in the delivery (before retail) of electricity to end users. It is generally considered to include medium-voltage (less than 50 kV) power lines, electrical substations and pole-mounted transformers, low-voltage (less than 1000 V) distribution wiring and sometimes electricity meters.

In the early days of electricity distribution, direct current DC generators were connected to loads at the same voltage. The generation, transmission and loads had to be of the same voltage because there was no way of changing DC voltage levels, other than inefficient motor-generator sets. Low DC voltages were used (on the order of 100 volts) since that was a practical voltage for incandescent lamps, which were then the primary electrical load. The low voltage also required less insulation to be safely distributed within buildings.

The losses in a cable are proportional to the square of the current, the length of the cable, and the resistivity of the material, and are inversely proportional to cross-sectional area. Early transmission networks were already using copper, which is one of the best economically feasible conductors for this application. To reduce the current and copper required for a given quantity of power transmitted would require a higher transmission voltage, but no convenient efficient method existed to change the voltage level of DC power circuits. To keep losses to an economically practical level the Edison DC system needed thick cables and local generators. Early DC generating plants needed to be within about 1.5 miles (2.4 km) of the farthest customer to avoid the need for excessively large and expensive conductors.

The adoption of alternating current (AC) for electricity generation following the War of Currents dramatically changed the situation. Power transformers, installed at power stations, could be used to raise the voltage from the generators, and transformers at local substations reduced it to supply loads. Increasing the voltage reduced the current in the transmission and distribution lines and hence the size of conductors required and distribution losses incurred. This made it more economical to distribute power over long distances. Generators (such as hydroelectric sites) could be located far from the loads.

In North America, early distribution systems used a voltage of 2200 volts corner-grounded delta. Over time, this was gradually increased to 2400 volts. As cities grew, most 2400 volt systems were upgraded to 2400/4160 volt, three-phase systems. In three phase networks that permit connections between phase and neutral, both the phase-to-phase voltage (4160, in this example) and the phase-to-neutral voltage are given; if only one value is shown, the network does not serve single-phase loads connected phase-to-neutral. Some city and suburban distribution systems continue to use this range of voltages, but most have been converted to 7200/12470Y, 7620/13200Y, 14400/24940Y, and 19920/34500Y.

European systems used 3300 volts to ground, in support of the 220/380Y volt power systems used in those countries. In the UK, urban systems progressed to 6.6 kV and then 11 kV (phase to phase), the most common distribution voltage. North American and European power distribution systems also differ in that North American systems tend to have a greater number of low-voltage, step-down transformers located close to customers' premises. For example, in the US a pole-mounted transformer in a suburban setting may supply 1-3 houses, whereas in the UK a typical urban or suburban low-voltage substation would normally be rated between 315kVA and 1000kVA (1MVA) and supply a whole neighbourhood.

This is because the higher voltage used in Europe (415V vs 230V) may be carried over a greater distance with acceptable power loss. An advantage of the North American setup is that failure or maintenance on a single transformer will only affect a few customers. Advantages of the UK setup are that the transformers may be fewer, larger and more efficient, and due to diversity there need be less spare capacity in the transformers, reducing power wastage. In North American city areas with many customers per unit area, network distribution will be used, with multiple transformers and low-voltage busses interconnected over several city blocks.

Rural Electrification systems, in contrast to urban systems, tend to use higher voltages because of the longer distances covered by those distribution lines. 7200, 12470, 25000, and 34500 volt distribution is common in the United States; 11 kV and 33 kV are common in the UK, New Zealand and Australia; 11 kV and 22 kV are common in South Africa. Other voltages are occasionally used. In New Zealand, Australia, Saskatchewan, Canada, and South Africa, single wire earth return systems (SWER) are used to electrify remote rural areas.

While power electronics now allow for conversion between DC voltage levels, AC is still used in distribution due to the economy, efficiency and reliability of transformers. High-voltage DC is used for transmission of large blocks of power over long distances, or for interconnecting adjacent AC networks, but not for distribution to customers.

Distribution Network Configurations

Distribution networks are typically of two types, radial or interconnected. A radial network leaves the station and passes through the network area with no normal connection to any other supply. This is typical of long rural lines with isolated load areas. An interconnected network is generally found in more urban areas and will have multiple connections to other points of supply.

These points of connection are normally open but allow various configurations by the operating utility by closing and opening switches. Operation of these switches may be by remote control from a control centre or by a lineman. The benefit of the interconnected model is that in the event of a fault or required maintenance a small area of network can be isolated and the remainder kept on supply.

Within these networks there may be a mix of overhead line construction utilizing traditional utility poles and wires and, increasingly, underground construction with cables and indoor or cabinet substations. However, underground distribution is significantly more expensive than overhead construction. In part to reduce this cost, underground power lines are sometimes co-located with other utility lines in what are called Common utility ducts. Distribution feeders emanating from a substation are generally controlled by a circuit breaker which will open when a fault is detected. Automatic Circuit Reclosers may be installed to further segregate the feeder thus minimizing the impact of faults.

Long feeders experience voltage drop requiring capacitors or voltage regulators to be installed. Characteristics of the supply given to customers are generally mandated by contract between the supplier and customer. Variables of the supply include:

- AC or DC - Virtually all public electricity supplies are AC today. Users of large amounts of DC power such as some electric railways, telephone exchanges and industrial processes such as aluminium smelting usually either operate their own or have adjacent dedicated generating equipment, or use rectifiers to derive DC from the public AC supply
- Voltage, including tolerance (usually +10 or -15 percentage)
- Frequency, commonly 50 and 60 Hz, 16-2/3 Hz for some railways and, in a few older industrial and mining locations, 25 Hz.
- Phase configuration (single phase, polyphase including two phase and three phase)
- Maximum demand (usually measured as the largest amount of power delivered within a 15 or 30 minute period during a billing period)
- Load Factor, expressed as a ratio of average load to peak load over a period of time. Load factor indicates the degree of effective utilization of equipment (and capital investment) of distribution line or system.
- Power factor of connected load
- Earthing arrangements - TT, TN-S, TN-C-S or TN-C
- Maximum prospective short circuit current
- Maximum level and frequency of occurrence of transients

Modern Distribution Systems

The modern distribution system begins as the primary circuit leaves the sub-station and ends as the secondary service enters the customer's meter socket. A variety of methods, materials, and equipment are used among the various utility companies, but the end result is similar. First, the energy leaves the sub-station in a primary circuit, usually with all three phases.

The most common type of primary is known as a wye configuration (so named because of the shape of a "Y".) The wye configuration includes 3 phases (represented by the three outer parts of the "Y") and a neutral (represented by the centre of the "Y".) The neutral is grounded both at the substation and at every power pole. In a typical 12470Y/7200 volt system, the pole mount transformer's primary winding is rated for 7200 volts and is connected across one phase of power and the neutral.

The primary and secondary (low voltage) neutrals are bonded (connected) together to provide a path to blow the primary fuse if any fault occurs that allows primary voltage to enter the secondary lines. An example of this type of fault would be a primary phase falling across the secondary lines. Another example would be some type of fault in the transformer itself. The other type of primary configuration is known as delta. This method is older and less common. Delta is so named because of the shape of the Greek letter delta, a triangle. Delta has only 3 phases and no neutral. In delta there is only a single voltage, between two phases (phase to phase), while in wye there are two voltages, between two phases and between a phase and neutral (phase to neutral). Wye primary is safer because if one phase becomes grounded, that is, makes connection to the ground through a person, tree, or other object, it should trip out the fused cutout similar to a household circuit breaker tripping.

In delta, if a phase makes connection to ground it will continue to function normally. It takes two or three phases to make connection to ground before the fused cutouts will open the circuit. The voltage for this configuration is usually 4800 volts. Transformers are sometimes used to step down from 7200 or 7600 volts to 4800 volts or to step up from 4800 volts to 7200 or 7600 volts. When the voltage is stepped up, a neutral is created by bonding one leg of the 7200/7600 side to ground.

This is commonly used to power single phase underground services or whole housing developments that are built in 4800 volt delta distribution areas. Step downs are used in areas that have been upgraded to a 7200/12500Y or 7600/13200Y and the power company chooses to leave a section as a 4800 volt setup. Sometimes power companies choose to leave sections of a distribution grid as 4800 volts because this setup is less likely to trip fuses or reclosers in heavily wooded areas where trees come into contact with lines.

Economic and Political

Traditionally the electricity industry has been a publicly owned institution but starting in the 1970s nations began the process of deregulation and privatisation, leading to electricity markets. A major focus of these was the elimination of the former so called *natural monopoly* of generation, transmission, and distribution. As a consequence, electricity has become more of a commodity. The separation has also led to the development of new terminology to describe the business units, *e.g.* line company, wires business and network company.

ALTERNATING-CURRENT SHOCK CIRCUITS

The previous discussion was based on shock circuits which use a battery as a voltage source. It is usually better, in practice as well as in theory, to use an a-c shock source. All of the foregoing principles apply to a-c as well as to d-c shocks, and since the properties of a-c circuits have not yet been fully described, d-c circuits were used in the preceding discussion. However, there are certain properties peculiar to a-c circuits which are relevant to shock boxes.

The animal, in this circuit, would get an a-c shock. Current would flow through him just as in the d-c circuit, except that the direction of current flow would keep reversing itself each time the switch is thrown. Now suppose you were to connect a zero-centre voltmeter across the grids. In a voltmeter, when the current flows one way, the pointer moves to the right. When it flows the other way, the pointer moves left. In the usual meter, a leftward movement cannot be read, but in a zero-centre meter it can. When the reversing switch is in one position, the meter will read + 115, and when it is reversed, the pointer will swing over until it reads – 115 volts.

A periodic repetition of the switching will result in the temporal voltage pattern. The a-c voltage represented by this figure would be described as a square-wave shock of 115-volt amplitude, with a fundamental frequency equal to half the number of reversals per second. Square-wave shocks are used extensively to stimulate nerve tissue in physiological preparations.

The shock from the circuit is a sine-wave shock, because the voltage delivered to wall sockets is sinosoidal. In almost all parts of the United States, the line power is held to a frequency of 60 cycles per second, and to a voltage of approximately 115 volts.

For the sine wave, the term "fundamental frequency" has the same meaning as for the square wave. It is the number of cycles occurring per second. For the sine wave, the "fundamental frequency" is the only frequency, and so the term "fundamental" is usually dropped. The square wave can be said to contain many (theoretically an infinity of) harmonic frequencies in addition to the fundamental, and for this reason the term fundamental is used in describing the number of complete cycles per second.

The simplest measure of voltage of a sine wave is the so-called peak to-trough amplitude. But there is another more common measure of voltage that requires some explanation. If two light bulbs were lit, one by a *square-wave* of 115 volts (230 volts peak-to-trough) and the other by direct current at 115 volts, the two bulbs would be equally bright because the direction of current flow through a light bulb does not affect its brightness. If the direct current were now left across one bulb and a very low frequency sine-wave voltage of 230 volts peak-to-trough put across the other, the one lit by the sine wave would always be either dimmer or, at the very top and bottom of the cycle, equally bright.

The average brightness thus would be less for the sine wave than for the direct current or for a square wave of equal peak-to trough voltage. If the frequency were increased to, say, 60 cycles per second, the changes in voltage would be so rapid that the bulb would not follow them; its brightness would remain fairly constant at a level lower than the brightness of the d-c-powered bulb. To light a bulb with a sine wave and make it just as bright as one lit by 115 volts d-c, the magnitude of the sine wave would have to be increased. It turns out that the necessary peak-to-trough voltage of the sine wave is $2 \times 230 = 324$ volts.

The power delivered by a sine wave of 324 volts, peak-to-trough, equals the power of direct current at 115 volts. Since it is generally more useful to specify power than peak-to trough voltage, a-c voltages are usually rated in terms of their equivalent d-c voltages. Therefore, to say that the wall socket voltage is 115 volts a-c means it will light a bulb with the same brightness as a 115-volt d–c source.

Its peak-to-trough voltage is about 324 volts. Whenever the voltage of a sine wave is given, it should be understood to mean the equivalent d–c voltage, not the peak-to-trough voltage. (This equivalent voltage is usually called the rms, or root-mean-square voltage, because it is actually the square root of the average of the square of the instantaneous voltages, integrated over one complete cycle.)

The shock circuit is a constant-voltage circuit and delivers a 115–volt, 60–cycle shock. For a lower voltage shock, a voltage divider may be added. This divider works in exactly the same way as it would for a d-c supply. To make the circuit deliver more than 115 volts is not so easy. If it were direct current we could just add some more batteries in series, but the only source of alternating current usually available is the wall socket, at 115 volts.

A simple and efficient device that gives higher or lower voltages than 115 volts is called the transformer. To make the circuit into a constant-current shocker, a very high resistance must be added in series with the source and the animal. Such a circuit definitely requires a transformer to obtain higher source voltages.

DIRECT CURRENT

Direct current (DC) is the unidirectional flow of electric charge. Direct current is produced by such sources as batteries, thermocouples, solar cells, and commutator-type electric machines of the dynamo type. Direct current may flow in a conductor such as a wire, but can also be through semiconductors, insulators, or even through a vacuum as in electron or ion beams. In direct current, the electric charges flow in a constant direction, distinguishing it from alternating current (AC). A term formerly used for *direct current* was Galvanic current.

TYPES OF DIRECT CURRENT

Direct current may be obtained from an alternating current supply by use of a current-switching arrangement called a rectifier, which contains electronic elements (usually) or electromechanical elements (historically) that allow current to flow only in one direction. Direct current may be made into alternating current with an inverter or a motor-generator set. The first commercial electric power transmission (developed by Thomas Edison in the late nineteenth century) used direct current.

Because of the advantage of alternating current over direct current in transforming and transmission, electric power distribution today is nearly all alternating current. For applications requiring direct current, such as third rail power systems, alternating current is distributed to a substation, which utilizes a rectifier to convert the power to direct current. Direct current is used to charge batteries, and in nearly all electronic systems as the power supply.

Very large quantities of direct-current power are used in production of aluminum and other electrochemical processes. Direct current is used for some railway propulsion, especially in urban areas. High voltage direct current is used to transmit large amounts of power from remote generation sites or to interconnect alternating current power grids.

VARIOUS DEFINITIONS

Within electrical engineering, the term *DC* is used to refer to power systems that use only one polarity of voltage or current, and to refer to the constant, zero-frequency, or slowly varying local mean value of a voltage or current. For example, the voltage across a DC voltage source is constant as is the current through a DC current source. The DC solution of an electric circuit is the solution where all voltages and currents are constant. It can be shown that any stationary voltage or current waveform can be decomposed into a sum of a DC component and a zero-mean time-varying component; the DC component is defined to be the expected value, or the average value of the voltage or current over all time.

Although DC stands for "Direct *Current*", DC sometimes refers to "constant polarity." With this definition, DC voltages can vary in time, such as the raw output of a rectifier or the fluctuating voice signal on a telephone line. Some forms of DC (such as that produced by a voltage regulator) have almost no variations in voltage, but may still have variations in output power and current.

APPLICATIONS

Direct-current installations usually have different types of sockets, switches, and fixtures, mostly due to the low voltages used, from those suitable for alternating current. It is usually important with a direct-current appliance not to reverse polarity unless the device has a diode bridge to correct for this (most battery-powered devices do not). DC is commonly found in many low-voltage applications, especially where these are powered by batteries, which can produce only DC, or solar power systems, since solar cells can produce only DC.

Most automotive applications use DC, although the alternator is an AC device which uses a rectifier to produce DC. Most electronic circuits require a DC power supply. Applications using fuel cells (mixing hydrogen and oxygen together with a catalyst to produce electricity and water as byproducts) also produce only DC. Many telephones connect to a twisted pair of wires, and internally separate the AC component of the voltage between the two wires (the audio signal) from the DC component of the voltage between the two wires (used to power the phone).

Telephone exchange communication equipment, such as DSLAM, uses standard –48V DC power supply. The negative polarity is achieved by grounding the positive terminal of power supply system and the battery bank. This is done to prevent electrolysis depositions. An electrified third rail can be used to power both underground (subway) and overground trains.

BIASING (ELECTRONICS)

Biasing in electronics is the method of establishing predetermined voltages and/or currents at various points of a circuit to set an appropriate *operating point*. The operating point of a device, also known as *bias point* or *quiescent point* (or simply *Q-point*), is the DC voltage and/or current which, when applied to a device, causes it to operate in a certain desired fashion. The term is normally used in connection with devices such as transistors and diodes which are used in amplification or rectification.

IMPORTANCE IN LINEAR CIRCUITS

Linear circuits involving transistors typically require specific DC voltages and currents to operate correctly, which can be achieved using a biasing circuit. As an example of the need for careful biasing, consider a transistor amplifier. In linear amplifiers, a small input signal gives larger output signal without any change in shape (low distortion): the input signal causes the output signal to vary up and down about the Q-point in a manner strictly proportional to the input. However, because a transistor is non-linear, the transistor amplifier only approximates linear operation.

For low distortion, the transistor must be biased so the output signal swing does not drive the transistor into a region of extremely non-linear operation. For a bipolar transistor amplifier, this requirement means that the transistor must stay in the *active* mode, and avoid cut-off or saturation. The same requirement applies to a MOSFET amplifier, although the terminology differs a little: the MOSFET must stay in the *active* mode (or saturation mode), and avoid cut-off or Ohmic operation (or triode mode). More detail follows.

BIPOLAR JUNCTION TRANSISTORS

For bipolar junction transistors the bias point is chosen to keep the transistor operating in the active mode, using a variety of circuit techniques, establishing the Q-point DC voltage and current. A small signal is then applied on top of the Q-point bias voltage, thereby either modulating or switching the current, depending on the purpose of the circuit. The quiescent point of operation is typically near the middle of DC *load line*. The process of obtaining certain DC collector current at a certain DC collector voltage by setting up operating point is called biasing.

After establishing the operating point, when input signal is applied, the output signal should not move the transistor either to saturation or to cut-off. However, this unwanted shift still might occur, due to the following reasons:

- Parameters of transistors depend on junction temperature. As junction temperature increases, leakage current due to minority charge carriers (I_{CBO}) increases. As I_{CBO} increases, I_{CEO} also increases, causing an increase in collector current I_C. This produces heat at the collector junction. This process repeats, and, finally, Q-point may shift into the saturation region. Sometimes, the excess heat produced at the junction may even burn the transistor. This is known as thermal runaway.
- When a transistor is replaced by another of the same type, the Q-point may shift, due to changes in parameters of the transistor, such as *current gain* (â) which varies slightly for each unique transistor.

To avoid a shift of Q-point, bias-stabilization is necessary. Various biasing circuits can be used for this purpose.

COMPLEX SWITCHING CIRCUITS

All of the subsequent examples and problems will be typical of those which come up repeatedly in research. Many of the examples deal with learning experiments on animals because complex switching circuitry is even more extensively used in research on rats and pigeons in Skinner boxes than in the study of human behaviour. Probably this is a result of the fact that by verbal instruction it is possible to throw into operation a human subject's own built-in, logical machinery, thus replacing a lot of relays.

This chapter contains many problems, and it is important to solve them all. Solving a problem means drawing the solution on real paper. It often *seems* possible to read a problem and arrive at a solution without actually writing anything down. However, thinking up a way of solving a problem is easy. Solving it is the hard part.

Problem. John has a ball that weighs 3 pounds. It is hollow, and made of rubber weighing 63 pounds per cubic foot. The ball just barely floats in water. How thick are its walls?

Solution. Use division.

HOLDING RELAYS

There is one particular relay circuit that is used very extensively. It is called the holding, or lockup, circuit, and is used when a circuit element must remain activated after the activating circuit is turned off. In essence, a relay is made to close as soon as a switch is operated, hold itself closed even after the switch is opened, and release only when a second switch is operated. The basic circuit is illustrated in the following example.

Suppose you are given the problem of designing a circuit to measure the length of time it takes a rat to run from point *X* in a straight alley to another point *Y* farther down the alley. First, some device must be used to sense when the rat passes each of the two points. In solving the present problem, we will use the method mentioned in the previous chapter of hinging a section of the floor at *X* and another at *Y*, so that the rat's weight will momentarily operate a SPDT microswitch at each of the two points. The clock commonly used for this type of timing consists of an electric motor driving the hands through an electric clutch.

The motor runs constantly, but the hands turn only so long as the clutch is engaged. The clutch operates as long as a small current is being passed through it, and disengages as soon as the current is turned off. To time a rat, then, a circuit is needed which will cause a current to begin to flow through the clutch when one SPDT switch (at *X*) is momentarily operated, and stop fl owing when another (at *Y*) is operated. Neither microswitch alone could operate the clutch in the desired way because, during the time the rat is between *X* and *Y*, both switches are open so that the clutch would not be engaged.

Suppose, instead, that the clutch is turned on and off through a relay. Now, if the relay could be made to close when switch *X* is closed, hold itself closed even after switch *X* is opened, and release when switch *Y* is operated, the problem would be almost solved. (The clutch could then be turned on and off by a pair of contacts on the relay.) The clutch connected through a relay, and the relay connected directly to one of the microswitches. This circuit is no improvement over connecting the clutch directly through the microswitch.

As soon as *X* is closed by the rat's running over that section of the alley, the relay closes, allowing current to flow in the clutch circuit. However, as soon as the rat leaves that section of the alley, the microswitch will open again, the relay will thus be released, and the clutch is turned off. These contacts are connected in parallel with the normally open contacts of the microswitch. The action of the circuit is as follows. At first, the microswitch and the relay are open so that no current flows anywhere (except, of course, through the clock motor). (Be sure that it is clear from the diagram that no current can flow. Try to find a closed path from one side of a power source to the other side of the same source. If there is no such path, then there can be no current flow, but as soon as there is such a path, there must be current flow.)

Now the rat steps on the hinged alley section, closing the normally open contacts of the microswitch. This furnishes a path for current to flow through the relay coil (dashed arrows), and the resulting magnetic field will cause the relay armature to pull down, closing the two pairs of normally open contacts. The top pair activates the clutch. The lower pair provides a second path for the current to flow through the relay coil (dotted arrows). This time, when the rat leaves the section and the microswitch opens again, the relay will stay closed because there is still a path allowing current to flow through the coil. With the microswitch open, current will flow only along the dotted lines. Under these conditions, the relay is, in a sense, holding itself closed and once it has been closed, microswitch *X* no longer has any effect at all.

The clock will start running when the rat first crosses the hinged section of the alley and will keep running indefinitely, as the circuit is now drawn. Even if the rat were to retrace and go back and forth several times over this alley section, the clock would be unaffected.

The relay is your 115-volt a–c relay, and switch X is one of the microswitches. (Observe the same precautions as were discussed in the previous chapter: remember that 115 volts can be painful.) If the circuit is connected up correctly, the light should go on when you first press the microswitch and stay on after you let go. Now the bulb is lit (the clock is running). How do you turn it off? There are several ways. First, you can unscrew the bulb.

This is the equivalent of putting a switch in the clutch circuit and opening it. Suppose that this switch were actually the second microswitch *Y* in the alley, connected. That is, the normally closed pair is connected in series with the clutch. When the rat hits section *X*, the relay closes and the clock starts. When he hits section *Y*, the clock stops. If at that moment the clock could be read very quickly, the circuit problem would be solved. But as soon as the rat leaves section *Y*, the clock will start up again. Prove this by unscrewing the bulb and then putting it back again.

If the section *Y* switch could actually turn off the power to the *relay*, then the relay would stay open even after the section is returned to its normally closed condition, because the only things keeping the relay closed are its own contacts. In the circuit on your board, turn off the light by pulling out the wall plug. Now when the plug is connected again, the light will remain off and the whole circuit is restored to its initial condition.

On your board, connect up the normally closed contacts of the second microswitch. This switch will now do what you were doing when you momentarily disconnected the wall plug. If the circuit is working properly, the light will turn on when switch *X* is momentarily pressed, and stay on until switch *Y* is momentarily pressed.

This circuit, the holding relay circuit, is very fundamental to switching circuit design. It is used as it has been presented here or with only slight modifications in a very large proportion of the switching circuits found in behavioural research. For this reason, it is worthwhile for you to draw the circuit diagram over and over again until it becomes almost automatic. Be sure you understand its principles of operation. This will save you considerable time when you are solving many of the remaining problems in this book.

If the circuit were used in a straight alley way, it would permit the measurement of the total time between the first crossing of *X* and the first crossing of Y (*i.e.*, stepping on Y again would not restart the clock). Consider instead the problem of designing a circuit to measure the total time the animal spends in the portion of the alley between X and Y. In other words, if he runs out of this portion, past Y, and then turns around and runs back into it once more, the clock should start again.

Since this is probably the most difficult problem yet presented, it may be a good one with which to illustrate the sort of process used by people skilled at circuit design to solve relatively complicated problems.

Many of the processes involved in arriving at solutions to this kind of problem are difficult to verbalize in a coherent way. If prodded, people who are skilled at circuit design frequently will say that they knew by intuition where to start and what things to try. Intuition is probably an excellent word to describe what happens. Most people using such intuition will readily agree that it is developed through a combination of some intelligence and a lot of experience. Here, an attempt will be made to enumerate some of the submerged logic involved.

The first step in arriving at a solution is to state the circuit requirements as explicitly as possible. In this particular problem, the circuit must:

- Start the clock any time the animal (*a*) crosses X going towards Y or (*b*) crosses Y going towards X.
- Stop the clock any time he (*a*) crosses X going away from Y or (*b*) crosses Y going away from X.

(For any problem there are a number of solutions. For example, this problem may be easily solved, in theory at least, by hinging the entire section between X and Y and resting it on a microswitch. In practice, it is difficult to make a large section of alley operate that way. The solutions presented herein are chosen both because they are practical and because they illustrate various important aspects of circuit design. If you arrive at a solution different from one in the text-and this happens very often—that does not necessarily mean that yours is incorrect.)

Examination of the apparatus requirements listed above indicates that the circuit must sense the direction in which the rat is traveling as well as his presence when he is at point X or point Y. If there were two short, hinged floor sections at X, and if the circuit could determine the order in which these two sections were traversed, it would indicate which way the animal was going when he went by X, and either of the sections might also indicate his presence at X. In other words, it may be that if there were four hinged floor sections, each with a microswitch, two at X and two more at Y, it would be possible to solve the problem. A guess about what kinds of circuit elements might solve the problem can be considered the second step in the solution.

The third step might then be to see whether or not such a collection of elements really can be made to work. To do this, it is probably most efficient to begin by representing, on a large sheet of paper, each of the possible elements without any particular regard for how they will be connected up.

The clock will run so long as there is current flowing through its clutch coil, and it becomes clear and considering the original problem that the clutch must be operated through something that is not yet represented. This is true because, although the clutch must remain activated when the rat is, say, halfway between X and Y, there is nothing drawn yet that will act that way. As in the preceding problem, a holding relay circuit is strongly indicated. A relay is thus drawn in with at least enough contacts to hold itself, and connected so as to operate the clutch. Now the clutch no longer need be considered. The problem is to get the relay to be on all the time the rat is between X and Y. If this is done, the clock will necessarily do the same.

Figuring out additional connections requires consideration of the step-by-step operation of the circuit. Let the rat come out of the start box and begin to walk up the alley. The clutch is to engage, that is, the relay is to close, when the rat passes X. Therefore, either microswitch X_1, or X_2 must close the relay. Try X_1 first. With this circuit, when the rat walks past X and stands between X and Y, the clock keeps running (the relay is connected to hold itself on). Now, if the rat turns around and crosses X again, the clock are supposed to go off. Since X_2 is left over, try its normally closed contacts to turn off the clock.

This will stop the clock all right when the rat recrosses X_2 going towards the start box, but two other things are wrong. First, if he goes a little farther and crosses X_1 too, the clock will go back on again. Second, after he leaves the start box in the first place, he will start the clock at X_1, but stop it again as he goes by X_2. Therefore, try reversing the roles of X_1 and X_2 to see if that works any better. Now, when the rat crosses X_1 going towards Y, nothing happens. When he crosses X_2, the clock starts. If he then turns around and recrosses X_2 going the wrong way, nothing happens.

But as soon as he crosses X_1 going the wrong way, the clock stops. When he again turns around and recrosses X going towards Y, the clock starts once more. So far, then, this circuit is operating according to plan. All that remains is to make the switches at Y operate in a similar, but *opposite*, manner. In other words, it would seem plausible, as a first guess, to make Y1 start the clock and Y_2 stop it. (Crossing Y towards X should do the same thing as crossing X towards Y.) The circuit is connected this way and will solve the problem. In this circuit, either Y_1 or X_2 will start the clock, and either X_1 or Y_2 will stop it.

The clock will always be running when the rat is between X and Y because the rat must have most recently closed either X_2 or Y_1 to get there. And any time the rat leaves the section between X and Y, the clock must be stopped because the rat must have most recently opened either X_1 or Y_2. The next problem illustrates another modification of a holding relay circuit. A relay that is turned on and off repeatedly at a fairly fast rate will buzz. In the holding relay circuit, the relay holds itself on. If instead a relay is connected in such a way that it turns itself *off*, then it must buzz.

Before the switch is closed, the relay is open, so that the normally closed contacts of the relay are closed. As soon as the switch is closed, current will flow through the relay contacts and through the relay coil, causing the relay armature to pull in. But when the armature pulls in far enough to open the normally closed contacts, the circuit through the relay coil is broken, so the armature is released. This allows the contacts to close again, and the cycle automatically repeats itself.

Connect up this circuit on your board, using the sensitive relay (not the 115-volt one). Begin by using as low a voltage across the relay as is convenient (say 3 volts), and increase it, if necessary, until the relay buzzes. Be sure that you understand which terminals are connected to which parts of the relay (*i.e.*, which terminals on the body of the relay are internally connected to the relay coil) before you connect up the battery. If you cannot tell about some terminal by looking at the relay, check it with an ohmmeter to see to which part it is connected. Practically all buzzers commercially available operate on exactly the same principle as the one you have constructed. The only important difference between your buzzer and a doorbell buzzer is that the latter uses a much cheaper relay.

DC TRANSISTOR CIRCUITS

Most of us deal with electronic products on a routine basis and have some experience with personal computers. A basic component for the integrated circuits found in these electronics and computers is the active, three terminal device known as the transistor. Understanding the transistor is essential before an engineer can start an electronic circuit design.

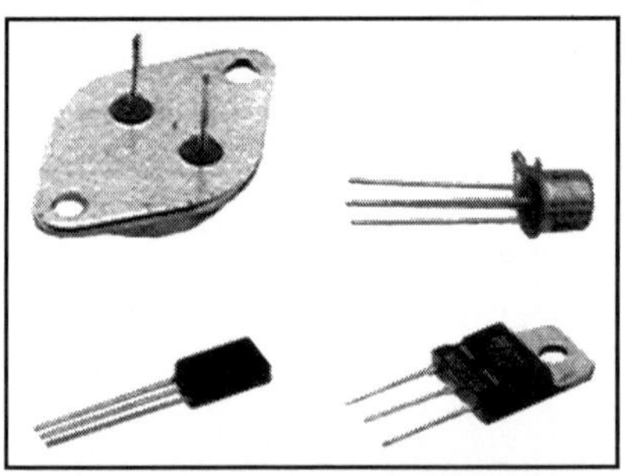

Fig. Various types of Transistor.

In Below Figure depicts various kinds of transistors commercially available. There are two basic types of transistors: bipolar junction transistors (BJTs) and field-effect transistor (FETs). Here, we consider only the BJTs, which were the first of the two and are still used today. Our objective is to present enough detail about the BJT to enable us to apply the techniques developed in this chapter to analyse dc transistor circuits. There two types of BJTs: npn and pnp, with their circuit symbols as shown in. Each type has three terminals, designated as emitter (E), base (B) and collector (C). For the npn transistor, the currents and voltages of the transistor are specified as in. Applying KCL to gives

$$I E = I B + I C$$

where IE, IC, and IB are emitter, collector, and base currents, respectively.

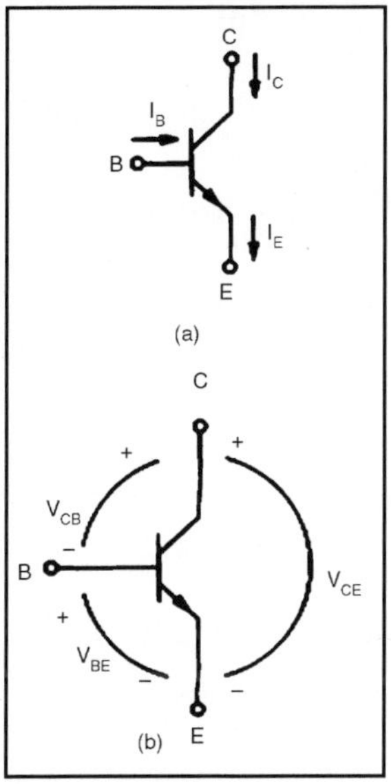

Fig. The terminal Variable of an npn Transistor:

a) Currents, b) Voltages.

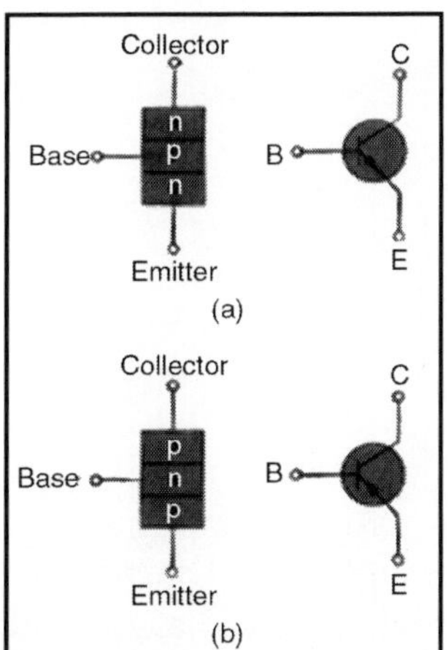

Fig. Two Types of BJTs and their Circuit Symbols: a) npn, b) pnp.

Similarly, applying KVL to gives,

$$V CE + V EB + V BC$$

Where VCE, VEB, and VBC are collector-emitter, emitter-base, and base-collector voltages. The BJT can operate in one of three modes: active, cutoff, and saturation. When transistors operate in the active mode, typically V ≈ 0.7V,

$$I C = aI E$$

Where α is called the common-base current gain. In Equation, α denotes the fraction of electrons injected by the emitter that are colleted by the collector. Also,

$$I C = bI B$$

Where β is known as the common-emitter current gain. The α and β are characteristic properties of a given transistor and assume constant values for that transistor. Typically, α takes values in the range of 0.98 to 0.999, while β takes values in the range 50 to 1000 from Equation to Equation, it is evident that,

$$I E = (1 + b) I B$$

and

$$b = a\ 1\ a$$

These equations show that, in the active mode, the BJT can be modeled as a dependent current-controlled current source. Thus, in circuit analysis, the dc equivalent model in may be used to replace the npn transistor in. Since β in Equation is large, a small base current controls large currents in the output circuit. Consequently, the bipolar transistor can be serve as an amplifier, producing both current gain and voltage gain. Such amplifiers can be used to furnish a considerable amount of power to transducers such as loudspeakers or control motors.

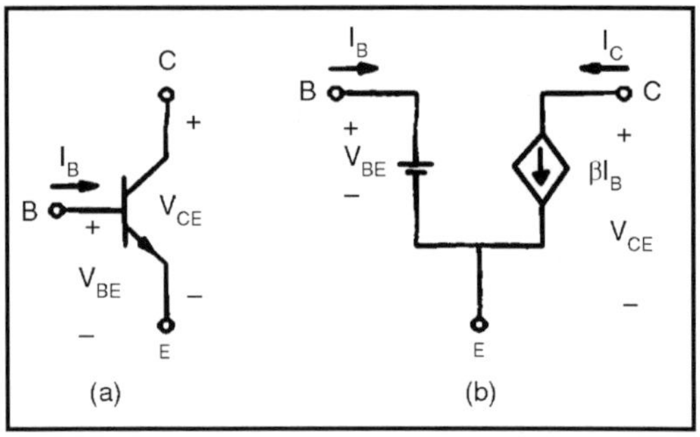

It should be observed in the following examples that one cannot directly analyse transistor circuits using nodal analysis because of the potential difference between the terminals of the transistor. Only when the transistor is replaced by its equivalent model, we can apply nodal analysis.

9

Digital Memory and Circuit Device

DIGITAL MEMORY DEVICES

LOOK-UP TABLES

Having learned about digital memory devices in the last chapter, we know that it is possible to store binary data within solid-state devices. Those storage "cells" within solid-state memory devices are easily addressed by driving the "address" lines of the device with the proper binary value(s). Suppose we had a ROM memory circuit written, or programmed, with certain data, such that the address lines of the ROM served as inputs and the data lines of the ROM served as outputs, generating the characteristic response of a particular logic function. Theoretically, we could programme this ROM chip to emulate whatever logic function we wanted without having to alter any wire connections or gates.

Consider the following example of a 4 × 2 bit ROM memory (a very small memory!) programmed with the functionality of a half adder:

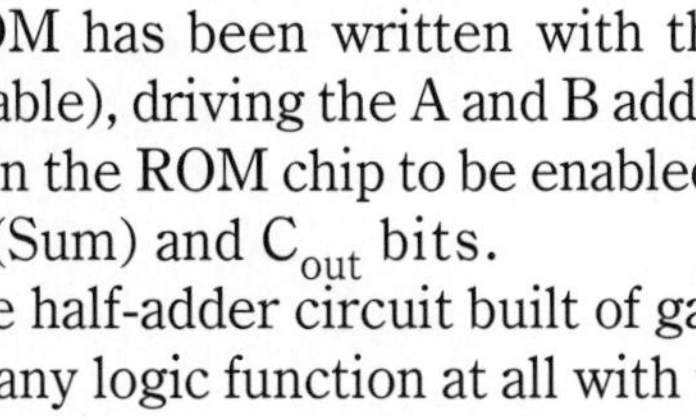

Address		Data	
A	B	C_{out}	Σ
0	0	0	0
0	1	0	1
1	0	0	1
1	1	1	0

If this ROM has been written with the above data (representing a half-adder's truth table), driving the A and B address inputs will cause the respective memory cells in the ROM chip to be enabled, thus outputting the corresponding data as the Ó (Sum) and C_{out} bits.

Unlike the half-adder circuit built of gates or relays, this device can be set up to perform any logic function at all with two inputs and two outputs, not just the half-adder function. To change the logic function, all we would need to do is write a different table of data to another ROM chip. We could even use an EPROM chip which could be re-written at will, giving the ultimate flexibility in function.

It is vitally important to recognize the significance of this principle as applied to digital circuitry. Whereas the half-adder built from gates or relays *processes* the input bits to arrive at a specific output, the ROM simply *remembers* what the outputs should be for any given combination of inputs.

This is not much different from the "times tables" memorized in grade school: rather than having to calculate the product of 5 times 6 (5 + 5 + 5 + 5 + 5 + 5 = 30), school-children are taught to remember that 5 × 6 = 30, and then expected to recall this product from memory as needed. Likewise, rather than the logic function depending on the functional arrangement of hard-wired gates or relays (hardware), it depends solely on the data written into the memory (software).

Such a simple application, with definite outputs for every input, is called a *look-up table*, because the memory device simply "looks up" what the output(s) should to be for any given combination of inputs states.

This application of a memory device to perform logical functions is significant for several reasons:

- Software is much easier to change than hardware.
- Software can be archived on various kinds of memory media (disk, tape), thus providing an easy way to document and manipulate the function in a "virtual" form; hardware can only be "archived" abstractly in the form of some kind of graphical drawing.
- Software can be copied from one memory device (such as the EPROM chip) to another, allowing the ability for one device to "learn" its function from another device.
- Software such as the logic function example can be designed to perform functions that would be extremely difficult to emulate with discrete logic gates (or relays!).

The usefulness of a look-up table becomes more and more evident with increasing complexity of function. Suppose we wanted to build a 4-bit adder circuit using a ROM. We'd require a ROM with 8 address lines (two 4-bit numbers to be added together), plus 4 data lines (for the signed output):

With 256 addressable memory locations in this ROM chip, we would have a fair amount of programming to do, telling it what binary output to generate for each and every combination of binary inputs. We would also run the risk of making a mistake in our programming and have it output an incorrect sum, if we weren't careful. However, the flexibility of being able to configure this function (or any function) through software alone generally outweighs that costs.

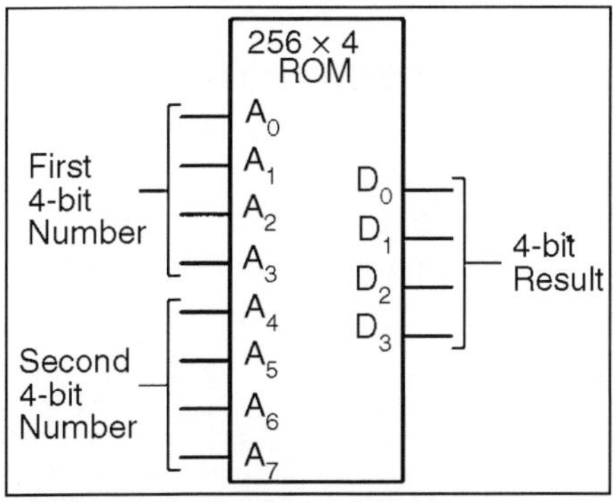

Consider some of the advanced functions we could implement with the above "adder." We know that when we add two sets of numbers in 2's complement signed notation, we risk having the answer overflow. For instance, if we try to add 0111 (decimal 7) to 0110 (decimal 6) with only a 4-bit number field, the answer we'll get is 1001 (decimal -7) instead of the correct value, 13 (7 + 6), which cannot be expressed using 4 signed bits. If we wanted to, we could avoid the strange answers given in overflow conditions by programming this look-up table circuit to output something else in conditions where we know overflow will occur (that is, in any case where the real sum would exceed +7 or -8). One alternative might be to programme the ROM to output the quantity 0111 (the maximum positive value that can be represented with 4 signed bits), or any other value that we determined to be more appropriate for the application than the typical overflowed "error" value that a regular adder circuit would output. It's all up to the programmer to decide what he or she wants this circuit to do, because we are no longer limited by the constraints of logic gate functions.

The possibilities don't stop at customised logic functions, either. By adding more address lines to the 256 × 4 ROM chip, we can expand the look-up table to include multiple functions:

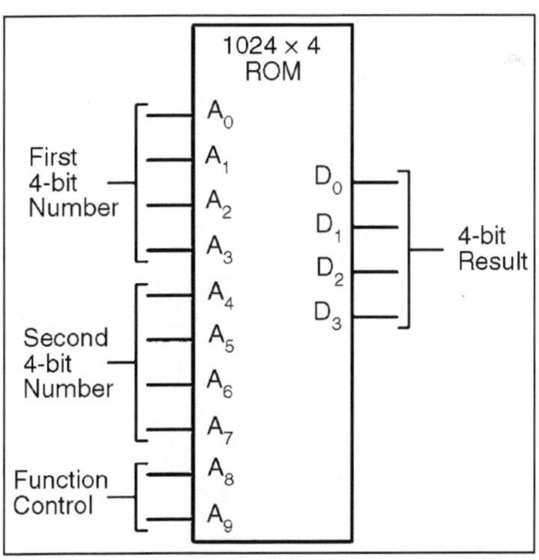

With two more address lines, the ROM chip will have 4 times as many addresses as before (1024 instead of 256). This ROM could be programmed so that when A8 and A9 were both low, the output data represented the *sum* of the two 4-bit binary numbers input on address lines A0 through A7, just as we had with the previous 256 × 4 ROM circuit.

For the addresses A8=1 and A9=0, it could be programmed to output the *difference* (subtraction) between the first 4-bit binary number (A0 through A3) and the second binary number (A4 through A7).

For the addresses A8=0 and A9=1, we could programme the ROM to output the difference (subtraction) of the two numbers in reverse order (second - first rather than first - second), and finally, for the addresses A8=1 and A9=1, the ROM could be programmed to compare the two inputs and output an indication of equality or inequality.

What we will have then is a device that can perform four different arithmetical operations on 4-bit binary numbers, all by "looking up" the answers programmed into it.

If we had used a ROM chip with more than two additional address lines, we could programme it with a wider variety of functions to perform on the two 4-bit inputs.

There are a number of operations peculiar to binary data (such as parity check or Exclusive-ORing of bits) that we might find useful to have programmed in such a look-up table.

Devices such as this, which can perform a variety of arithmetical tasks as dictated by a binary input code, are known as*Arithmetic Logic Units* (ALUs), and they comprise one of the essential components of computer technology. Although modern ALUs are more often constructed from very complex combinational logic (gate) circuits for reasons of speed, it should be comforting to know that the exact same functionality may be duplicated with a "dumb" ROM chip programmed with the appropriate look-up table(s).

In fact, this exact approach was used by IBM engineers in 1959 with the development of the IBM 1401 and 1620 computers, which used look-up tables to perform addition, rather than binary adder circuitry. The machine was fondly known as the "CADET," which stood for "Can't Add, Doesn't Even Try."

DIGITAL COMPUTING SYSTEM

In 1832, Samuel F. B. Morse used the very same idea to propose that telegram *words* be coded into *binary addresses* or *binary codes* that could be transmitted over telegraph lines and decoded at the receiving end to unravel the telegram. Morse abandoned his scheme, illustrated in Figure, as too complicated and, in 1838, proposed his fabled Morse code for coding letters (instead of words) into objects (dots, dashes, spaces) capable of a threefold difference onely (sic).

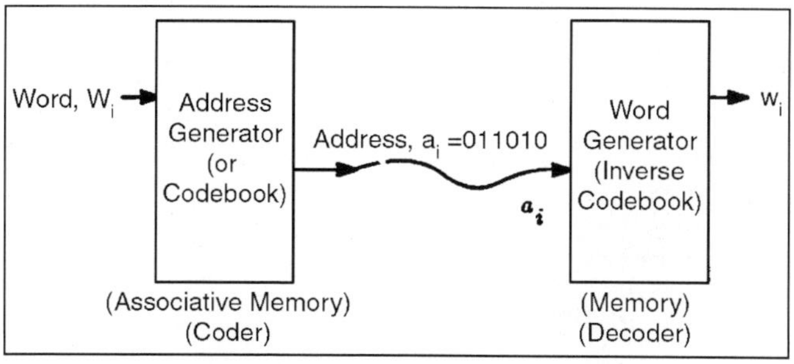

Fig. Generalized Coder-Decoder

The basic idea of Figure is used today in cryptographic systems, where the "address a_i" is an encyphered version of a messagew_i ; in vector quantizers, where the "address a_i" is the address of a close approximation to data w_i ; in coded satellite transmissions, where the "address a_i" is a data word w_i plus parity check bits for detecting and correcting errors; in digital audio systems, where the "address a_i" is a stretch of digitized and coded music; and in computer memories, where a_i is an address (a coded version of a word of memory) and w_i is a word in memory.

In this chapter we study three fundamental questions in the construction of binary addresses or binary codes. First, what are plausible schemes for mapping symbols (such as words, letters, computer instructions, voltages, pressures, etc.) into binary codes? Second, what are plausible schemes for coding likely symbols with short binary words and unlikely symbols with long words in order to minimize the number of binary digits (bits) required to represent a message? Third, what are plausible schemes for "coding" binary words into longer binary words that contain "redundant bits" that may be used to detect and correct errors? These are not new questions. They have occupied the minds of many great thinkers. Sir Francis recognized that arbitrary messages had binary representations. Alan Turing, Alonzo Church, and Kurt Goedel studied binary codes for computations in their study of computable numbers and algorithms. Claude Shannon, R. C. Bose, Irving Reed, Richard Hamming, and many others have studied error control codes. Shan- non, David Huffman, and many others have studied the problem of efficiently coding information.

EXPLANATION

Did you know that everything a computer does is based on ones and zeroes? It's hard to imagine, because you hear people talking about the absolutely gargantuan (huge) numbers that computers "crunch". But all those huge numbers - they're just made up of ones and zeros. It's kind of like the computer is made up of a bunch of lightswitches, and each lightswitch controls just one lightbulb. On or Off. One or Zero. But if you took *all* of those lightbulbs together, and said "Let's make each sequence of On-and-Off represent a different number!" Well then, you could get some pretty large numbers.

Sequence

There are four different ways we could flip those switches:

- Both Off
- First Off, Second On
- First On, Second Off
- Both On

Binary Code takes each of those combinations and assigns a number to it, like this:

- Both Off = 0
- First Off, Second On = 1
- First On, Second Off = 2
- Both On = 3

Intermediate Explanation

Another way of thinking about it is this: let's give each lightbulb a point value. We'll say the first lightbulb is worth two points, and the second one is worth one point.

Now take a look at the combinations:

- Both Off = 0 + 0 = 0
- First Off, Second On = 0 + 1 = 1
- First On, Second Off = 2 + 0 = 2
- Both On = 2 + 1 = 3

If we added in another lightbulb, we would make it worth twice as many points as the 2 pointer. Then, if all the bulbs were on, the point value would be 4 + 2 + 1 = 7. And if we added in *another* bulb, we would make it worth 8 points (twice as much as four). Now if they are all turned on, that's worth 8 + 4 + 2 + 1 = 15. As you can see, it's going to take a lot of lightbulbs to make a really big number! Finally, even though we're giving point values to each of those lightbulbs, when we write them down we still only write them as ones and zeros. One means On, and Zero means Off.

So let's say we had 8 lightbulbs, and they were set up like this:

Off Off On On Off On Off Off.

The point values of those eight bulbs are:

0 + 0 + 32 + 16 + 0 + 4 + 0 + 0 (remember – we only give points if they're turned on!)

And that adds up to 52. So we would say the sequence of bulbs is worth 52. But how do we write it?

We write it like this: 00110100

So now we can say $00110100_{\text{binary}} = 52$.

8421 BCD CODE

In the 8421 Binary Coded Decimal (BCD) representation each decimal digit is converted to its 4-bit pure binary equivalent.

For example: $57_{dec} = 0101\ 0111_{bcd}$

Addition is analogous to decimal addition with normal binary addition taking place from right to left. For example,

6	0110	BCD for 6	42	0100 0010	BCD for 42
+3	0011	BCD for 3	+27	0010 0111	BCD for 27
	1001	BCD for 9		0110 1001	BCD for 69

Where the result of any addition exceeds 9(1001) then six (0110) must be added to the sum to account for the six invalid BCD codes that are available with a 4-bit number.

This is illustrated in the example below,

8	1001	BCD for 8
+7	0111	BCD for 7
	1111	exceeds 9 (1001) so
	0110	add six (0110)
	0001 0101	BCD for 15

Note that in the last example the 1 that carried forward from the first group of 4 bits has made a new 4-bit number and so represents the "1" in "15". In the examples above the BCD numbers are split at every 4-bit boundary to make reading them easier. This is not necessary when writing a BCD number down. This coding is an example of a binary coded (each decimal number maps to four bits) weighted (each bit represents a number, 1, 2, 4, etc.) code.

GRAY CODE

Gray coding is an important code and is used for its speed, it is also relatively free from errors. In pure binary coding or 8421 BCD then counting from 7 (0111) to 8 (1000) requires 4 bits to be changed simultaneously. If this does not happen then various numbers could be momentarily generated during the transition so creating spurious numbers which could be read.

Gray coding avoids this since only one bit changes between subsequent numbers. To construct the code there are two simple rules. First start with all 0s and then proceed by changing the least significant bit (lsb) which will bring about a new state.

The first 16 Gray coded numbers are indicated below:

Decimal	Gray Code
0	0000
1	0001
2	0011

3	0010
4	0110
5	0111
6	0101
7	100
8	1100
9	1101
10	1111
11	1110
12	1010
13	1011
14	1001
15	1000

To convert a Gray-coded number to binary then follow this method:

- The binary number and the Gray-coded number will have the same number of bits.
- The binary MSB (left-hand bit) and Gray code MSB will always be the same.
- To get the binary next-to-MSB (*i.e.* next digit to the right) add the binary MSB and the gray code next-to-MSB. Re cord the sum, ignoring any carry.
- Continue in this manner right through to the end.

Gray coding is a non-BCD, non-weighted reflected binary code.

4221 BCD CODE

The 4221 BCD code is another binary coded decimal code where each bit is weighted by 4, 2, 2 and 1 respectively. Unlike BCD coding there are no invalid representations. The decimal numbers 0 to 9 have the following 4221 equivalents,

Decimal	4221	1's complement
0	0000	1111
1	0001	1110
2	0010	1101
3	0011	1100
4	1000	0111
5	0111	1000
6	1100	0011
7	1101	0010
8	1110	0001
9	1111	0000

the 1's complement of a 4221 representation is important in decimal arithmetic.

In forming the code remember the following rules:

- Below decimal 5 use the right-most bit representing 2 first.
- Above decimal 5 use the left-most bit representing 2 first.
- Decimal 5 = 2+2+1 and not 4+1.

ROMS CONTROL SYSTEMS

A very common application for look-up table ROMs is in control systems where a custom mathematical function needs to be represented. Such an application is found in computer-controlled fuel injection systems for automobile engines, where the proper air/fuel mixture ratio for efficient and clean operation changes with several environmental and operational variables.

Tests performed on engines in research labouratories determine what these ideal ratios are for varying conditions of engine load, ambient air temperature, and barometric air pressure.

The variables are measured with sensor transducers, their analog outputs converted to digital signals with A/D circuitry, and those parallel digital signals used as address inputs to a high-capacity ROM chip programmed to output the optimum digital value for air/fuel ratio for any of these given conditions.

Sometimes, ROMs are used to provide one-dimensional look-up table functions, for "correcting" digitized signal values so that they more accurately represent their real-world significance.

An example of such a device is a *thermocouple transmitter*, which measures the millivoltage signal generated by a junction of dissimilar metals and outputs a signal which is supposed to *directly* correspond to that junction temperature. Unfortunately, thermocouple junctions do not have perfectly linear temperature/voltage responses, and so the raw voltage signal is not perfectly proportional to temperature.

By digitizing the voltage signal (A/D conversion) and sending that digital value to the address of a ROM programmed with the necessary correction values, the ROM's programming could eliminate some of the non-linearity of the thermocouple's temperature-to-millivoltage relationship, so that the final output of the device would be more accurate. The popular instrumentation term for such a look-up table is a digital *characterizer*.

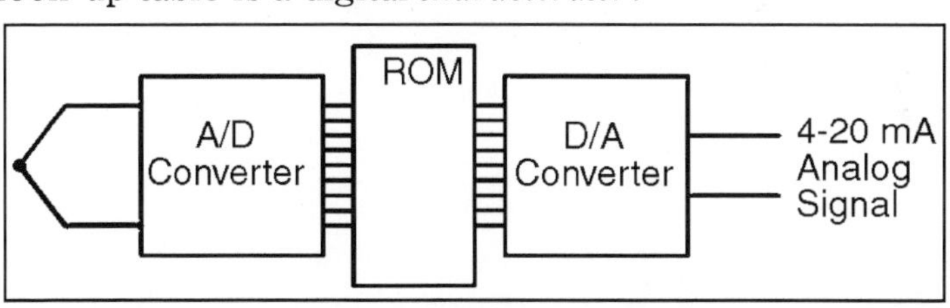

Another application for look-up tables is in special code translation. A 128 × 8 ROM, for instance, could be used to translate 7-bit ASCII code to 8-bit EBCDIC code:

Again, all that is required is for the ROM chip to be properly programmed with the necessary data so that each valid ASCII input will produce a corresponding EBCDIC output code.

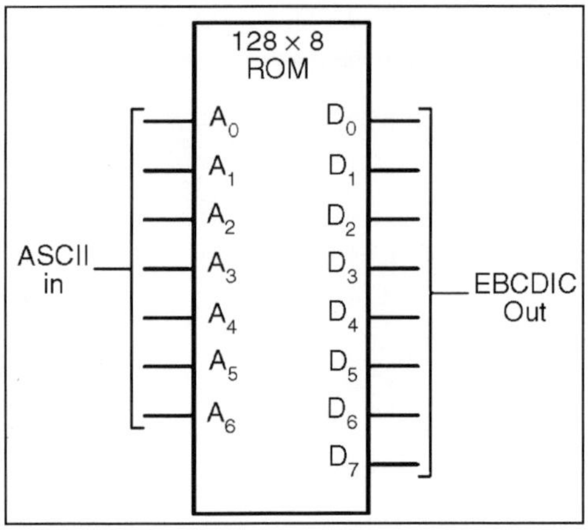

FINITE-STATE MACHINES

Feedback is a fascinating engineering principle. It can turn a rather simple device or process into something substantially more complex. We've seen the effects of feedback intentionally integrated into circuit designs with some rather astounding effects:

- Comparator + negative feedback→ controllable-gain amplifier
- Comparator + positive feedback→ comparator with hysteresis
- Combinational logic + positive feedback → multivibrator
- In the field of process instrumentation, feedback is used to transform a simple measurement system into something capable of control:
- Measurement system + negative feedback→closed-loop control system

Feedback, both positive and negative, has the tendency to add whole new dynamics to the operation of a device or system.

Sometimes, these new dynamics find useful application, while other times they are merely interesting. With look-up tables programmed into memory devices, feedback from the data outputs back to the address inputs creates a whole new type of device.

The *Finite State Machine*, or *FSM*:

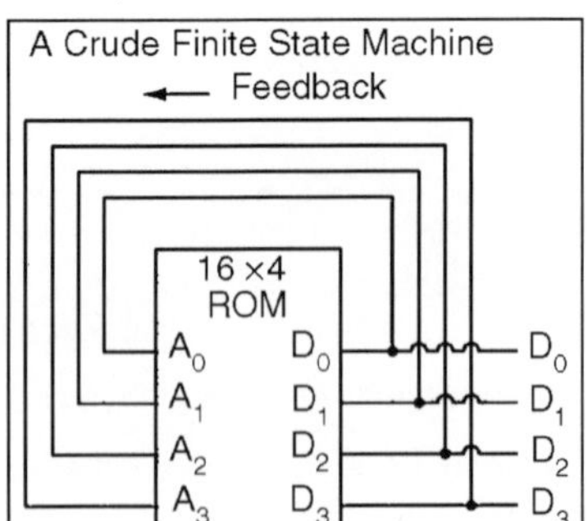

The above circuit illustrates the basic idea: the data stored at each address becomes the next storage location that the ROM gets addressed to. The result is a specific sequence of binary numbers (following the sequence programmed into the ROM) at the output, over time. To avoid signal timing problems, though, we need to connect the data outputs back to the address inputs through a 4-bit D-type flip-flop, so that the sequence takes place step by step to the beat of a controlled clock pulse:

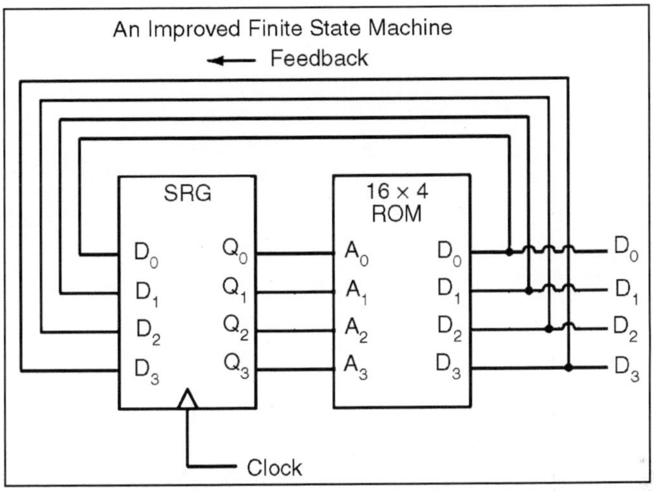

An analogy for the workings of such a device might be an array of post-office boxes, each one with an identifying number on the door (the address), and each one containing a piece of paper with the address of another P.O. box written on it (the data). A person, opening the first P.O. box, would find in it the address of the next P.O. box to open. By storing a particular pattern of addresses in the P.O. boxes, we can dictate the sequence in which each box gets opened, and therefore the sequence of which paper gets read.

Having 16 addressable memory locations in the ROM, this Finite State Machine would have 16 different stable "states" in which it could latch. In each of those states, the identity of the next state would be programmed in to the ROM, awaiting the signal of the next clock pulse to be fed back to the ROM as an address. One useful application of such an FSM would be to generate an arbitrary count sequence, such as Grey Code:

Address → *Data*			*Grey Code count sequence*:
0000	→	0001	0 0000
0001	→	0011	1 0001
0010	→	0110	2 0011
0011	→	0010	3 0010
0100	→	1100	4 0110
0101	→	0100	5 0111
0110	→	0111	6 0101

0111	→	0101	7 0100
1000	→	0000	8 1100
1001	→	1000	9 1101
1010	→	1011	10 1111
1011	→	1001	11 1110
1100	→	1101	12 1010
1101	→	1111	13 1011
1110	→	1010	14 1001
1111	→	1110	15 1000

Try to follow the Grey Code count sequence as the FSM would do it: sta rting at 0000, follow the data stored at that address (0001) to the next address, and so on (0011), and so on (0010), and so on (0110), etc. The result, for the programme table shown, is that the sequence of addressing jumps around from address to address in what looks like a haphazard fashion, but when you check each address that is accessed, you will find that it follows the correct order for 4-bit Grey code. When the FSM arrives at its last programmed state (address 1000), the data stored there is 0000, which starts the whole sequence over again at address 0000 in step with the next clock pulse.

We could expand on the capabilities of the above circuit by using a ROM with more address lines, and adding more programming data:

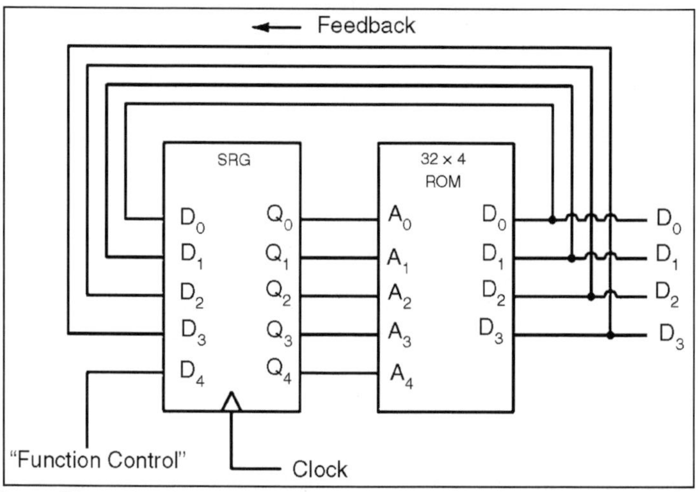

Now, just like the look-up table adder circuit that we turned into an Arithmetic Logic Unit (+, –, ×,/functions) by utilizing more address lines as "function control" inputs, this FSM counter can be used to generate more than one count sequence, a different sequence programmed for the four feedback bits (A0 through A3) for each of the two function control line input combinations (A4 = 0 or 1).

Address	→	*Data*	*Address*	→	*Data*
00000	→	0001	10000	→	0001

00001	→	0010	10001	→	0011
00010	→	0011	10010	→	0110
00011	→	0100	10011	→	0010
00100	→	0101	10100	→	1100
0010	→	0110	10101	→	0100
00110	→	0111	10110	→	0111
0011	→	1000	10111	→	0101
01000	→	1001	11000	→	0000
01001	→	1010	11001	→	1000
01010	→	1011	11010	→	1011
01011	→	1100	11011	→	1001
01100	→	1101	11100	→	1101
01101	→	1110	11101	→	1111
01110	→	1111	11110	→	1010
01111	→	0000	11111	→	1110

If A4 is 0, the FSM counts in binary; if A4 is 1, the FSM counts in Grey Code. In either case, the counting sequence is arbitrary: determined by the whim of the programmer. For that matter, the counting sequence doesn't even have to have 16 steps, as the programmer may decide to have the sequence recycle to 0000 at any one of the steps at all. It is a completely flexible counting device, the behaviour strictly determined by the software (programming) in the ROM. We can expand on the capabilities of the FSM even more by utilizing a ROM chip with additional address input and data output lines.

Take the following circuit, for example:

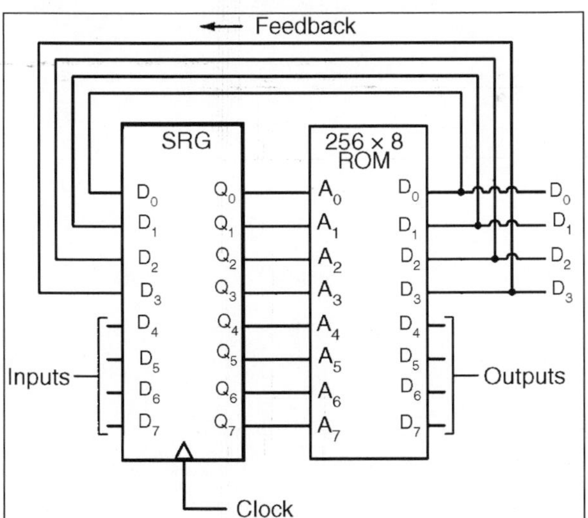

Here, the D0 through D3 data outputs are used exclusively for feedback to the A0 through A3 address lines. Date output lines D4 through D7 can be programmed to output something other than the FSM's "state" value. Being that four data output bits are being fed back to four address bits, this is still a 16-state device. However, having the output data come from other data output lines gives the programmer more freedom to configure functions than before.

In other words, this device can do far more than just count! The programmed output of this FSM is dependent not only upon the state of the feedback address lines (A0 through A3), but also the states of the input lines (A4 through A7). The D-type flip/flop's clock signal input does not have to come from a pulse generator, either. To make things more interesting, the flip/flop could be wired up to clock on some external event, so that the FSM goes to the next state only when an input signal tells it to. Now we have a device that better fulfills the meaning of the word "programmable." The data written to the ROM is a programme in the truest sense: the outputs follow a pre-established order based on the inputs to the device and which "step" the device is on in its sequence. This is very close to the operating design of the *Turing Machine*, a theoretical computing device invented by Alan Turing, mathematically proven to be able to solve any known arithmetic problem, given enough memory capacity.

Bibliography

Arunima Baruah: *Computer Networking in Libraries*, Kalpaz Publication, New Delhi, 2011.

Ben-Ze'Ev: *Love Online: Emotions on the Internet*, Cambridge University Press, Delhi, 2001.

Carol Anne Strange: *Start And Run An Internet Business*, Rupa Publication, Delhi, 2010.

Chris Bates: *Web Programming: Building Internet Applications*, Wiley Publication, London, 2006.

Christopher Allen: *Oracle: OCP Building Internet Applications I & II Exam Guide (With CD)*, Tata McGraw-Hill, Delhi, 2004.

Diane Barrett and Todd King: *Computer Networking Illuminated*, Jones and Bartlett Learning, London, 2008.

Fallows: *Pro JSF and Ajax: Building Rich Internet Components*, Springer Publication, London, 2005.

Harsh Marwah: *Computer System Architecture*, Anmol Publication, Delhi, 2010.

Irene E McDermott: *The Librarian's Internet Survival Guide : Strategies for the High-Tech Reference Desk*, Ess Ess Publication, Delhi, 2008.

James Kempf: *Wireless Internet Security: Architecture and Protocols*, Cambridge University Press, Delhi, 2001.

K.L. James: *The Internet : A User's Guide*, PHI Learning, Delhi, 2000.

K.M.Shrivastava: *News Agencies : From Pigeon to Internet*, Sterling

Publication, London, 2006.

Kizza: *Computer Network Security*, Springer Publication, London, 2001.

LEE: *Mobile Cellular Telecommunications: Analog and Digital Systems*, Tata McGraw-Hill, Delhi, 2006.

M. Usha and T. S. Srikanth: *Computer System Architecture and Organization*, Wiley Publication, London, 2012.

Milos Ercegovac, Tomas Lang, Jaime h. Moreno: *Introduction to Digital Systems*, Wiley Publication, London, 2000.

Miron Abramovici, Melvin A. Breuer & Arthur D. Friedman: *Digital Systems Testing and Testable Design*, Jaico Publication, Jaipur, 2003.

Nandan Kamath: *Law Relating to Computers Internet and E-Commerce*, Universal Law Publishers, Delhi, 2012.

NAVABI: *Verilog Digital System Design, Second Edition (SIE)*, Tata McGraw-Hill, Delhi, 2008.

Nicholas Carr: *The Shallows: How the Internet is Changing the Way We Think, Read and Remember*, Atlantic Publication, Delhi, 2010.

Nina Verma: *Social Engineering : A Means to Violate a Computer System*, Global Vision Publication, Delhi, 2011.

Nishit Mathur: *Computer Systems and Applications*, A.P.H. Publishing Corporation, Delhi, 2010.

Pradeep Pokhriyal: *Computer Networking*, APH Publication, Delhi, 2012.

Rajeev Ranjan: *Computer System Organization and Architecture*, Centrum Press, Delhi, 2011.

Rajneesh Narayan: *Computer System Architecture*, Centrum Press, Delhi, 2010.

S.S. Khandare: *Computer Systems and Programming In C*, S. Chand Publisher, Delhi, 2010.

Samir M. Chaudhari and Ankita H. Pandya: *Computer Networking and the Internet*, Shree Niwas Publications, Delhi, 2011.

Satyapriya Bhattacharjee: *A Textbook of Objective Computer Networking*, Wisdom Press, Delhi, 2012.

Index